Contributions to Management Science

The series *Contributions to Management Science* contains research publications in all fields of business and management science. These publications are primarily monographs and multiple author works containing new research results, and also feature selected conference-based publications are also considered. The focus of the series lies in presenting the development of latest theoretical and empirical research across different viewpoints.

This book series is indexed in Scopus.

Elena Borin • Mara Cerquetti • Marta Crispí • Judith Urbano
Editors

Cultural Leadership in Transition Tourism

Developing Innovative and Sustainable Models

Springer

Editors
Elena Borin
Link Campus University
Rome, Italy

Mara Cerquetti
University of Macerata
Macerata, Italy

Marta Crispí
Faculty of Humanities
Universitat Internacional de Catalunya
Barcelona, Spain

Judith Urbano
Faculty of Humanities
Universitat Internacional de Catalunya
Barcelona, Spain

ISSN 1431-1941 ISSN 2197-716X (electronic)
Contributions to Management Science
ISBN 978-3-031-14120-1 ISBN 978-3-031-14121-8 (eBook)
https://doi.org/10.1007/978-3-031-14121-8

This Springer imprint is published by the registered company Springer Nature Switzerland AG
The registered company address is: Gewerbestrasse 11, 6330 Cham, Switzerland

Contents

Cultural Organisations as Drivers for Change in Tourism

Elena Borin, Mara Cerquetti, Marta Crispí, and Judith Urbano

Abstract In the context of the broader changes in tourism approaches, cultural tourism is a distinct example of the potential paradigm shift in the tourism industry. Culture is indeed closely related to territory, identity, traditions and, therefore, inhabitants and communities. The aim of this book is to encourage discussion on the concept of cultural leadership in this transition phase for tourism and thereby lay alternative paths of development for cultural tourism destinations at a time of profound change and dialogue in the post-pandemic period. Also, within the context of increasing attention and discussion on post-pandemic tourism innovation, the presented contributions aim to trigger reflection on the potential to design more specific and appropriate strategies and measures for cultural tourism destinations and on the role cultural organisations (public, private and civic) might play in more sustainable models of tourism development for cities and regions. Within this framework, the volume aims to present a multidisciplinary approach that includes contributions from academics and professionals in the field of cultural tourism management, urban planning and sustainable development, as well as heritage, museums and cultural and creative industries.

E. Borin (✉)
Department of Health and Life Sciences, Link Campus University, Rome, Italy
e-mail: e.borin@unilink.it

M. Cerquetti
Department of Education, Cultural Heritage and Tourism, University of Macerata, Macerata, Italy
e-mail: mara.cerquetti@unimc.it

M. Crispí · J. Urbano
Faculty of Humanities, Universitat Internacional de Catalunya, Barcelona, Spain
e-mail: mcrispi@uic.cat; jurbano@uic.cat

E. Borin et al. (eds.), *Cultural Leadership in Transition Tourism*, Contributions to Management Science, https://doi.org/10.1007/978-3-031-14121-8_1

1 Time for a Change: A Moment of Reflection on Past and Emerging Paths of Development in Tourism

Over the last 20 years, the growing pressure of tourism flows on cities and regions all over the world has led to the questioning of current tourism development models, especially in relation to sustainability issues (Veiga et al., 2018). Although tourism has become a significant source of economic development, the potentially negative impact of tourism in social, environmental and cultural terms has increasingly been raised as a serious problem for the long term. Its negative effects, not only on the survival of local ecosystems but also on the well-being and quality of life of local people, the tourist experience and the life-cycle duration of a tourism destination have been amply discussed. Concepts such as overtourism, created in the 1980s (Jungk, 1980; Krippendorf, 1986), and sustainable tourism development have become buzzwords, once again highlighting the growing awareness communities and global actors have of the themes of sustainability and quality in the field of tourism (Veiga et al., 2018).

Consequently, the search for models that promote sustainable tourism development has become a pressing issue (Benner, 2019). This is particularly important not only for wildlife destinations (Newsome, 2020) but also for cultural destinations (García-Hernández et al., 2017; Martínez, 2018) where the environmental threat is linked to a risk of social and cultural unsustainability (Bertocchi & Visentin, 2019; Novy & Colomb, 2016; Seraphin et al., 2018) and a failure to preserve both tangible and intangible cultural heritage (Artal-Tur et al., 2018; De Luca et al., 2020; Díaz-Parra & Jover, 2021). On the one hand, the crisis caused by the COVID-19 pandemic highlighted the unsustainability of the pre-crisis tourism model and underscored the need for far-reaching structural changes and degrowth strategies (Romagosa, 2020). On the other hand, it also made clear the importance of tourism promotion for sustaining local economies. Thus, there is increasing demand for designing new models and approaches to place tourism at the core of recovery strategies but also to make it more respectful of local identities, inhabitants and sociocultural ecosystems.

Several theoretical models have been devised to study overtourism (Capocchi et al., 2020; Volgger, 2018) and to reflect upon and measure the carrying capacity of tourism destinations (Capocchi et al., 2019; Muler Gonzalez et al., 2018; Koens et al., 2018; Perkumiene & Pranskuniene, 2019). While the limit capacity of natural ecosystems has been widely investigated, the study of the limit capacity of "cultural ecosystems" (Barker, 2020; Bonet & Donato, 2011; Borin & Donato, 2015; Borin, 2016; De Bernard et al., 2021) is still evolving (Martínez, 2018). Indeed, although, in addition to quantitative indicators, discussions and investigations into the limit capacity of tourism destinations increasingly take into account qualitative indicators that aim to capture sociocultural dimensions such as CCVS (carrying capacity value stretch) (Mansfeld & Jonas, 2006), the cultural dimensions of sustainable tourism development are rather underexplored, especially in terms of cultural cities and regions that mainly base their tourism model on cultural assets (Capocchi et al., 2019). When it comes to professions and policy development, the debate also seems to be concerned with the integration of different dimensions in generic terms while not giving due consideration to the specificities of cultural destinations. The guidelines developed by important organisations such as UNWTO (2018) or resulting from reports commissioned by the European Union (Peeters et al., 2018) mainly encourage strategies and measures for handling visitor flows in urban destinations that are rooted in time-based and space-based dispersal of tourists and coordinated communication strategies. Their recommendations and overall focus are on the need to engage not only tourism authorities but also residents and locals, alongside different stakeholders, in the design and implementation of tourism strategies. In the post-COVID-19 era, relationships with local communities are even more critical for the increasing value of proximity tourism. Indeed, in the uncertain and insecure context of the pandemic and the post-pandemic phase, tourists are more likely to explore their local regions and visit nearby destinations in response to international restrictions and their reduced purchasing power (Carr, 2020).

Moreover, in the context of the broader changes in tourism approaches, cultural tourism is a distinct example of the potential paradigm shift in the tourism industry. Culture is indeed closely related to territory, identity, traditions and, therefore, inhabitants and communities. For instance, historic buildings and built heritage play a significant part in the culture of a society, as they include and represent aspects of its history. The apparent dichotomy between preservation and enhancement of cultural assets has been one of the core topics of debate on cultural tourism (González, 2018; Seraphin et al., 2018). Indeed, it is crucial to maintain intrinsic heritage and cultural values even when cultural assets are accommodated to new socio-economic demands and uses, such as those of tourism. Cultural and creative actors have also been at the forefront of lobbying for a more holistic approach to development. One important example is the movement whose aim is to include culture as one of the pillars of the sustainability paradigm (Jagodzińska et al., 2015; Soini & Dessein, 2016), as a contribution to the ongoing debate on culture, sustainable development and cultural ecosystems (Borin & Donato, 2015; Holden, 2015). Also, in this context, the concept of cultural ecosystems seems particularly relevant. Indeed, it highlights a need to reinterpret culture as an ecology rather than as an economy and brings to the fore the interdependence and the relationships between cultural and creative actors and other stakeholders in sustainable local development. Therefore, cultural destinations could be interpreted as "cultural tourism ecosystems", i.e. local territorial ecosystems based on strong cultural assets in which civic, public and private stakeholders are linked by strong common cultural values and cultural identity. In cultural tourism ecosystems, cultural actors can potentially play a steering role, by driving actions and strategies of tourism development and by connecting the different stakeholders (from cultural and other sectors) operating in the territory (Borin, 2016). In other words, cultural organisations of different kinds could become the leaders of tourism development and promote a successful transition towards more sustainable models of cultural tourism. The cultural leadership concept (Sutherland & Gosling, 2010) could build upon concepts in tourism such as destination leadership (Beritelli & Laesser, 2014; Hristov & Zehrer, 2015; Pechlaner et al., 2019) to guide this moment of transformation of the tourism sector.

2 Aim and Content of the Book

The aim of this book is to encourage discussion on the concept of cultural leadership in this transition phase for tourism and thereby lay alternative paths of development for cultural tourism destinations at a time of profound change and dialogue in the post-pandemic period. Also, within the context of increasing attention and discussion on post-pandemic tourism innovation, it aims to trigger reflection on the potential to design more specific and appropriate strategies and measures for cultural tourism destinations and on the role cultural organisations (public, private and civic) might play in more sustainable models of tourism development for cities and regions.

Could cultural and creative actors lead a new approach to tourism development that could also bring more sustainable tourism practices, involving local communities and respecting local social and cultural identities and attributes?

The book's theoretical and empirical contributions from international researchers present different approaches to this topic. The chapters reflect on the past and current status of the cultural and creative sector as it relates to tourism and, more specifically, on cultural tourism management, governance and strategies to bring new ideas to the fore and provide food for thought for academic and cultural tourism professionals.

Within this framework, the volume aims to present a multidisciplinary approach that includes contributions from academics and professionals in the field of cultural tourism management, urban planning and sustainable development, as well as heritage, museums and cultural and creative industries.

The chapters are organised into four main thematic sections and focus on analyses of significant case studies, qualitative and quantitative investigations, along with theoretical contributions. A broad range of perspectives and approaches is presented, providing a broad representation of investigations in different cultural contexts and countries.

The book's objectives are twofold. First, it aims to present an overview of the best practices that link cultural and creative industries (CCIs), cultural heritage and tourism. For this, it contains a selection of works focused on innovative examples of development in the cultural and creative field which aim to encourage reflections on how they might open new paths towards more sustainable forms of tourism growth. Second, the book aims to explore cultural tourism more in depth, analysing it from various perspectives and angles, including cultural studies, economics, management, governance and leadership. The objective here is to gather information on the role of cultural and creative actors in different cultural tourism scenarios and to encourage debate on how innovative leadership from cultural organisations might promote more diverse and more sustainable tourism development models in cultural destinations. This could lead to the development of guidelines for a potential transition to new models of cultural tourism, in which cultural and creative organisations are considered the prospective leaders of a paradigm shift.

The target readers of the volume are academics and researchers in the field of cultural management and sustainable tourism development but also professionals operating in the fields of culture and tourism. The book also aims to provide insights that might benefit the broader community of stakeholders in cultural tourism destinations, since it represents a plurality of voices that could reflect the perspectives of a wider and multifold audience.

3 Structure of the Book

This volume is divided into four parts. Part I provides a theoretical and empirical introduction discussing the core issues and challenges cultural ecosystems face in this transition phase for tourism. The section stems from the interpretation of CCIs as levers for local and tourism development and aims to set the basis for further discussion on the relationship between cultural activities and local tourism enhancement. Borin and Donato "Cultural Ecosystem Approaches as Key for New Development Paths" analyse the shift from traditional cultural ecosystems towards culture-led ecosystems, highlighting the need for stronger cultural leadership to navigate the new circumstances. By discussing a broad set of crucial examples, the authors present four main emerging trends: exploration of new online and digital possibilities, increased attention on audience development and engagement, closer links with sustainability and sustainable development issues and the growing interdependence between CCIs and other sectors, including tourism. Chapter "Towards Sustainable Innovation in Tourism. The Role of Cultural Heritage and Heritage Communities" (Cerquetti and Romagnoli) focuses on the current extended concept of cultural heritage—open and inclusive, dynamic and relational—and the involvement of communities in heritage recognition, safeguard and enhancement. In this conceptual paper, the role of heritage communities arises as an emerging actor in cultural ecosystems as being crucial to triggering and leading sustainable innovation in tourism. Chapter "Cultural and Creative Industries as Activators and Attractors for Contemporary Culture-Driven Nomadism" (Bertola, Iannilli, Spagnoli and Vandi) moves from the theoretical approach to case studies. First, the authors present an interpretative framework methodology—the design-orienting scenario approach—based on the intersection of two interpretative variables: types of communities involved (inner/outer) and actions in the territory (reimagining/resignifying). Then, they discuss four example cases from Italy that are representative of current changes: Farm Cultural Park, a visionary project which reimagines the city of Favara, Sicily, and its community through the lenses of art and architecture; Lenz Rifrazioni, a creative hub in Parma, which revitalised an abandoned industrial neighbourhood through dialogue between marginal communities and multidisciplinary artistic residencies; Dolomiti Contemporanee, an extended lab for contemporary art and culture in disused public spaces in the Dolomite area with a high-value potential; and Lottozero, a co-working space for textile design, art and culture based in the textile district of Prato, which helps promote circular economy practices involving local and international actors from various sectors and disciplines. Saleh and Ost (Chapter "Cultural and Creative Industries in the Brussels

Region. Challenges and Opportunities of the Ecosystem") close the first part of the book with an in-depth case study. The authors focus on the CCI ecosystem in the Brussels region, by mapping the reskilling/upskilling programmes through a survey. After preliminary desk research on the 1170 CCIs in the 19 municipalities of the Brussels region, 34 questionnaires were collected and analysed. The research suggests that innovative cultural projects and initiatives need critical entrepreneurial skills development. The key training components are business and operational model innovation and sustainability, i.e. sustainable working conditions, viable projects and sustainable impacts. Furthermore, ethical and cooperative governance, solid partnerships and disruptive technologies are identified as subsidiary ingredients for successful holistic upskilling for CCIs.

Part II explores the role of cultural heritage more in detail by focusing on the role of emerging typologies of heritage for developing new sustainable tourism paths. The first two chapters of this part analyse the contribution of industrial heritage. Investigating new opportunities for cultural tourism development in the post-pandemic era, Garofano, Riviezzo and Napolitano (Chapter "Industrial Tourism as a New Opportunity for Tourism Development in the Post-Pandemic Era") provide a theoretical framework of the industrial tourism phenomenon, analysing its various definitions and manifestations in addition to data on its impact on visitors. The authors discuss a number of best practices and experiences in industrial tourism in Italy, distinguishing them into three typologies: industrial archaeology, corporate museums and factory tours. Rossato (Chapter "Documentation Activities for Historic Railways: Detection of Potential and Weaknesses for a Sustainable Cultural Heritage Resource") goes into this topic by presenting a possible methodology for documenting and enhancing historic railways as socio-economic drivers and tourism resources. The research focuses on two regions in Asia and South America facing development issues in a low-income scenario, that is, the Darjeeling Himalayan Railway (India) and the Curitiba-Paranaguá historic railway (Brazil). Making use of geospatial and non-geospatial data, the chapter provides administrators and stakeholders with tools for managing railway sites as tourism resources. The analysis in this chapter leads the way for Chapter "Digital Documentation for Enhancement and Conservation of Minor or Inaccessible Heritage Sites" (Maietti), which explores the role of the digital documentation of cultural heritage in improving its conservation and accessibility. The author presents the case studies of Stellata di Bondeno and Mesola, both in the province of Ferrara, in the north-east of Italy. The last chapter in part II (Ocón) broadens the discussion by investigating the latest trends in digitalisation as applied to cultural heritage tourism, particularly in contexts of crisis like the COVID-19 pandemic, where in-person opportunities to engage with cultural heritage assets are significantly reduced. The chapter argues that, although largely unintentional initially, the push for these new technologies during the pandemic has generated more sustainable and equal access paths to cultural heritage assets, particularly for visitors with limited means or mobility. The discussion takes place through an analysis of case studies from Asia which applied different kinds of technologies: 360-degree technology, 3D modelling and photogrammetry, virtual exhibitions, podcasts and augmented reality (AR).

Part III disentangles the relationship between cultural assets, cities and territories while focusing on the integrated enhancement of place-specific cultural assets (e.g. tangible and intangible cultural heritage, museums, local products, etc.) for promoting sustainable tourism development in both traditional and emerging destinations. In this framework, Crispí and Urbano (Chapter "Girona Episcopal: An Analysis of a New Cultural Brand") discuss Girona Episcopal as a new cultural brand which strengthens Girona's attractiveness by linking together the buildings and collections in its powerful bishopric, i.e. Girona Cathedral, the church of Sant Feliu and the Museu d'Art de Girona. The brand is analysed by means of the cultural heritage branding and brand orientation models. The research results summarise the strengths, weaknesses, opportunities and threats of the project in a SWOT analysis. The need to innovate the urban cultural offer is also the focus of Chapter "Multidimensional Value Creation Through Cultural Programmes? Challenges for Veszprém-Balaton ECoC 2023" (Cerquetti, Lőrincz and Raffay-Danyi). In this chapter, the authors analyse the case of Veszprém-Balaton, European Capital of Culture (ECoC) in 2023, with the aim of enriching the international debate on value creation through cultural programmes. The research findings reveal the need to improve the levels of involvement and networking and highlight the role of process innovations, in addition to product innovations, for ensuring long-term sustainable development. Aiming to contribute to the theoretical debate on overtourism in UNESCO destinations, the authors of Chapter "Partnerships as a Tool for Eliminating Tourist Oversaturation in UNESCO Destinations: Case Study of Slovakia" (Kvasnová and Marciš) adopt an exploratory case study approach to investigate the methods and strategies used to eliminate oversaturation in the UNESCO town of Banská Štiavnica, Slovakia. The research results suggest a conceptual framework for eliminating overtourism in UNESCO destinations, based on institutional, product, process, managerial and marketing innovations. Finally, in Chapter "Sustanaiblity and Commodification? The Role of Cultural Assets in the Development of New Paths for Tourism in Catalonia", Barbieri and del Mármol question culture-based development strategies by highlighting the paradox in the opportunity for making development processes more sustainable and, at the same time, for reproducing economic and cultural inequalities. The chapter addresses this issue by analysing cultural policies in Catalonia and, specifically, the relationship between cultural heritage enhancement and tourism promotion in the Catalan Pyrenees.

Part IV goes further into the analysis by exploring cultural leadership in transition tourism beyond the cultural sector. The objective of this part is to understand what stimuli we can get from the current tourism debate to help us reinterpret tourism and culture in a more sustainable way. Savych and Shkoda (Chapter "Impact of Cultural Relationship Marketing on Business Activity") examine the impact of cultural relationship marketing on business activity, focusing on automotive part distributors in Ukraine. The research findings prove the importance of the cultural heritage of a country where a company is headquartered, its tourist attractiveness and food traditions for building strong relationships with stakeholders. Molderez and Perera (Chapter "Inclusive Tourism, a Cultural Paradigm Shift in Approaching Tourism Development") discuss inclusive tourism as a cultural paradigm shift in tourism development. By adopting the Base of the Pyramid (BoP) as an inclusive business model, the authors discuss the involvement of the local community as a change agent and a developer for a greater understanding of local culture, with positive impacts on the conservation of natural and cultural heritage. The research is supported by an illustrative case study, the Hiriwadunna Village Tour in Sri Lanka. Finally, Nogueira, Lins de Moraes, Marques and Correia (Chapter "Towards a Demand Profile of Wwoofers in Portugal: Ecological, Rural and Voluntary Principles in Tourism") move from tourism supply to tourism demand by analysing Worldwide Opportunities on Organic Farm (WWOOF) tourists as an emerging cultural change in tourism based on ecological, rural and voluntary principles. The research focuses the analysis of ecotourism practices on the case of Portugal and identifies three different tourist profiles: sustainable gappers, sustainable escapers and sustainable lifers.

In conclusion, the book provides multiple perspectives, theoretical reflections and case studies which help understand how CCIs can lead the current transition phase in tourism. Explorative research presented in this volume shows that cultural ecosystems have changed over the last 20 years, particularly in their relationship with local communities, through the use of digital tools and by strengthening connections with other sectors.

Local communities are emerging as an important new actor that can and should be involved in decision-making processes. Cooperation with local communities could be the basis for redesigning our approach to tourism, and cultural and creative organisations could potentially stimulate the creation of co-designing and participatory practices. Moreover, the involvement of communities in decisions relating to destinations could help identify strategies to decrease tourism pressures in overcrowded cultural cities and redirect tourism flows towards potential new cultural destinations. From this point of view, the new approach to cultural heritage and the increasing focus on industrial and minor heritage could be an opportunity to prevent abandonment and depopulation and promote socio-economic regeneration and sustainable local development, even in rural and remote areas (Cerquetti & Cutrini, 2021). Digital technologies could help assess, preserve and enhance cultural heritage and increase its availability to visitors with limited means or mobility. Thus, the use of technologies and digital tools could be interpreted as a means to achieve the objectives of social sustainability and accessibility, subject, of course, to a critical analysis of their potential contribution. Finally, several chapters present cases of an increasing intersection of culture, tourism and other fields (industrial sectors, natural and environmental preservation, organic farming, automotive sector, etc.) which points to a need to work on the connections, rather than the separations, between domains of our economies and to enhance the mutual benefits and adopt inclusive and holistic approaches when designing tourist destinations.

In short, the challenges in developing more sustainable models of tourism are manifold: cultural and creative organisations could play leading roles in this transition phase, if they embark on a process of change that requires different levels of innovation, including infrastructures (e.g. digital technologies for ensuring cultural heritage conservation and accessibility), capacity building (e.g. training in key areas such as entrepreneurship and sustainability) and solid cross-sectoral partnerships between the actors involved.

References

Artal-Tur, A., Briones-Peñalver, A. J., & Villena-Navarro, M. (2018). Tourism, cultural activities and sustainability in the Spanish Mediterranean regions: A probit approach. *Tourism & Management Studies, 14*(1), 7–18.

Barker, V. (2020). The democratic development potential of a cultural ecosystem approach. *Journal of Law, Social Justice and Global Development, 24*, 1–14.

Benner, M. (2019). *From overtourism to sustainability: A research agenda for qualitative tourism development in the Adriatic*. Heidelberg University. MPRA Paper No. 92213. Retrieved June 7, 2022, from https://mpra.ub.uni-muenchen.de/92213/1/MPRA_paper_92213.pdf.

Beritelli, P., & Laesser, C. (2014). Getting the cash-cow directors on board–an alternative view on financing DMOs. *Journal of Destination Marketing & Management, 2*(4), 213–220.

Bertocchi, D., & Visentin, F. (2019). "The overwhelmed city": Physical and social over-capacities of global tourism in Venice. *Sustainability, 11*(24), 6937.

Bonet, L., & Donato, F. (2011). The financial crisis and its impact on the current models of governance and management of the cultural sector in Europe. *ENCATC Journal of cultural management and policy, 1*(1), 4–11.

Borin, E. (2016). *Public-private Partnership in the Cultural Sector: A comparative analysis of European models*. PIE Peter Lang.

Borin, E., & Donato, F. (2015). Unlocking the potential of IC in Italian cultural ecosystems. *Journal of Intellectual Capital, 16*(2), 285–304.

Capocchi, A., Vallone, C., Pierotti, M., & Amaduzzi, A. (2019). Overtourism: A literature review to assess implications and future perspectives. *Sustainability, 11*(12), 3303.

Capocchi, A., Vallone, C., Amaduzzi, A., & Pierotti, M. (2020). Is 'overtourism' a new issue in tourism development or just a new term for an already known phenomenon? *Current Issues in Tourism, 23*(18), 2235–2239.

Carr, A. (2020). COVID-19, indigenous peoples and tourism: A view from New Zealand. *Tourism Geographies, 22*(3), 491–502.

Cerquetti, M., & Cutrini, E. (2021). The role of social ties for culture-led development in inner areas. The case of the 2016–2017 Central Italy earthquake. *European Planning Studies, 29*(3), 556–579.

De Bernard, M., Comunian, R., & Gross, J. (2021). Cultural and creative ecosystems: A review of theories and methods, towards a new research agenda. *Cultural Trends*, 1–22.

De Luca, G., Shirvani Dastgerdi, A., Francini, C., & Liberatore, G. (2020). Sustainable cultural heritage planning and management of overtourism in art cities: Lessons from atlas world heritage. *Sustainability, 12*(9), 3929.

Díaz-Parra, I., & Jover, J. (2021). Overtourism, place alienation and the right to the city: Insights from the historic Centre of Seville, Spain. *Journal of Sustainable Tourism, 29*(2–3), 158–175.

García-Hernández, M., de la Calle-Vaquero, M., & Yubero, C. (2017). Cultural heritage and urban tourism: Historic City Centres under pressure. *Sustainability, 9*, 1346.

González, A. T. (2018). Venecia: el problema del "overtourism" y el impacto de los cruceros. *Investigaciones Regionales/Journal of Regional Research, 42*, 35–51.

Holden, J. (2015). *The ecology of culture*. Arts and Humanities Research Council.

Hristov, D., & Zehrer, A. (2015). The destination paradigm continuum revisited: DMOs serving as leadership networks. *Tourism Review, 70*(2), 116–131.

Koens, K., Postma, A., & Papp, B. (2018). Is Overtourism overused? Understanding the impact of tourism in a City context. *Sustainability, 10*, 4384.

Krippendorf, J. (1986). Die Ferienmenschen. *München: dtv Verlag.*
Jagodzińska, K., Sanetra-Szeliga, J., Purchla, J., Van Balen, K., Thys, C., Vandesande, A., & Van der Auwera, S. (2015). *Cultural heritage counts for Europe*. Full Report.
Jungk, R. (1980). Wie viele Touristen pro Kektar Strand? *Geo, 10*, 154–156.
Mansfeld, Y., & Jonas, A. (2006). Evaluating the socio-cultural carrying capacity of rural tourism communities: A 'value stretch' approach. *Tijdschrift voor Economische en Sociale Geografie, 97*(5), 583–601.
Martínez, P. J. (2018). The new holistic paradigm and the sustainability of historic cities in Spain: An approach based on the world heritage cities. *Sustainability, 10*, 2301.
Muler Gonzalez, V., Coromina, L., & Galí, N. (2018). Overtourism: Residents' perceptions of tourism impact as an indicator of resident social carrying capacity-case study of a Spanish heritage town. *Tourism Review, 73*, 277–296.
Newsome, D. (2020). The collapse of tourism and its impact on wildlife tourism destinations. *Journal of Tourism Futures, 7*(3), 295–302.
Novy, J., & Colomb, C. (2016). Urban tourism and its discontents: An introduction. In C. Colomb & J. Novy (Eds.), *Protest and resistance in the Tourist City* (pp. 1–30). Routledge.
Pechlaner, H., Innerhofer, E., & Erschbamer, G. (Eds.). (2019). *Overtourism: Tourism management and solutions*. Routledge.
Peeters, P., Gössling, S., Klijs, J., Milano, C., Novelli, M., Dijkmans, C., et al. (2018). *Research for TRAN committee–Overtourism: Impact and possible policy responses*. European Parliament, Policy Department for Structural and Cohesion Policies.
Perkumiene, D., & Pranskuniene, R. (2019). Overtourism: Between the right to travel and residents' rights. *Sustainability, 11*, 2138.
Romagosa, F. (2020). The COVID-19 crisis: Opportunities for sustainable and proximity tourism. *Tourism Geographies, 22*(3), 690–694. https://doi.org/10.1080/14616688.2020.1763447
Seraphin, H., Sheeran, P., & Pilato, M. (2018). Over-tourism and the fall of Venice as a destination. *Journal of Destination Marketing & Management, 9*, 374–376.
Soini, K., & Dessein, J. (2016). Culture-sustainability relation: Towards a conceptual framework. *Sustainability, 8*(2), 167.
Sutherland, I., & Gosling, J. (2010). Cultural leadership: Mobilizing culture from affordances to dwelling. *The Journal of Arts Management, Law, and Society, 40*(1), 6–26.
UNWTO. (2018). *Tourism and the sustainable development goals–good practices in the Americas*. World Tourism Organization.
Veiga, C., Santos, M. C., Águas, P., & Santos, J. A. C. (2018). Sustainability as a key driver to address challenges. *Worldwide Hospitality and Tourism Themes.*
Volgger, M. (2018). The end of tourism through localhood and overtourism? An exploration of current destination governance challenges. In H. Pechlaner (Ed.), *Tourismus Und Lebensraum. Perspektiven Für Alpine Destinationen*. Springer.

Part I
Cultural Ecosystems Facing Transition Tourism: Core Issues and Challenges

Cultural Ecosystem Approaches as Key for New Development Paths: A Reflection on Management and Governance Implications

Elena Borin and Fabio Donato

Abstract The purpose of this paper is to analyse some of the main trends of development of the cultural and creative sector from the perspective of cultural ecosystems.

The authors argue that the concept of cultural ecosystems could be relevant in reinterpreting the cultural and creative sector in a moment of profound change, leading to a different approach to the management and governance of the cultural and creative field as well as the way we interpret its connections with and contribution to society and sustainable development.

The study highlights four main trends which have recently emerged: the exploration of new online and digital possibilities, an increased attention to audience development and engagement, a closer connection with sustainability and sustainable development, and the growing interdependence between CCIs and other sectors, including tourism. These trends can be interpreted as a shift from the traditional cultural ecosystem approaches towards culture-led ecosystems where a stronger cultural leadership is required to navigate the new circumstances.

The research not only impacts the scientific debate on cultural ecosystems but also triggers deeper reflection, for practitioners and policy makers, on how to implement structural changes to adapt to the emerging scenarios.

The research could be further developed, implementing a quantitative approach and international and longitudinal dimensions to further investigate these topics and validating these preliminary research results.

E. Borin (✉)
Department of Health and Life Sciences, Link Campus University, Rome, Italy
e-mail: e.borin@unilink.it

F. Donato
Department of Economics and Management, University of Ferrara, Ferrara, Italy
e-mail: fabio.donato@unife.it

E. Borin et al. (eds.), *Cultural Leadership in Transition Tourism*, Contributions to Management Science, https://doi.org/10.1007/978-3-031-14121-8_2

1 Introduction

The increasing popularity of the concept of cultural and creative ecosystems seems to reflect a shift towards a more holistic and inclusive approach to the cultural and creative sector (Borin & Donato, 2015; Holden, 2015). It indeed stresses the links with the territory, diverse stakeholders from other sectors, as well as with local communities and their cultural and creative identities and values (Bonet & Négrier, 2018; Borin, 2017). Moreover, the concept of cultural and creative ecosystems has been linked to the reflection on the role culture can play in leading development in a sustainable way (Sutherland & Gosling, 2010).

During the pandemic period, the cultural and creative sector experienced a rupture of its traditional models (Radermecker, 2021; Tan & Tan, 2021), which highlighted the need to reinterpret its development paradigm in light of different ways to connect with audiences and stakeholders (Martínez Pino, 2018; Zbuchea et al., 2021) using innovative tools and digital technologies but also leveraging connections with other sectors, such as tourism (Artal-Tur et al., 2018). The concept of cultural ecosystems seems to be particularly promising for interpreting this moment of change; it seems therefore relevant to reflect on how it could lead to a diverse approach to both the management and governance of the cultural and creative sector and the way we interpret its connections with, and contribution to, society. However, given the significant changes experienced over the last 2 years, it seems necessary to first reflect on current and emerging experiences and trends and then address how the concept of cultural ecosystem could evolve to meet the increasing challenges of the future and its implications in terms of management and governance in the field.

This paper investigates this topic by means of a theoretical and empirical investigation, based on the analysis of the trends emerging in the pandemic context through multiple examples of cultural and creative organisations using new solutions during the pandemic restrictions. The paper examines how these solutions point to a new cultural ecosystem concept and how these experiences could impact the management and governance of cultural and creative organisations in this new ecosystem framework.

Based on this analysis, the authors attempt to answer the following main research questions:

- How has the concept of cultural ecosystem evolved over the last years?
- What is the potential of this cultural and creative ecosystems concept to lead to a reinterpretation of the cultural and creative sector?
- What are the implications in terms of management and governance for the organisations in this field?

This contribution is divided into six main sections. The second section provides a literature review on the development of cultural ecosystems and on the link between culture and the sustainable development debate, also reflecting on the pandemic impact. Section 3 explains the research design and methodology, while Sect. 4

illustrates the research results. In Sect. 5, we critically discuss the results of the analysis. Section 6 draws some concluding remarks, highlighting the scientific and practice contribution of the research as well as its limitations and further developments.

2 Cultural and Creative Ecosystems: Theoretical Premises

Over the last 20 years, the concept of 'ecosystem', stemming from the biology field, has become increasingly popular and adapted to several contexts and domains (Costanza et al., 1997). Among them, it has been increasingly popular in management and business studies (Basole, 2009; Iansiti & Levien, 2004; Peltoniemi, 2006; Stam, 2015), thus adapting the original features of biology ecosystems to a business context. Among the many definitions, Moore (1996) described business ecosystems as interconnected systems of different stakeholders (customers, suppliers, financing entities, trade associations, trade unions, NGOs, governmental bodies and other interacting stakeholders) whose activities are mutually supportive. Business ecosystems are not centrally directed but rather influenced by members and guided in their development by a collective consciousness created and distributed inside the ecosystem (Kinder et al., 2021). A whole trend of business literature focused on entrepreneurial ecosystems (Isenberg, 2014; Jansen et al., 2017; Spigel, 2017; Spigel & Harrison, 2018; Stam, 2015). These studies focused not only on defining entrepreneurial ecosystems but also on identifying the factors that influence their success. Isenberg (2014) and Stam (2015) argue that entrepreneurial ecosystems are dynamic, living networks that encompass several diverse but interdependent actors that can cooperate to trigger virtuous entrepreneurial processes. In the representation of entrepreneurial ecosystems, scholars underline the link with the territorial context, not only pointing out that each ecosystem is 'site-specific' and represents the general macro- and meso-environment of a territory (Aage & Belussi, 2008; Cohendet et al., 2010; Drake, 2003) but also identifying ecosystem factors that are related to the socio-economic structure of their location (Isenberg, 2010, and 2011a, b; Blanke & Chiesa, 2013). For instance, the model proposed by Isenberg (2011a, b) categorises six key dimensions of entrepreneurial ecosystems (policy, finance, culture, supports, human capital, and markets) that basically describe the socio-economic elements of the territory. Also, the prominent model designed by Stam in 2015 identifies key factors of successful entrepreneurial ecosystems that are framework conditions and systemic conditions, both related to territorial resources. Framework conditions are physical infrastructure, formal institutions, demand for new products (goods and services), and enabling culture; systemic conditions are factors which lead to entrepreneurial activities such as leadership, knowledge, entrepreneurship networks, finance, support services, and talent (Jansen et al., 2017).

The ecosystem concept has been early embraced by several sectors and risen at the centre of European industrial development strategies (EC, 2020a) including the cultural and creative sector. Notwithstanding the fact that this field has been a late adopter, the ecosystem framework has been central in the latest discussions on culture and creativity. Indeed, understanding the sector's richness and diversity can be aided by using the ecosystem metaphor. In the cultural and creative sector, ecosystem discussion has been developed in relation to the concept of cultural ecosystems (Holden, 2015; Borin & Donato, 2015), cultural and creative ecosystems (Barker, 2018, 2020; Henry & Barker, 2016), and entrepreneurial cultural ecosystems (Borin & Jolivet, 2021). The discussion on cultural ecosystems keeps adding new elements of reflection, and researchers and practitioners are still adapting it to the peculiarities of this field. Among the first studies, the work of Holden (2004 and 2015) stands out for its attempt at a definition of cultural ecosystems on the basis of the 'ecology of culture' approach. His model—which identifies three main spheres of interaction, namely, 'commercial culture', 'publicly funded culture', and 'home-made culture'—was criticised for limiting the domains of cultural ecology to an analysis of the relations among just cultural and creative industries. It was later updated by other studies that interpret cultural ecosystems based on the connections and contribution of cultural and creative organisations to a territory and its stakeholders (Borin & Donato, 2015; Barker, 2020). A significant stream of research focuses on entrepreneurial ecosystems in the cultural and creative sector and explores the motives of collaborations among the components of cultural ecosystems (Ballico, 2017; Borin, 2017; Lash & Urry, 1994; Protogerou et al., 2017). This group of works underlines that entrepreneurs in the cultural and creative sector decide to cluster mainly on the basis on their desire to collectively produce cultural values (Ballico, 2017; Lash & Urry, 1994) rather than for financial or policy incentives or for reasons of physical proximity with customers and competitors as in other ecosystems. Other studies have addressed issues of governance (Borin, 2017: Oakes, 2019), exploring the potential of multi-stakeholder (public, private, and civic) partnerships as tools for sustainable coordination and exchange between the stakeholders operating in a cultural entrepreneurial ecosystem (Borin, 2017). Indeed, in cultural ecosystems, cultural heritage institutions work alongside cultural and creative firms, governmental authorities, and local communities in cultural ecosystems, also launching partnerships with other sectors (Borin & Donato, 2015). Governance systems and management models for these ecosystems must reflect this approach to maximise the potential of the cultural sector as a whole; they are essential for fostering culture-led, long-term growth (Bonet & Donato, 2011). The necessity to increase and structure exchanges and links between diverse organisations is also one of the recurrent theoretical topics within these frameworks.

Another line of inquiry investigated the links between cultural entrepreneurial ecosystems and territory, particularly cities. On the one hand, they talked of their potential of cultural and creative ecosystems to be the guiding principles of processes of city revitalisation or reviving urban areas (Grandadam et al., 2013; He & Gebhardt, 2014). On the other hand, they asserted that the ecosystem approach is necessary to cultivate a 'fertile soil' in cities and territories where creativity might thrive through specialised procedures, policy tools, and relationships between important stakeholders that could improve, foster, and activate the potential of creative and cultural actors (Cohendet et al., 2010; Grandadam et al., 2013). The argument on cultural ecosystems and territory, on the other hand, appears to be predominantly centred on urban regeneration and development (Andres & Chapain, 2013; Comunian et al., 2010; Cooke & Lazzeretti, 2008; Pascual, 2016; Florida, 2005 and 2017; Duxbury, 2015; Kagan & Hahn, 2011). Nonetheless, there are prominent examples of cultural ecosystems flourishing in rural areas (Markusen, 2007; Borin & Paunovic, 2016; Borin & Jolivet, 2021).

The pandemic, however, seems to question the traditional approaches to cultural ecosystems and in particular its links with the territory, mainly due to the thriving of online access to cultural contents and requests for digital cultural material (Radermecker, 2021). In these increasingly digital spaces, new ecosystem actors could explore their roles in the ecosystem (Artal-Tur et al., 2018; Martínez Pino, 2018; Zbuchea et al., 2021), further expressing a period of rupture of cultural ecosystems' traditional structures (Radermecker, 2021; Tan & Tan, 2021). This highlights the necessity to reflect on the evolving concept of cultural ecosystem and reinterpret it, considering the several experiences and innovations that emerged during the pandemic.

This chapter aims at exploring this topic and the management and governance implications related to the implementation of new cultural ecosystem perspective in the cultural and creative field and in society.

3 Research Design and Methodology

The aim of this chapter is to investigate how emerging trends are impacting the management and governance of cultural and creative organisations and interpret them through an ecosystem perspective/understanding to identify management and governance practices for the sector in the future.

According to the approach delineated by Gustafsson (2017), we based our research on qualitative material and therefore secondary sources. This includes scientific articles, case studies, and other literature, as well as several examples of documentary material available from online sources (Gustafsson, 2017, p. 1). We decided to use secondary data related to significant case studies (e.g. websites, reports, press articles) and classify them into four main categories. The cases selected for the analysis present key examples of how the identified trend occurs in practice and have been chosen to satisfy two main criteria: (1) they support and make visible the reasoning at the basis of the theoretical categorisation, and (2) they provide evidence of the changes and emerging patterns cultural and creative organisations have been facing in recent years and during the pandemic period.

The results of this analysis are presented in the following sections, first focusing on the categorisation of data into the above-mentioned four main trends and then discussing the data on the basis of the ecosystem framework and management and governance lessons we can learn from them.

4 Research Results

As analysed in the previous sections, in recent years, the literature has highlighted how the CCI sector has undergone various processes of change and innovation. According to our analysis, these changes can be grouped into four main trends that will be examined in the following sections: (1) a strong push towards digitisation and online engagement, with the consequent creation of ad hoc initiatives and the experimentation of new strategies and digital tools, (2) a new understanding of the role of the audience in terms of audience engagement and development processes, (3) a growing attention to the role of culture and creativity in the sustainable development debate, and (4) an intensifying interdependence and interrelation between CCIs and other sectors of the economy and society.

The first trend we identified is the *exploration of new online and digital possibilities* (Giannini & Bowen, 2022; Kist, 2020; Orlandi, 2020; Rendell, 2021; Samaroudi et al., 2020). This trend is of course not a prerogative of the cultural and creative sector, but rather a shift of the whole economy, in line with the international policy strategies (e.g. the European digital strategy, EC, 2020b). In this overall framework, for several years now, cultural heritage institutions have been moving towards digital, not only implementing digital activities in training and education but also digitising their collections and contents or developing online communication and audience engagement initiatives. Among the many examples that can be named in recent years, we could mention the online education initiatives of Sotheby's (UK), the digitisation of the artworks of the *Rijksmuseum* in Amsterdam (the Netherlands),[1] or the streaming service of the *Wiener Staatsoper* in Vienna (Austria), launched already in 2014,[2] or innovative communication via social media implemented by the *Musée d'Orsay* in Paris[3] (France). However, perhaps the most challenging aspect of digitisation has been going on at the strategic, partnership development and organisational level. Indeed, some cultural institutions have created departments and teams dedicated to developing digital strategies. Among them,

[1]The Rijksmuseum di Amsterdam in the Netherlands completely digitised its collections and made them downloadable for free from its website. More information is available at https://www.rijksmuseum.nl/en/rijksstudio?ii=0&p=0&from=2022-01-09T16%3A45%3A42.4486724Z, accessed 10 June 2022.

[2]The Wiener Staatsoper, after a successful trial season in 2014, launched a programme of 40 world-wide live streams of its opera and ballet performances in 2015, which still successfully continues and has further developed during the pandemic. More information is available at https://www.wiener-staatsoper.at/, accessed 10 June 2022.

[3]The Musée d'Orsay worked with artist Jean-Philippe Delhomme to establish an unusual social media communication for over 2 years starting from the beginning of 2020. Every Monday for the entire year, Delhomme was given full artistic authority over the museum's official Instagram account to do anything he liked. Delhomme's illustrations reimagined the museum's historic masterpieces, as well as putting prominent nineteenth-century artists in a contemporary context by giving them a voice on social media.

the case of Qatar Museums[4] stands out, a governmental organisation that oversees several museums and numerous heritage sites in Qatar. Qatar Museums has recently embarked on a digital transformation journey and is a shining example of how to create a digital strategy from scratch. In February 2020, the digital transformation included the creation of an in-house human-centred digital experience team to bundle digital competences and functions and oversee the multi-constituent organisation's digital evolution and long-term planning. Innovation, digital goods, digital contents, digital productions, research, and analytics are all under the purview of the digital experience team. In developing their strategy, it was critical to enlist the help of international specialists and consultants, especially for training and upskilling local team members. Working with management and organisation technologies like Trello, Monday, Microsoft Teams, Planner, and ClickUP to support working across time zones was regarded as necessary when organising their international team. Similarly, the National Gallery Singapore (China)[5] from its creation has aimed to become a globally recognised, forward-thinking art museum with an avant-garde technological strategy. The Gallery elaborated a digital strategy developed in collaboration with an IT and Digital Advisory Board that aimed to reach several objectives related to digitisation: ensuring the National Gallery stays ahead of the curve in terms of technology advancement; adoption of new technology and digital initiatives that enhance visitor experience as well as productivity; strategic development and implementation of the IT master plan and digital strategy; strategic partnerships and opportunities; identifying, monitoring, and addressing risks related to the overall IT and digital strategy; and continuous assessment of technical choices and planning. Data analytics played an important role in the museum's progress, with the digital department identifying a variety of indicators to analyse developments and evaluate the effectiveness of implementing technology tools. Furthermore, in autumn 2020, the Gallery teamed up with the renowned integrated marketing communications network *Mullen Lowe* to create an innovative digital ecosystem, optimise the Gallery's digital solutions, and improve visitor experiences dramatically. The results were promising, with the launching of *#GalleryAnywhere*, an online collections search gateway that allows anybody to explore and search for artworks, library, and archival content from any location on Earth.

The year 2021, moreover, saw a further acceleration of online activity for the entire cultural and creative industries sector, with important innovative ideas also brought about by the development in the use of big data, AI (artificial intelligence), and new blockchain technologies. For instance, big data and artificial intelligence were at the centre of a collaboration between the *Istituto Musei di Bologna* (Bologna Museums Institute) and *ENEA* (National Agency for New Technologies, Energy and

[4]More information about Qatar Museums is available at https://qm.org.qa/en/, accessed 10 June 2022.

[5]The National Gallery Singapore, which first opened its doors in 2015, is a visual arts organisation that houses the world's biggest public collection of Singapore and Southeast Asian modern art. More information about the gallery is available at https://www.nationalgallery.sg/, accessed 10 June 2022.

Sustainable Economic) in Italy for the project *ShareArt* aiming at understanding visitors' experience in municipal museums.[6] Cameras were installed in panels inside the museum, focusing on the visitor's face. They use artificial intelligence to detect, among other things, where visitors are most interested in a work of art, and the innovative system is able to 'measure' the visitor's experience by sharing different data obtained by recording and examining certain indicators and using a 'big data' application capable of extracting information by examining large amounts of different data. The year 2021 saw a boom in NFTs (non-fungible tokens) and the entry of several cultural institutions (that are usually reluctant to experiment) into the metaverse. NFTs, initially only popular in the art market sector, have become part of the experience of other cultural and creative industries over the last 2 years, from the music industry with the creation of Kingships and their signing of a contract with Universal to museums. Aside from digitally native institutions like the Museum of Crypto Art,[7] more traditional cultural institutions like the König Galerie[8] and Sotheby's[9] have entered the metaverse and built digital structures to house artworks. Following some high-profile NFT purchases and 'The Ethereal Aether', a virtual exhibition of NFT art held in November 2021, also Russia's State Hermitage Museum (Russia), is set to unveil its own 'digital avatar' in the metaverse, entitled 'Celestial Hermitage'.[10] In 2021, the British Museum (UK) partnered with NFT Platform *La Collection* to digitise more than 200 artworks by Katsushika Hokusai.[11] The artworks will be made available across five scarcity levels, according to common NFT sales practices. The British Museum will make no further primary sales on these artworks after 31 January 2022; after that date, all NFT sales will be via the secondary market.

The strong acceleration of digitalisation influenced the second of the trends we identified, that of a *new approach to the concept of audience development and engagement* (Mak et al., 2021; Radermecker, 2021; Zagrebelnaia, 2021); this trend is also a process that has been ongoing for several years but has accelerated considerably during the pandemic. The pandemic in fact brought about new ways and online tools for engagement and pushed cultural organisations to use these tools

[6]More information about the project is available at https://innovationorigins.com/en/museum-bologna-in-italy-uses-big-data-to-view-visitors/ (in English) or https://ict.enea.it/event/beni-culturali-al-via-il-progetto-shareart-per-misurare-il-gradimento-delle-opere-darte-con-lintelligenza-artificiale/ (in Italian), accessed 10 June 2022.

[7]More information is available at https://museumofcryptoart.com/, accessed 10 June 2022.

[8]More information is available at https://www.koeniggalerie.com/exhibitions/32598/koenig-galerie-app/, accessed 10 June 2022.

[9]More information on Sotheby's metaverse is available at https://metaverse.sothebys.com/hackatao-queens-and-kings, accessed 10 June 2022.

[10]The Celestial Hermitage and the exhibition held in the virtual space of the Hermitage can be visited at https://celestialhermitage.ru/en/, accessed 10 June 2022.

[11]More information is available at https://lacollection.io/?gclid=CjwKCAiA866PBhAYEiwANkIneA31a260cNTxmTvTjerWFfE7d2Digt-G4l-7e7dGx5m847ej8nHc2BoCn94QAvD_BwE, accessed 10 June 2022.

to enhance the role of the audience in the co-creation of contents, initiatives, and strategies. There have been various stages of development of this process. During the first lockdowns, the audience was still perceived as a passive actor, a recipient of the communication processes of cultural heritage institutions. In those periods, the main concern of cultural institutions was to maintain a relationship with their audiences through the sharing of repertoire works, previously recorded or digitised, and the 'displacement' of traditional online activities. This is the case, for example, of the Royal Opera House (London, UK),[12] one of the most famous opera houses in the world, which immediately took care to make its operas and performances available on its online platform and social channels, thus enabling it to maintain a relationship with its audience despite the impossibility of a live performance. Similarly, the London Transport Museum (London, UK), famous for its guided tours through the subterranean capital of the UK,[13] during the first lockdown decided to introduce online guided tours. They became a huge success since the museum managed to increase the content and the value of the tours by displaying London stations and other areas with severely limited access, where public onsite tours had previously been impossible. As a result, the museum was able to entice repeat visitors as well as attract new ones, maintaining and even increasing one of their main sources of revenues. Furthermore, the Museums + Heritage Awards 2021 have shortlisted the Hidden London tours in the category 'The Best Use of Digital UK'.

In a second phase, cultural institutions tried to create a more interactive connection with their audiences, entering their daily lives and creating initiatives and formats that could also attract different audiences. An example of this phase is the initiative of Los Angeles County Museum of Art (LACMA) (USA). At a time when interpersonal and working relationships were mainly mediated through video conferences, LACMA made it possible to download its most famous works of art in a format suitable for video conferences.[14] This initiative, so simple yet so pervasive in the new reality in which people were living, was a huge success and invited new audiences to interpret works of art not only as distant contents but also as part of their own life experience and everyday life. At the same time, many cultural institutions have tried to create online content linked to their traditional activities but inviting viewers to take a more active role in the enjoyment of the content, either on a more playful level tending towards online gamification or with experiences more oriented towards the development of a critical spirit and co-creation of content through digital modes. An example of gamification is the Newark Museum of Art in New Jersey, USA, which devised a series of online interactive formats to call for a more active

[12]The Royal Opera House launched in March 2020 a free programme of broadcasts and online contents, accessible also via Facebook and YouTube. More information on this programme is available at https://www.roh.org.uk/news/the-royal-opera-house-launches-a-programme-of-free-online-content-for-the-culturally-curious-at-home, accessed 10 June 2022.

[13]The guided tours are still offered and accessible from the website of the London Transport Museum: https://www.ltmuseum.co.uk/hidden-london/virtual-tours, accessed 10 June 2022.

[14]The programme and free downloads are accessible at https://www.lacma.org/videobackgrounds, accessed 10 June 2022.

participation of its visitors. Among them is the immersive escape room game 'Escape from the Ballantine Mansion',[15] which is based on a historic house that is now a museum wing. As further development of the relationship with audiences, we can consider the example of the 'Invisible Archaeology'[16] exhibition at the Egyptian Museum of Turin (Italy), which was designed to be enjoyed both in-person and fully online. The exhibition attempts to encourage critical participation and partial co-creation of content; when enjoyed online, the exhibition allowed visitors to actively create the exhibition itinerary and choose the content of the exhibition, thus simulating the curatorial experience. In a way, these experiences could be considered among the 'preparatory' activities for the next phase of audience development in which citizens and communities are increasingly involved in defining the development of initiatives and policies and in co-designing a new post-pandemic scenario. Among them, we can mention the launch of the New European Bauhaus,[17] promoted by the European Commission in 2020. 'The New European Bauhaus is a creative and interdisciplinary initiative, convening a space of encounter to design future ways of living, situated at the crossroads between art, culture, social inclusion, science and technology. It brings the Green Deal to our living places and calls for a collective effort to imagine and build a future that is sustainable, inclusive, and beautiful for our minds and for our souls' (Source: NEB website). The first year of the initiative (October 2020–June 2021) was entirely dedicated to a co-design phase in which potential stakeholders were invited to engage in dialogue and present via an online platform while the European Commission was 'active listening'. This co-design phase helped the initiative to gather contributions and ideas on how to shape beautiful, sustainable, and inclusive living spaces which formed the basis of the subsequent implementation phase that translated into proposals for action with a policy document, a Commission Communication. In parallel, the initial group of partners was sent to work on widening the scope of the new European Bauhaus and building an open community that articulates international cooperation with local action. This example shows us how audiences are interpreted in a broader way, not merely as visitors/participants in cultural activities but as stakeholders of an enlarged and hybrid cultural ecosystem (online and in-person) overcoming the traditional physical and geographical interaction barriers of analogue systems.

A third trend in the development of the cultural and creative sector during these last years, accelerated once again in the pandemic period, is the one that has led to the *progressive rethinking on its link with the theme of sustainability and sustainable development* (Dell'Ovo et al., 2021; Roigé et al., 2021). The theme of the contribution of the cultural and creative field to sustainable development had already been

[15] The initiative, along with several other online activities proposed by the museum, is available at https://www.newarkmuseumart.org/escape-ballantine-house, accessed 10 June 2022.

[16] More information is available at https://museoegizio.it/en/explore/exhibitions/archeologia-invisibile/, accessed 10 June 2022.

[17] More information is available at https://europa.eu/new-european-bauhaus/about/about-initiative_en, accessed 10 June 2022.

discussed and analysed for years (CHCfE, 2015; Soini & Dessein, 2016), especially in reference to the recognition of culture as the fourth pillar of sustainable development. It has also been at the centre of important policy strategies of development, such as the European Green Deal (EC, 2019) and the New European Bauhaus Initiative (EC, 2021). However, during the pandemic, we have witnessed an evolution of this relationship that has led to the repetition of the traditional question of not only how the cultural and creative sector can contribute to sustainable development but also how the sector is pursuing sustainability approaches in its practices in line with the goals of the 2030 Agenda. Regarding the first question, it is emblematic how the climate discourse has been greatly enriched by reflections on the role of culture in the 2030 Agenda. This is visible, for instance, in the developments of the United Nations Climate Change Conference. In the United Nations Climate Change Conference (COP21), held in Paris in 2015, cultural events were part of the programme of initiatives accompanying the meeting, but the role of culture didn't specifically enter the climate debate. During the recent COP26 meeting held in 2021 in Glasgow (UK),[18] although there is still a substantial absence of artists at the table, we can note the presence of several debates and round tables reflecting on the contribution of this sector to sustainable development, especially in terms of awareness-raising. An interesting example was the session entitled 'How Can Art and Culture Accelerate a Just Transition?', promoted by the non-profit association Little Sun, founded by the famous contemporary artist Anish Kapoor, which not only illustrated the cultural projects in support of the climate promoted by the association but also established an important dialogue with artists, curators, and environmentalists on the subject.

Regarding the second point, the last 2 years have witnessed a growing awareness by cultural organisations of the importance of promoting cultural practices that meet the criteria of sustainability and including performance indicators related to these aspects in annual reports and financial statements. The Thyssen-Bornemisza Museum in Madrid (Spain), for example, has proposed a sustainability report, where it shows its commitment to the 2030 Agenda and indicates its accountability of its progress in implementing sustainable economic, social, and environmental management practices.[19] In 2021, also the IVAM (Instituto Valenciano de Arte Moderno (Valencian Institute of Modern Art, Spain)) drew up a multi-annual Sustainability Plan 2021–2023[20] that included a series of strategic lines in this area. In this sustainability plan, a table sets out specific sustainability goals for the museum and then indicates to which SDGs each of the institutional sustainability goals contributes. This makes it possible to see which SDGs the IVAM will contribute the most through this plan. At a macro level, in the UK, 'environmental

[18] More information is available at https://ukcop26.org/, accessed 10 June 2022.

[19] The full Sustainability Report of the Thyssen-Bornemisza Museum is available at www.thyssenbornemisza.org

[20] The full sustainability plan is available at https://www.ivam.es/es/ivam-obert-portal-de-transparencia/informacion-corporativa/, accessed 10 June 2022.

responsibility' is one of the four investment principles the Arts Council England (ACE) 2020–30 Strategy[21] is built around; from 2021–2022 onwards, all NOPs in the country, including cultural and creative organisations, are invited to submit an environmental report. Also in the UK, the 'Framework for Environmental Action for Museums' was a pioneering handbook developed by the British organisation Julie's Bicycle[22] in collaboration with ACE with the objective to encourage sustainability reflections inside museums but also to implement management control systems that could assess and account for sustainability approaches inside museums. The framework is structured into 12 key areas showing different levels of environmental practice, from least to most ambitious.

Finally, a fourth trend strongly emphasised during the pandemic is that of *rethinking the connections between the cultural and creative sector and other sectors of the economy and society.* Over the last 2 years, a stronger cooperation between culture and the digital industries has been created, as already highlighted in the previous paragraphs. However, this was not the only connection that was highlighted during the pandemic. There was also an increase of partnerships between the cultural and creative sector, and several industries related to a whole range of products, from clothing and cosmetics to technology, food, and beverage items, have been established in terms of IP licenses; museums license the cultural IP of their collections to diversify their revenue streams in a moment of 2020 when they recorded a drop of visitors and related income losses.[23] A very strong link has also been established between culture and creativity and the field of health and well-being (Tan & Tan, 2021; Tubadji, 2021). This discourse emerged in the demand for digital contents by people in lockdown periods, considering them crucial to ensuring psychological well-being. This need could be related to already existing practices of art therapy, but the pandemic period led to a further investigation of its potential and triggered the creation of actual policies and programmes aiming at reinforcing the connection between culture, creativity, and health. Of particular interest is the Arts in Health programme, developed in Australia by the Women's & Children's Hospital Foundation.[24] This Foundation, whose statutory aim is to promote a better quality of life for hospital patients, has included a programme of exhibitions organised by famous artists which can be visited by its patients and medical staff

[21] More information about the Arts Council England 2020–30 Strategy is available at https://www.artscouncil.org.uk/publication/our-strategy-2020-2030, accessed 10 June 2022.

[22] Julie's Bicycle is a not for profit located in London mobilising the arts and culture to take action on the climate and ecological crisis. The 'Framework for Environmental Action for Museums' elaborated by Julie's Bicycle is available at https://juliesbicycle.com/

[23] Visitor figures for the world's top 100 art museums dropped by 77% in 2020, resulting in significant decreases in revenue streams (e.g. the Louvre and the Metropolitan Museum of Art have reported losses of $110 million and $150 million, respectively).

[24] More information on this programme is available at https://wchfoundation.org.au/arts-in-health/, accessed 10 June 2022.

especially during the pandemic. Similarly, the Arts in Hospital Centre Glasgow[25] has partnered with several local bodies, including NHS Greater Glasgow and Clyde and Glasgow University, to create a range of online artistic and creative content and resources in response to the COVID-19 pandemic for use by patients, their families, and medical staff. The lockdown also further strengthened the collaboration between the cultural and creative sector and the educational sector; during the lockdown, many cultural institutions tried their hand at producing educational content for various audiences (Choi et al., 2020; Tranta et al., 2021; Puspasari et al., 2020). Significant is the experience of the Barnes Foundation in Philadelphia (USA)[26] which implemented, as a response to the lockdown, a rich programme of online art history education aimed primarily at adults but also at students of different ages. This initiative received very positive feedback from participants and was also a huge success from a financial point of view, proving to be an important revenue stream in the foundation's budget.[27] Also, the already mentioned National Gallery Singapore (China) developed during the pandemic an online education programme for young children, called 'Small Big Dreamers',[28] through its Keppel Centre for Arts Education. It consisted of content from the museum's collections combined with a series of activities that could be carried out in every home with simple materials and specifically proposed for teachers and families as art education activities for young children.

Finally, the traditional link between the cultural and creative sector and the tourism sector was further reflected upon during the pandemic (Flew & Kirkwood, 2021; Tully, 2020). In particular, during the periods of timid reopening between pandemic peaks, cultural institutions proved to be interesting drivers of initiatives for a more sustainable (and safe) redistribution of tourism in peripheral and rural areas and for the creation of programmes aimed at non-traditional tourist groups and which more closely meet the inclusion needs of a society changed by the experience of the last 2 years. This was particularly evident in the initiatives implemented in two Italian cities 'afflicted' by overtourism in the pre-pandemic period, such as Florence and Venice. In Florence, interesting is the programme *Terre degli Uffizi*[29] promoted for the period 2021–2026 by the Uffizi Museum in Florence (one of the most visited

[25] More information on this programme is available at https://www.artinhospital.com/, accessed 10 June 2022.

[26] The initiative is still available at https://www.barnesfoundation.org/classes, accessed 10 June 2022.

[27] About one-third of the previous year's adult education revenue was reached already within the first month of shifting to online. Overall, the earnings have reached more than $650,000 since its launch last March—more than double the revenue that in-person classes had generated in 2019. Increase was reached not only in revenue but also in outreach, gaining more than 2800 students from 39 states and 6 countries. Alongside students, the Barnes has also gained contribution from international curators and art historians.

[28] More information is available at https://www.nationalgallery.sg/see-do/programme-detail/347658918/smallbigdreamersathome, accessed 10 June 2022.

[29] More information is available at https://www.uffizi.it/en/terre-degli-uffizi, accessed 10 June 2022.

museums in Italy and in the world) which aims at exhibiting works from the museum's collection in exhibitions organised in minor museums located in the Tuscan territory outside the traditional tourist destinations. The initiative aims at enhancing the whole area around the city of Florence, in connection with two other initiatives such as *Uffizi Diffusi* and *Piccoli Grandi Musei*, both part of a long-term strategy to enhance the Tuscan territory and its cultural heritage and to foster a widespread and more sustainable tourism development in the area. In Venice, the 'Hidden Jewels of Venice'[30] initiative, promoted by the *Fondazione Venezia Servizi*, aims to involve minor cultural sites outside the traditional tourism circuits in the creation and promotion of unusual itineraries that relieve the pressure on the city's most overcrowded and visited places, thus proposing a more sustainable tourism model.

5 Discussion: Managerial Implication of the Ecosystem Approach to CCS and the Shift from Cultural Ecosystems to Culture-Led Ecosystems

The previous section illustrated significant experiments in the cultural and creative field that point to the need to rethink/revisit the traditional cultural ecosystem concept. They indeed reflect a reinterpretation of the sector as an interconnected, hybrid cultural ecosystem where culture and creativity are the leverages for different approaches to societal development. New cultural ecosystems are no longer strongly bound to a specific local dimension but open to stakeholders from the whole world that could interact through digital technologies. The contribution of audiences in the ecosystem is facilitated by online interactions and encouraged at different levels in the co-designing of contents, programmes, and policies. In this emerging cultural ecosystem, cultural organisations closely collaborate with digital industries, health organisations, education institutions, and tourism to guide innovative developments. In this ecosystem, culture not only does contribute to awareness-raising for sustainable development but also actively incorporates sustainability measures in cultural practices.

This emerging ecosystem is therefore not simply a cultural ecosystem but rather a 'culture-led ecosystem', where cultural and creative organisations could potentially be leaders in a paradigm shift impacting sustainable development processes. As a result, cultural and creative organisations are required to adapt to their changing role and reflect on how to change their management and governance practices to fruitfully perform this task. The examples we provided gave us some interesting insights on this issue.

First, the exploration of online and digital technologies, big data, NFT, and metaverse during these last 2 years indicates that the cultural and creative industries

[30] More information is available at https://www.gioiellinascostidivenezia.it/en/, accessed 10 June 2022.

are increasingly suitable to operate in a hybrid environment, physical but all the time more digital, that transcends the geographical limits and operates globally. This requires changes related to the organisational structures of cultural organisations. On the one hand, there is an increasing need for specialised staff to develop and manage these tools that can be either developed in-house, outsourced, or implemented through partnerships. Developing them in-house implies not simply the creation of related roles and jobs inside the cultural organisations (such as in the cases of Qatar museums or the National Gallery Singapore). It also requires training and education of the cultural organisations' staff who are going to use these technologies (curators, education department staff, communication offices, etc.) to empower them in liaising with experts and encourage them to take responsibility in using these tools. If outsourced or implemented through partnerships with external organisations (such as in the case of the Hermitage Museum or the British Museum), there is a need to correctly structure these collaborations as long-term tools for the common advancement of the diverse actors involved. In both cases, there is a need for integration of competences and digital infrastructures (i.e. specific software and working platforms) that could enable the different actors to effectively work together as an ecosystem of shared competences, experiences, and projects. It must be noted that these innovations could prove particularly difficult for small- and middle-sized cultural organisations, which lack sufficient organisational structures both for developing new competences and for playing a strong role in negotiating digitisation partnerships. The role of larger cultural institutions is therefore crucial in including smaller organisations in the emerging ecosystems and legitimising their role in appropriate cooperation schemes.

Second, the emerging approaches to audience engagement indicate that citizens and communities could be not simply stakeholders of the cultural ecosystem but become active partners of cultural organisations in steering an emerging culture-led environment. To overcome geographical and communication barriers and implement participation and co-creation mechanisms, the use of digital tools is necessary. This implies that cultural organisations include these tools from the beginning in the design of their participatory processes. In turn, once again, this entails creating the digital infrastructures that enable these interactions to take place and the staff or partnerships necessary to develop and manage them, including both large, small, and middle cultural institutions.

Third, rethinking the growing links between the cultural and creative sector and other sectors of the economy (e.g. education, health, and tourism) is paving the way for expanding the traditional cultural ecosystems through a broader ecosystem approach that is even more multisectoral and multi-stakeholder. This, however, indicates the necessity to introduce changes in the governance of both cultural organisations and other field organisations to fully develop the potential of these alliances. On the one hand, cultural organisations could benefit from having representatives of other fields and organisations in their governance boards, since they could bring significant inputs in terms of strategic and innovative developments. On the other hand, representatives of cultural organisations could also be included in the boards of collaborating sectors, bringing their point of view and ideas to contribute to their strategic development. Furthermore, the creation of governance structures that reflect on development from a multisectoral perspective could be an interesting possibility to explore. The tourism sector could be particularly affected by this paradigmatic shift. Indeed, whereas the creation of territorial governance structures is already a common practice for local and tourism development and enhancement (as the literature on creative cities, regional regeneration, and destination governance highlighted), these structures are usually quite limited in terms of sectors involved and usually promoted mainly by local authorities or DMOs (Destination Management Organisations). In the new approach we are proposing, cultural organisation could take the lead in a deeper governance reflection, not only working alongside tourism organisations but also becoming the pivot for broader alliances and the inclusion of other economic and social partners. This will require again an inclusive, collaborative approach uniting culture, tourism, and territories for overcoming a rooted mindset that interprets culture as instrumental to tourism and sees tourism mainly in its economic dimension and not in its identity-creating capacity.

Finally, the increasing attention to the link between culture and sustainability points to the necessity to include cultural leaders as part of the discussion on sustainability and to further reflect on how to make the cultural sector more sustainable and account for sustainability practices inside cultural organisations. This second point is particularly relevant: it should not be limited just to introducing sustainability practices in cultural organisations but also implies a restructuring of the managerial approaches and a better accountability of sustainability in the reporting mechanisms of cultural organisations. The attempts to develop sustainability reports and include sustainability indicators developed by pioneering cultural organisations, such as the IVAM or Thyssen-Bornemisza museums, are paving the way for introducing new management control approaches that are planning, measuring, and reporting the sustainability impact of cultural organisations. The process of introducing different performance measurement systems in cultural organisations needs to be coupled with specific accounting, reporting, and management competence in their staff and, more generally, with a mindset shift in the traditional accounting practices of cultural organisations.

6 Conclusions

The aim of this chapter is to analyse the concept of cultural ecosystem in its latest developments and to discuss its potential for reinterpreting the cultural and creative sector in a moment of profound change as that we have experience over the last 2 years. The paper also examines the implications in terms of management and governance of the sector, based on an analysis of multiple case studies of emerging innovative approaches.

From the literature review, it emerges that the ecosystem concept has become increasingly popular, not just in natural sciences but also in the economic and management field, with specific references to business and entrepreneurial ecosystems. In the cultural and creative field, the concept of ecosystem has been used in a stream of studies related to the ecology of culture approach that paved the way to the cultural ecosystem concept. Cultural ecosystems are ecosystems of cooperation developed at the meso-territorial level, implemented mainly through multi-stakeholder partnerships between public authorities, cultural organisations, and related sectors. They have a strong territorial link, created because of a shared cultural identity, and they consider citizens and communities as important ecosystem components. The cultural ecosystem approach is strictly linked to the studies of creative cities and destination governance approaches.

By means of a series of in-depth examples, we highlighted that over the last 2 years, many changes have occurred in the cultural and creative sector, concerning the use of digital technologies and online tools, audience participation, the synergies with other sectors, and the role of cultural organisations in the sustainable development discourse. These trends indicate that the cultural ecosystem concept should be updated and evolve towards a 'culture-led' ecosystem approach in which cultural organisations could be the leaders in processes of change. This implies a restructuring of the management and governance models of the cultural and creative field, which aimed at enabling to operate in a hybrid (digital-physical) environment and enhancing the role of peoples and audiences, the long-term cooperation with other sectors, and the implementation of sustainability approaches.

This research provides first insights on an ongoing complex phenomenon, contributing to the literature on cultural ecosystems and identifying interesting tools for professionals and practitioners in the cultural and creative sector and beyond. However, more research with a quantitative approach and international and longitudinal dimension could be useful in further exploring these topics and validating the research results.

Additionally, further developments of this research could explore the implications of the culture-led ecosystem concept in terms of policy initiatives, funding, and the potential influence that this approach might have in implementing cooperation with other sectors.

References

Aage, T., & Belussi, F. (2008). From fashion to design: Creative networks in industrial districts. *Industry and Innovation, 15*(5), 475–491.

Andres, L., & Chapain, C. (2013). The integration of cultural and creative industries into local and regional development strategies in Birmingham and Marseille: Towards an inclusive and collaborative governance? *Regional Studies, 47*(2), 161–182.

Artal-Tur, A., Briones-Peñalver, A. J., & Villena-Navarro, M. (2018). Tourism, cultural activities and sustainability in the Spanish Mediterranean regions: A probit approach. *Tourism & Management Studies, 14*(1), 7–18.

Ballico, C. (2017). Another typical day in this typical town: Place as inspiration for music creation and creative expression. *Australian Geographer, 48*(3), 349–363.

Barker, V. J. (2018). *On the creative ecosystem: Investigating ecosystem approaches through the creative sector*. Retrieved June 10, 2021 from https://pure.coventry.ac.uk/ws/portalfiles/portal/41003377/Barker_2018_PhD.pdf

Barker, V. (2020). The democratic development potential of a cultural ecosystem approach. *Journal of Law, Social Justice and Global Development, 24*, 1–14. https://doi.org/10.31273/LGD.2019.2405

Basole, R. C. (2009). Visualization of interfirm relations in a converging mobile ecosystem. *Journal of Information Technology, 24*(2), 144–159.

Blanke, J., & Chiesa, T. (2013). *The travel & tourism competitiveness report 2013. WEF–the world economic forum*. Retrieved on February 27, 2020 from http://p285140.mittwaldserver.info/sites/default/files/upload/WEF_TT_Competitiveness_Report_2013.pdf

Bonet, L., & Donato, F. (2011). The financial crisis and its impact on the current models of governance and management of the cultural sector in Europe. *ENCATC Journal of cultural management and policy, 1*(1), 4–11.

Bonet, L., & Négrier, E. (2018). The participative turn in cultural policy: Paradigms, models, contexts. *Poetics, 66*, 64–73.

Borin, E. (2017). *Public-private Partnership in the Cultural Sector: A comparative analysis of European models*. PIE Peter Lang.

Borin, E., & Donato, F. (2015). Unlocking the potential of IC in Italian cultural ecosystems. *Journal of Intellectual Capital, 16*(2), 285–304.

Borin, E., & Jolivet, C. (2021). Entrepreneurial cultural ecosystems in rural contexts: Some insights from rural cultural centers in France. In P. Demartini et al. (Eds.), *Cultural initiatives for sustainable development* (pp. 321–341). Springer.

Borin, E., & Paunovic, I. (2016). The case of louvre-lens: Regional regeneration through cultural innovation. In H. Pechlaner, M. Gon, & M. Valeri (Eds.), *Innovazione, sostenibilità e competitiità. Teoria ed evidenze empiriche (pp. 67–76)*. Giappichelli Editore.

CHCfE Consortium. (2015). *Cultural heritage counts for Europe*. International Cultural Centre.

Choi, M., Tessler, H., & Kao, G. (2020). Arts and crafts as an educational strategy and coping mechanism for Republic of Korea and United States parents during the COVID-19 pandemic. *International Review of Education, 66*(5), 715–735.

Cohendet, P., Grandadam, D., & Simon, L. (2010). The anatomy of the creative city. *Industry and Innovation, 17*(1), 91–111.

Comunian, R., Chapain, C., & Clifton, N. (2010). Location, location, location: Exploring the complex relationship between creative industries and place. *Creative Industries Journal, 3*(1), 5–10.

Cooke, P. N., & Lazzeretti, L. (Eds.). (2008). *Creative cities, cultural clusters and local economic development*. Edward Elgar Publishing.

Costanza, R., d'Arge, R., de Groot, R., Farber, S., Grasso, M., Hannon, B., Limburg, K., Naeem, S., O'Neill, R. V., Paruedo, J., Raskin, R. G., Sutton, P., & van den Belt, M. (1997). The value of the world's ecosystem services and natural capital. *Nature, 15*(387), 253–260.

Dell'Ovo, M., Dell'Anna, F., Simonelli, R., & Sdino, L. (2021). Enhancing the cultural heritage through adaptive reuse. A multicriteria approach to evaluate the Castello Visconteo in Cusago (Italy). *Sustainability, 13*(8), 4440. https://doi.org/10.3390/su13084440

Drake, G. (2003). 'This place gives me space': Place and creativity in the creative industries. *Geoforum, 34*(4), 511–524.

Duxbury, N. (2015). European cities as cultural projects: Where is culture in urban sustainability policy? In *Culture and sustainability in European cities* (pp. 89–105). Routledge.

European Commission. (2019). *Communication from the commission to the European parliament, the European council, the council, the European economic and social committee and the committee of the regions–The European Green Deal*. Retrieved on June 10, 2021 from https://eur-lex.europa.eu/resource.html?uri=cellar:b828d165-1c22-11ea-8c1f-01aa75ed71a1.0002.02/DOC_1&format=PDF

European Commission. (2020a). *Communication from the commission to the European parliament, the European council, the council, the European economic and social committee and the committee of the regions–A New Industrial Strategy for Europe*. Retrieved October 11, 2021 from https://eur-lex.europa.eu/legal-content/EN/TXT/PDF/?uri=CELEX:52020DC0102&from=EN

European Commission. (2020b). *Shaping Europe's digital future.* Retrieved June 10, 2021 from https://ec.europa.eu/info/sites/default/files/communication-shaping-europes-digital-future-feb2020_en_4.pdf

European Commission. (2021). *Communication from the commission to the European parliament, the council, the European economic and social committee and the committee of the regions–New European Bauhaus: Beautiful, sustainable, together.* Retrieved June 10, 2021 from https://europa.eu/new-european-bauhaus/system/files/2021-09/COM%282021%29_573_EN_ACT.pdf

Flew, T., & Kirkwood, K. (2021). The impact of COVID-19 on cultural tourism: Art, culture and communication in four regional sites of Queensland, Australia. *Media International Australia, 178*(1), 16–20.

Florida, R. (2005). *Cities and the creative class*. Routledge.

Giannini, T., & Bowen, J. P. (2022). Museums and digital culture: From reality to Digitality in the age of COVID-19. *Heritage, 5*(1), 192–214.

Grandadam, D., Cohendet, P., & Simon, L. (2013). Places, spaces and the dynamics of creativity: The video game industry in Montreal. *Regional Studies, 47*(10), 1701–1714.

Gustafsson, J. (2017). *Single case studies vs. multiple case studies: A comparative study.*

He, J. L., & Gebhardt, H. (2014). Space of creative industries: A case study of spatial characteristics of creative clusters in Shanghai. *European Planning Studies, 22*(11), 2351–2368.

Henry, N., & Barker, V. (2016, October). Policy and the location of value creation: The creative microenterprise and the creative ecosystem. In *19th Uddevalla Symposium: Geography, Open Innovation, Diversity, and Entrepreneurship* (pp. 205–218).

Holden, J. (2004). Creative Reading: Young people, reading and public libraries. *Demos.*

Holden, J. (2015). *The ecology of culture.* Arts and Humanities Research Council.

Iansiti, M., & Levien, R. (2004). *Keystones and dominators: Framing operating and technology strategy in a business ecosystem* (pp. 1–82). Harvard Business School.

Isenberg, D. J. (2010). How to start an entrepreneurial revolution. *Harvard Business Review, 88*(6), 40–50.

Isenberg, D.J. (2011a). Introducing the entrepreneurship ecosystem: Four defining characteristics. *Forbes.* http://www.forbes.com/sites/danisenberg/2011/05/25/introducing-the-entrepreneurship-ecosystem-four-defining-characteristics/

Isenberg, D. (2011b). *The entrepreneurship ecosystem strategy as a new paradigm for economy policy: Principles for cultivating entrepreneurship.* Babson Entrepreneurship Ecosystem Project.

Isenberg, D. (2014). *What an entrepreneurship ecosystem actually is.* Harvard Business Review blog, Retrieved January 7, 2022 from https://hbr.org/2014/05/what-an-entrepreneurial-ecosystem-actually-is

Jansen, H. H., Denise van Blitterswijk, & Jurgen Nieuwenhuijsen. (2017). *SME Ecosystem in the sultanate of Oman, sensing and seizing opportunities in the tourism industry.* E. C. o. Entrepreneurship.

Kagan, S., & Hahn, J. (2011). Creative cities and (un) sustainability: From creative class to sustainable creative cities. *Culture and Local Governance*, 11–27.

Kinder, T., Stenvall, J., Six, F., & Memon, A. (2021). Relational leadership in collaborative governance ecosystems. *Public Management Review, 23*(11), 1612–1639.

Kist, C. (2020). Museums, challenging heritage and social media during COVID-19. *Museum and Society, 18*(3), 345–348.

Lash, S. U., & Urry, J. (1994). *Economies of signs and space.* Sage.

Mak, H. W., Fluharty, M., & Fancourt, D. (2021). Predictors and impact of arts engagement during the COVID-19 pandemic: Analyses of data from 19,384 adults in the COVID-19 social study. *Frontiers in psychology.* Advanced online publication. https://doi.org/10.3389/fpsyg.2021.626263.

Markusen, A. (2007). A consumption base theory of development: An application to the rural cultural economy. *Agricultural and Resource Economics Review, 36*(1), 9–23.

Martínez Pino, J. (2018). The new holistic paradigm and the sustainability of historic cities in Spain: An approach based on the world heritage cities. *Sustainability, 10*(7), 2301.
Moore, J. F. (1996). *The death of competition: Leadership & strategy in the age of business ecosystems*. Harper Business.
Oakes, T. (2019). Making cultural cities in China: Governance, state-building, and precarious creativity. *Environment and Planning A: Economy and Space, 51*(1), 178–186.
Orlandi, S. D. (2020). Museums web strategy at the Covid-19 emergency times. *DigitCult-Scientific Journal on Digital Cultures, 5*(1), 57–66.
Pascual, J. (2016). Culture as a pillar in sustainability: The best is yet to come. *Economia della Cultura, 26*(4), 557–572.
Peltoniemi, M. (2006). Preliminary theoretical framework for the study of business ecosystems. *E: CO, 8*(1), 10–19.
Protogerou, A., Kontolaimou, A., & Caloghirou, Y. (2017). Innovation in the European creative industries: A firm-level empirical approach. *Industry and Innovation, 24*(6), 587–612.
Puspasari, S., Suhandi, N., & Iman, J. N. (2020). Augmented reality development for supporting cultural education role in SMB II museum during Covid-19 pandemic. In *2020 Fifth International Conference on Informatics and Computing* (ICIC) (pp. 1–6). IEEE.
Radermecker, A. S. V. (2021). Art and culture in the COVID-19 era: For a consumer-oriented approach. *SN Business & Economics, 1*(1), 1–14.
Rendell, J. (2021). Staying in, rocking out: Online live music portal shows during the coronavirus pandemic. *Convergence, 27*(4), 1092–1111.
Roigé, X., Arrieta-Urtizberea, I., & Seguí, J. (2021). The sustainability of intangible heritage in the COVID-19 era–resilience, reinvention, and challenges in Spain. *Sustainability, 13*(11), 5796. https://doi.org/10.3390/su13115796
Samaroudi, M., Echavarria, K. R., & Perry, L. (2020). Heritage in lockdown: Digital provision of memory institutions in the UK and US of America during the COVID-19 pandemic. *Museum Management and Curatorship, 35*(4), 337–361.
Soini, K., & Dessein, J. (2016). Culture-sustainability relation: Towards a conceptual framework. *Sustainability, 8*(2), 167.
Spigel, B. (2017). The relational organization of entrepreneurial ecosystems. *Entrepreneurship Theory and Practice, 41*(1), 49–72.
Spigel, B., & Harrison, R. (2018). Toward a process theory of entrepreneurial ecosystems. *Strategic Entrepreneurship Journal, 12*(1), 151–168.
Stam, E. (2015). Entrepreneurial ecosystems and regional policy: A sympathetic critique. *European Planning Studies, 23*(9), 1759–1769.
Sutherland, I., & Gosling, J. (2010). Cultural leadership: Mobilizing culture from affordances to dwelling. *The Journal of Arts Management, Law, and Society, 40*(1), 6–26.
Tan, M. K. B., & Tan, C. M. (2021). Curating wellness during a pandemic in Singapore: COVID-19, museums, and digital imagination. *Public Health, 192*, 68–71.
Tranta, A., Alexandri, E., & Kyprianos, K. (2021). Young people and museums in the time of covid-19. *Museum Management and Curatorship, 36*(6), 632–648.
Tubadji, A. (2021). Culture and mental health resilience in times of COVID-19. *Journal of Population Economics, 34*(4), 1219–1259.
Tully, G. (2020). Are we living the future? Museums in the time of Covid-19. *Tourism facing a pandemic: From crisis to recovery*, 229–242.
Zagrebelnaia, A. (2021). State of the art fair post-covid: Audience development strategies in a post-digital context. *European Journal of Cultural Management and Policy, 11*(2), 34–51.
Zbuchea, A., Romanelli, M., & Bira, M. (2021). Through the public's lens: Are museums active members of society? An investigation during the COVID-19 pandemic. In P. Demartini, L. Marchegiani, M. Marchiori, & G. Schiuma (Eds.), *Cultural initiatives for sustainable development* (pp. 61–95). Springer.

Toward Sustainable Innovation in Tourism: The Role of Cultural Heritage and Heritage Communities

Mara Cerquetti and Annamaria Romagnoli

Abstract This chapter provides a theoretical contribution investigating how the current notion of cultural heritage and the emerging role of heritage communities can trigger and lead sustainable innovation in tourism. After tracing the evolution of international conventions on cultural heritage since the Second World War, the research discusses the implications of the paradigm shift introduced by the Faro Convention (2005) for heritage and tourism management. The analysis focuses on the current extended concept of cultural heritage—open and inclusive, dynamic and relational—and on the involvement of heritage communities in heritage recognition, safeguard and enhancement. Ten years after the Faro Convention entered into force, the research results suggest that the international framework currently provided by UNESCO and the Council of Europe can help to design heritage-aware destinations and develop heritage-aware tourism. In the context of the COVID-19 pandemic, this approach also opens up unprecedented potentialities for slow and proximity tourism.

1 Introduction: Grabbing Cultural Heritage, a Slippery Concept

If culture is a difficult concept, "at once utterly familiar, but also complex and hard to fully pin down" (O'Brien, 2014, 2; see also Rosenstein, 2016), it is even more difficult to define the meaning of cultural heritage. On the one hand, this concept is affected by the evolution of the meaning of culture; on the other, given its relational and interactive nature, it is defined and constantly reshaped by the relationships

M. Cerquetti (✉)
Department of Education, Cultural Heritage and Tourism, University of Macerata, Macerata, Italy
e-mail: mara.cerquetti@unimc.it

A. Romagnoli
Department of Political Sciences, Communication and International Relations, University of Macerata, Macerata, Italy
e-mail: a.romagnoli15@unimc.it

E. Borin et al. (eds.), *Cultural Leadership in Transition Tourism*, Contributions to Management Science, https://doi.org/10.1007/978-3-031-14121-8_3

people have with their environment. Cultural heritage is not an intrinsic property of objects, but a construction, an invention, "a matter of perception" (Schofield, 2016, 3). It is:

> something which is created by contemporary thought and social action; heritage, simply does not exist as a set of things [. . .] but rather as a multitude of approaches and perspectives that refer to how we think about and choose to manage those things. (Schofield, 2015, 198)

As a social practice (Poirrier, 2004), cultural heritage is not only varied and dynamic (Van Velthem et al., 2017) but also variable and uncertain. Its definition is always historically determined and cannot be equated with the much greater fixity to which, for example, a legal-administrative definition is bound (Galasso, 1991, 8). It is undergoing constant transformation: it has changed over time, is currently changing and will change in the future. It can mean anything and everything (Fairclough et al., 2008).

As argued by Bottesi (2018, 127), the notion of cultural heritage should be based on an epistemological uncertainty acting as a stimulus for the search for new certainties, in a virtuous cycle that does not reify ideologies and experiences, but promotes their continuous transformation according to the needs of society. This implies moving away from the perception of cultural heritage as a static, "immutable document" (Lattanzi, 2013, 5) and accepting its artificiality as the result of an ongoing process of selection, interpretation and transmission (Battilani, 2017).

In this context, a two-way process emerges for analysing the identification of cultural heritage, its determinants and effects. On the one hand, the notion of cultural heritage is culturally determined by human beings as members of a society; on the other, the notion of cultural heritage affects how society approaches cultural heritage identification and management. Following Smith (2006, 14), we can conclude that:

> the way people talk about, discuss and understand things, such as "heritage", have a material consequence that matters. In addition, not only is discourse "used" to do things by actors, but discourses also do things to actors and are productive independently of actors [. . .]. As such, discourses are "inherently positioned", and so the collection of ideas, concepts, and categorizations regarding heritage give rise to different ways of "seeing" the social practice of managing "heritage" according to the positions of social actors. (Smith, 2006, 14)

However incomplete and full of gaps they may be, legal definitions provide an example of this two-way process: they acknowledge cultural changes in society when they are mature but also provide a new framework for recognising and managing cultural heritage, thus triggering a continuous cycle (Fig. 1).

Sharing this approach, the present research aims to understand:

1. If and how the international legal framework has affected cultural heritage perception and management.
2. Its implications for innovation in tourism management.

Specifically, the extended idea of cultural heritage supported by UNESCO and the active role of heritage communities introduced by the Faro Convention may contribute to designing heritage-aware destinations and developing heritage-aware

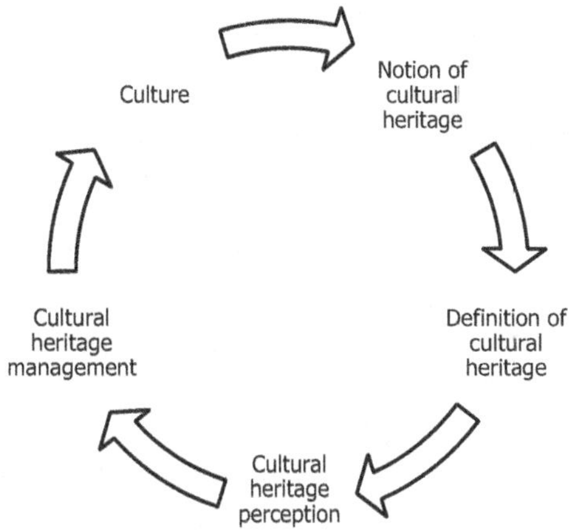

Fig. 1 The relationship between culture, cultural heritage and its management (own elaboration)

tourism. In the context of the COVID-19 pandemic, this approach can also bring about unprecedented challenges and opportunities for slow and proximity tourism.

The chapter is structured as follows: Sections 2 and 3 go over the evolution of the different approaches to cultural heritage of UNESCO and the Council of Europe's Conventions. Section 4 analyses the paradigm shift introduced by the current legal framework, and Sect. 5 discusses possible implications for sustainable innovation in tourism.

2 From Tangible to Intangible Heritage in UNESCO Conventions

The concept of cultural heritage started out at the centre of the international debate, mainly in relation to its protection. The discussion intensified after the Second World War, which had caused extensive damage to cultural heritage. Regulations for the defence of cultural heritage in cases of conflict already existed (Article 56 of the second Hague Convention on the laws and customs of war on land, 1899; Articles 27, 28, 56 of the fourth Hague Convention on the laws and customs of war on land, 1907; Roerich Pact, 1935), but the international community considered whether more effective protection measures were necessary. For this reason, under the aegis of UNESCO, the Convention for the Protection of Cultural Property in the Event of Armed Conflict was signed in the Hague in 1954.

One of the most interesting concepts introduced by the 1954 Convention was that of "cultural property". This is the first time the term is used in an international document. In previous papers, a very restricted category of cultural objects was considered worthy of protection, which can be summarised as "historical monuments", "works of art" and "works of science". The 1954 Convention formalised a broader term that included:

> (a) movable or immovable property of great importance to the cultural heritage of every people, such as monuments of architecture, art or history, whether religious or secular; archaeological sites; groups of buildings which, as a whole, are of historical or artistic interest; works of art; manuscripts, books and other objects of artistic, historical or archaeological interest; as well as scientific collections and important collections of books or archives or of reproductions of the property defined above;
>
> (b) buildings whose main and effective purpose is to preserve or exhibit the movable cultural property defined in sub-paragraph (a) such as museums, large libraries and depositories of archives, and refuges intended to shelter, in the event of armed conflict, the movable cultural property defined in sub-paragraph (a);
>
> (c) centers containing a large amount of cultural property as defined in sub-paragraphs (a) and (b), to be known as 'centers containing monuments'. (Art. 1)

The adoption of provisions for the protection of cultural assets in the event of armed conflict initiated reflections and an international regulatory process on the subject of cultural heritage that would change with the times. In the years following the 1954 Convention, the need arose to create a further system of protection to be applied in times of peace, which took into account the new threats brought by the social and economic evolution of modern civilisation. The international community's awareness stemmed from reflections made after certain historic events: the construction of the Aswan Dam, which would have led to the destruction of the Egyptian temples of Abu Simbel, if there had been no international intervention, and the 1966 floods which caused considerable damage to the cultural heritage of Venice and Florence and saw the mobilisation of many volunteers from all over the world. These episodes brought to light the existing limitations in the conservation and protection of cultural heritage, which were based exclusively on national law, and gave a strong input to the idea of an international instrument to fill these gaps (Albanesi, 2010).

In those years, an environmental consciousness began to spread. This initiated a process of awareness-raising of what would later become a global imperative: sustainable development (Senatore, 2013). In this context, certain phenomena such as industrialisation, pollution, anthropisation and urbanisation were considered risks, and the incontrovertible relationship between cultural goods and their surrounding environment was understood.

For this reason, in 1972, the UNESCO General Conference approved in Paris the Convention Concerning the Protection of the World Cultural and Natural Heritage. This document is the first legal recognition of the strong connection between culture and nature, as it established a common regime of conservation and protection for the most important human-made productions and the extraordinary works of nature (Francioni & Lenzerini, 2008). The definitions of "cultural heritage" and "natural heritage" offered by the 1972 Convention declared that only what is of "Outstanding Universal Value" was worthy of protection.

Therefore, UNESCO attached considerable importance to the traditional categories of art history, the great monuments and the great civilisations. This concept of cultural heritage mostly included European, Christian and monumental culture (Bortolotto, 2007).

During and after the debate that led to the drafting of the 1972 Convention, non-Western countries criticised this definition, because their cultural expressions were under-represented. UNESCO therefore had to change its paradigms, making adjustments that made it possible to overcome the "almost colonialist" vision, replacing aesthetic and artistic values with anthropological ones (Bortolotto, 2007). Since the 1990s, the meaning and the definition of cultural heritage have undergone changes and reconsiderations that have broadened the concept of cultural heritage and profoundly renewed the understanding and approach to heritage, now considered an anthropological whole.

This process of change (which actually began in the late 1960s and led to the 1989 UNESCO Recommendation on the Safeguarding of Traditional Culture and Folklore, the 2001 UNESCO Universal Declaration on Cultural Diversity and the Istanbul Declaration in 2002) culminated in the UNESCO Convention for the Safeguarding of the Intangible Cultural Heritage, approved in Paris in 2003. This Convention expanded the meaning of cultural heritage, to include intangible heritage, which is defined in Article 2:

> The "intangible cultural heritage" means the practices, representations, expressions, knowledge, skills—as well as the instruments, objects, artefacts and cultural spaces associated therewith—that communities, groups and, in some cases, individuals recognize as part of their cultural heritage. This intangible cultural heritage, transmitted from generation to generation, is constantly recreated by communities and groups in response to their environment, their interaction with nature and their history, and provides them with a sense of identity and continuity, thus promoting respect for cultural diversity and human creativity. For the purposes of this Convention, consideration will be given solely to such intangible cultural heritage as is compatible with existing international human rights instruments, as well as with the requirements of mutual respect among communities, groups and individuals, and of sustainable development.

Previous documents conceived cultural heritage as fixed, static and monumental, while the 2003 Convention extended this concept to consider it to be a living expression to be safeguarded in its evolution (Bortolotto, 2007); the living, localised, self-reproducing character of the intangible cultural heritage is therefore accentuated (Zagato, 2008).

Apparently, the main innovation of the definition of heritage would always be the passage from the tangible to the intangible, but, in reality, the Convention introduces other very important conceptual innovations. In fact, as time goes on, cultural expressions are no longer considered *objects*, but *processes* (Bortolotto, 2007): the focus thus shifts from heritage as a document to living heritage, from documentation to transmission, from products to processes, from objects to subjects and, more importantly, from the outstanding to the representative (Cerquetti, 2015, 253). Moreover, the Convention focused on individuals, not only as actors who make intangible heritage visible but also as having an active role in recognising what is representative (Cerquetti, 2015). However, as recently pointed out (Lázaro Ortiz &

Jiménez de Madariaga, 2022), despite the focus on communities and groups, "it is the Member States that dominate management processes" (Duvelle, 2017, 41).

The 2003 Convention introduced several concepts which were further developed by the subsequent Convention on the Protection and Promotion of the Diversity of Cultural Expressions, signed in Paris in 2005.

However, compared to previous documents, the focus of the 2005 Convention shifted from identity practices to activities that produce goods or objects of exchange and consumption capable of circulating and sharing cultural expressions. In fact, this was the first treaty in which explicit reference was made to the ambivalent nature of cultural—but also economic and commercial—goods, services and activities (Cunsolo, 2019).

In addition to the deep interest in commercial and economic aspects, the 2005 Convention is an important instrument that seeks to respond to the needs of society by addressing the increasing plurality that has resulted from globalisation (UNESCO, 2007).

3 Cultural Heritage and Heritage Communities within the Council of Europe's Conventions

In the years following the establishment of the Council of Europe (1949), when the Old Continent was still paying the price of the conflict, an awareness was reached: culture is at the heart of all relations between people and between nations, and a lack of awareness of that and other people's culture can lead to terrible consequences (Council of Europe, 2005a). Moreover, the reconstruction of cultural heritage after the Second World War had the support of a large portion of the population who, despite experiencing years of economic difficulties, retained a very strong link with their heritage (Guermandi, 2018). For these reasons, in 1954, the Council of Europe signed the European Cultural Convention in Paris. The Convention's aim of greater unity between countries was to be achieved by promoting dialogue, mutual understanding and mutual appreciation of the differences between Member States (Pavan-Woolfe, 2019). These actions would be pursued through the protection of the cultural heritage, defined for the first time as "Common European Heritage".

From an ideological point of view, the European Cultural Convention was absolutely innovative. The world was back from the exploitation of culture and cultural heritage for the aggrandisement of the nation state and totalitarian political regimes. In contrast, the 1954 Convention identified culture and cultural heritage as instruments for the enhancement of shared European values, thus overcoming the overemphasis on difference that had characterised the whole of the previous century, in favour of the commonality of the peoples of Europe (Pavan-Woolfe, 2019). The document considered free cultural expression and education in culture and cultures to be a source of enrichment for society and an antidote to nationalist regressions, with the goal of recognising common European heritage. Moreover, it encouraged a cooperation policy that would lead the Member States to "concerted action in promoting cultural activities of European interest".

Starting with the 1954 Convention, various legal texts have been drawn up to foster international cooperation and preserve cultural heritage as a bearer of European identity (Pavan-Woolfe, 2019).

For this reason, in 1969, the first European Convention on the Protection of the Archaeological Heritage was opened for signature in London by Member States. The document's main purpose was to define the measures to be adopted for the protection of archaeological areas; to regulate the circulation of archaeological assets, thus counteracting illicit trade; and, finally, to develop educational action for training and public awareness in these issues (Council of Europe, 1969). European archaeological heritage was considered a fundamental source of knowledge of civilisation, and the document asserts that, as it is the oldest resource in the history of Europe, the moral responsibility for the protection of this heritage should primarily be in the interest of the State under whose jurisdiction it falls but that it is also the concern of all European countries.

A few years later, attention shifted to architectural heritage, considered the most immediately visible part of cultural heritage and the part most affected by the devastation of war. Thus, in 1975, the Congress of Amsterdam, held at the end of the European Architectural Heritage Year, gave rise to the European Charter of Architectural Heritage. This document was very important as it emphasised the need to integrate conservation into urban planning, to preserve groups of historic buildings and their environment and to take heritage into account as part of economic and social life (Council of Europe, 1985).

This document is a charter; therefore, it only promoted general principles and recommendations, with no obligation to enact them (Smargiassi, 2012). Consequently, in 1985, the Convention for the Protection of the Architectural Heritage of Europe was signed in Granada. This treaty provided a definition for architectural heritage and restated the definition of integrated conservation introduced in the 1975 Charter. The aim of the Granada Convention was therefore to promote protection of monuments not as stand-alone objects, but rather as ones that have a relationship with their physical and social surroundings. The Council of Europe document encouraged a greater understanding between the protection of the architectural heritage and urban planning, thus establishing good practices of safeguard that were immediately considered to be a guarantee of more responsible development and improved quality of life (Pavan-Woolfe, 2019).

Following the conceptual outline of the Granada Convention, the Council of Europe thought it was necessary to update the 1969 London Convention, and, in 1992, a new European Convention for the Protection of the Archaeological Heritage was signed in Valletta. In fact, since 1969, the issues involved in protecting and promoting Europe's archaeological heritage had changed considerably, mainly because of rapid anthropisation and the consequent extensive urban planning interventions. The Valletta Convention required States that ratified to involve archaeologists in the entire planning process, thus ensuring a multidisciplinary collaboration between urban planners and regional planners (Council of Europe, 1992). The

Valletta Convention is ultimately more than one simple archaeological equivalent of the Granada Convention since it represents an important step from an approach focused on objects to one focused on activity (Fojut, 2009).

In the mid-1980s, the landscape approach was widely accepted, as was the concept of integrated conservation. In 2000, the European Landscape Convention was opened for signature in Florence by the Member States of the Council of Europe. Imitating the Granada and Valletta Conventions, the Florence treaty also introduced conceptual innovations. It went beyond the sectoral approach to the landscape as a function of an integrated and cross-cutting vision and considered the landscape as a "collective good extended to the whole territory" (Della Torre et al., 2016). It also introduced the perceptual factor of the population in recognition of the landscape and encouraged the involvement of communities in landscape policies. Finally, it extended the very concept of landscape to embrace not only the outstanding but also the everyday and even degraded landscapes. The European Landscape Convention, therefore, highlighted the need for conservation of landscape heritage as an integral part and characterising element of the cultural context and also the need to create a dialogue between social reality and development and planning strategies (Pavan-Woolfe, 2019).

Following the destruction of cultural goods during the Yugoslav Wars of the 1990s, the need to implement more stringent policies for the protection of cultural heritage became more evident. The Council of Europe reflected on the new meaning that cultural heritage had assumed. It was a source of identity and differentiation in a context in which the risk of uniformity, brought by globalisation, was very real (Directorate of Culture, Cultural and Natural Heritage, 2001). It was thought that a new document should be drawn up to consider this concept of cultural heritage while taking into account the international attention on intangible heritage. In 2005, the Convention on the Value of Cultural Heritage for Society was drawn up in Faro, Portugal.

One of the most interesting innovations of the Convention is that of cultural heritage as:

> a group of resources inherited from the past which people identify, independently of ownership, as a reflection and expression of their constantly evolving values, beliefs, knowledge and traditions. It includes all aspects of the environment resulting from the interaction between people and places through time.

The use of the word "resources" radically changes the concept of cultural heritage, which had always been defined as "assets". The Faro Convention clarifies that cultural heritage is not static, unchanged, but it exists to be used and, in parallel, recognises that it can bring benefits (Fairclough, 2009).

Furthermore, the new definition particularly focuses on the interactive nature of cultural heritage, recognising that it is defined, and redefined, by human actions, which is why it cannot be seen as static and immutable (Council of Europe, 2005b). In fact, another fundamental word is "people" because it introduces a collective profile into the notion of cultural heritage (Carmosino, 2013), indicating that it is a collective and shared phenomenon (Pinton, 2019). Therefore, people are placed at the centre of the cultural process, and the Faro Convention defines another interesting concept in this regard, that of heritage community:

> a heritage community consists of people who value specific aspects of cultural heritage which they wish, within the framework of public action, to sustain and transmit to future generations.

These are diverse, open and spontaneous groups formed by people who share common goals (Pinton, 2019) and who are not necessarily united by parameters such as citizenship, ethnicity, profession, social class or religion (Dolff-Bonekämper, 2009).

The Faro Convention also establishes innovative principles. In fact, it introduces the right to cultural heritage as a human right and bestows capital importance on democratic participation in cultural heritage, that is, in every activity in the "process of identification, study, interpretation, protection, conservation and presentation of the cultural heritage; public reflection and debate on the opportunities and challenges which the cultural heritage represents" (Article 12).

This Convention supports the potential of cultural heritage as a resource with multidimensional value, which can be used for sustainable development and for improving quality of life (Montella & Petraroia, 2016).

To achieve these principles, it is necessary to involve all actors as part of a process of democratic governance and shared responsibility.

4 A Paradigm Shift in Cultural Heritage Management

The international framework provided by UNESCO and the Council of Europe at the beginning of the twenty-first century has significantly broadened the concept of cultural heritage and has reshaped its relationship with human beings.

If the 2003 UNESCO Convention finally recognised the intangible cultural heritage (Vecco, 2007, 2010), the Faro Convention went so far as to move beyond the distinction between tangible and intangible to provide a holistic definition of cultural heritage (Thérond, 2009, 10). Compared to the UNESCO definition, the latter innovatively introduced the environmental dimension, by recognising "the inseparability of human and natural influences in landscape and also the inherited context of cultural environment" (Vícha, 2014, 32). This definition not only includes ordinary and everyday places that are significant for society (Schofield, 2015, 199) but also focuses on society's continuous and changing relationship with the environment.

As far as human beings are concerned, as already argued in Sect. 2, UNESCO has recognised cultural expressions as processes, pointing out the "living" nature of the intangible cultural heritage, which is continuously created and recreated by communities, groups or individuals. Going further, as already stated, the Faro Convention introduced the collective profile of cultural heritage, by using "the word people instead of individuals (everyone, persons...)" (Zagato, 2015, 144). Indeed, the Faro

Convention's main innovation is the introduction of the concept of heritage community defined as a variable geometry concept (Vícha, 2014, 32), resulting from a shared interest in a subject or having a geographical basis. It can be transnational and multidimensional, with no reference to specific ethnic groups or fixed communities (Van der Auwera & Schramme, 2011, 72). For this reason, it would be better to talk about heritage communities, in the plural, to include diverse communities that not only have an interest or a stake but also should and can have a voice (Schofield, 2016, 1).

This emerging framework enacts the paradigm shift suggested by Loulanski in Loulanski, 2006: from monuments to people, from objects to functions and from preservation per se to purposeful preservation, sustainable use and development. Specifically, the focus on heritage communities supports a people-centred approach in which people are:

1. *Performers*, such as when giving life to intangible cultural heritage.
2. *Recipients*, since cultural heritage is *for* people.
3. *Decision-makers*, because of the increasing involvement of citizens in the ongoing process of defining and managing cultural heritage—as a right and shared responsibility.

People have a right not only to cultural heritage and cultural life but also to heritage-building (Schofield, 2016, 6): they can benefit from cultural heritage and contribute to its enrichment (Faro Convention, Article 4).

Montella (2016, 15) effectively summarised the innovations introduced by the Faro Convention as a complete and profound reversal: of authority (from top to bottom), of the object (from the exceptional to everything), of value (from value itself to use value) and, therefore, of purpose (from "museification" to enhancement).

The implementation of this approach would lead to a multidimensional innovation of the current culture of protection, involving (1) the *object*, which can no longer consist of things, but rather contexts; (2) the *subject*, which cannot be the State alone, but must involve civil society and other stakeholders; and (3) the *process*, by replacing administrative constraints with civic bonds (Covatta & Cammelli, 2013, 295).

Moreover, the approach promoted by the Faro Convention overcomes the false dichotomy between cultural heritage *protection* and *enhancement*, highlighting the virtuous relationship between the two functions (Petraroia, 2010; Golinelli, 2015; Saviano et al., 2018; Cerquetti et al., 2019). This paradigm shift is based on the assumption that enhancement can help protection, not only by attracting more resources for cultural heritage preservation but also by increasing awareness of the value of cultural heritage among more people. If people recognise and appreciate the value of cultural heritage, they may be more willing to preserve it for future generations, by being actively involved in ensuring it is maintained and safeguarded over time. In addition, more resources could be submitted for cultural heritage safeguard, both directly, through the willingness of citizens to pay for it (e.g. donations, tickets, memberships, etc.), and indirectly, through an increase in public and private expenditure on cultural heritage, since policymakers, sponsors and other stakeholders would be more willing to invest in cultural heritage if it created value for a large number of people (Cerquetti & Ferrara, 2018). This virtuous relationship can trigger the heritage cycle, whereby an increasing understanding of cultural heritage and its environment "leads to people valuing it more and as a consequence caring for it better. An environment cared for will be enjoyed, and enjoyment normally brings a thirst to learn more, thus completing the cycle" (Thurley, 2005, 26).

As suggested by the Faro Convention, this process can boost cultural diversity, contemporary creativity, peaceful coexistence, quality of life and human and sustainable development, thus creating multi-stakeholder and multidimensional value. Montella (2009) identifies three different types of use value that can be drawn from cultural heritage:

1. *Presentation value*, an information-based value, corresponding to the communication of the meanings of cultural heritage, i.e. authentic, aesthetic, symbolic, social, historical and documentary value (Throsby, 2001). In the current context, the cultural value should be perceived not only by a limited cluster of users but also by all potential users.
2. *Landscape value*, a systemic value concerning the safeguard of the environment and territory through environmental policies and city planning (Golinelli, 2015). A system of preventive and planned conservation and active conservation, rather than end-of-pipe interventions, protective restrictions and bans, would promote quality of life and well-being, including environment and landscape quality.
3. *Production value*, referring to market uses flowing from cultural heritage and profit for businesses operating in different sectors, e.g. restoration, publishing, tourism, construction, real estate business, performing arts, etc. (Greffe, 2009). Moreover, when stratified through time in a specific context, cultural heritage influences value creation and competitive advantage, especially for "made in" firms.

In summary, we can visualise the current paradigm shift(s) concerning subject, process, object, value and objective as in Fig. 2:

The paradigm shift designed by UNESCO and the Council of Europe has consequences for heritage identification and management. On the one hand, as the meaning of cultural heritage has expanded, cultural heritage is everywhere, and everything or almost everything can become heritage (Heinich, 2009; Harrison, 2013). Particularly in times of crisis, heritage becomes a way of living in the present and finding an anchor point, which leads to a risk of commodification of culture and heritagisation of local inhabitants, such as in eco-museums (Hartog, 2020). On the other hand, communities with a stake in cultural heritage are increasing both in number and in nature. This scenario points to the need to integrate different needs, concerns and interests, through the cooperation between professionals, policymakers and experts from different disciplinary fields, e.g. scholars and heritage officials, as guides and facilitators, and local people, as experts with strong ties to a place (Schofield, 2015, 208).

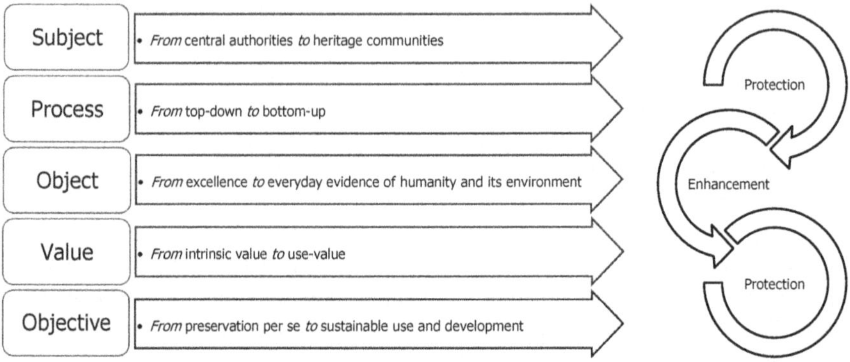

Fig. 2 The paradigm shift triggered by the current international legal framework (own elaboration)

Audiences play an increasingly active role in the multitude of communities involved in heritage matters. According to the model suggested by Bollwerk (2015) for measuring levels of engagement in digital public archaeology, audiences have been changing from crowds to communities. The relationship with them can move away from *publicity*, meaning dissemination of information, to *participation*, that is, their active involvement in projects, *collaboration*, where they have a voice in the definition and improvement of high-level objectives and in the design, and *co-creation*, where they are equal partners in the design and implementation of a project. As argued by Sokka et al. (2021, 12), the co-creative governance of cultural heritage "implies a hybrid definition of cultural heritage and a higher rate of citizen participation, where citizens contribute knowledge and other resources to solve problems efficiently".

5 Challenges and Opportunities for Sustainable Innovation in Tourism

Ten years after the Faro Convention entered into force (2011), the signing and ratification process is still ongoing.[1] Currently, it is not easy to predict how the Member States will execute the Faro paradigm shift. As already stated (Montella, 2016), its principles may remain mere "words in freedom" with no effect on cultural heritage protection and enhancement. Another risk to take into account is the politically correct ideological orientation that confuses due respect for all people and their differences with "exasperated formalisms", "sugary hypocrisies" and "linguistic conformism", sweetening even difficult and unpleasant realities and ignoring real conflicts (Volpe, 2020, 197).

[1] https://www.coe.int/en/web/conventions/full-list/-/conventions/treaty/199/signatures?p_auth=dg2WfyCT (accessed June 13, 2022)

Aside from any legal and official actions that may follow, there is no doubt that a new approach in thinking and managing cultural heritage has already gained ground, with recognition that cultural heritage exists *thanks to* and *for* people. Finally, in addition to the contribution of cultural heritage to community-building, another significant factor to emerge is the importance of communities in heritage-building. Many bottom-up experiences promoted in the light of Faro or only thanks to its suggestion or evocative power are confirmation of this cultural revolution (Consiglio & Riitano, 2015; D'Alessandro, 2015; Faircoulgh et al., 2015; Tasso, 2017; Cerreta & Giovene di Girasole, 2020; Volpe, 2020). First, the Council of Europe has promoted heritage walks, which are "*cultural routes* aimed at rediscovering territory and its history through direct contact with witnesses: people who lived and live in these regions holding a strong memory of them" (Tasso, 2017, 748). In addition, there are many other examples of "marriage between heritage and citizens" (Volpe, 2019, 107), cases of "civic activism" or "active citizenship" (Cerquetti et al., 2019) and "stories of spontaneous and necessary innovation" (Zabatino, 2015, 25), such as those started in the field of archaeology (Möller, 2019). They are expressions of social innovation through which citizens move from being recipients of a traditional, centralist public intervention to being protagonists and co-designers of interventions (Consiglio & Zabatino, 2015, 72), as a consequence of an unmet social need and the need to identify alternative management models.[2]

Even though the current debate is overladen with rhetoric of community,[3] the role of heritage communities in triggering and leading sustainability in tourism cannot be ignored. Indeed, heritage communities can be agents of innovation. Their active participation can boost their awareness of the value of cultural heritage and its environment, their willingness to preserve and transmit it to other people, which helps them design more authentic experiences and bring more awareness to tourists of cultural heritage and its environment (Fig. 3).

In order to be effective, this approach should be holistic and should integrate all the social, cultural and economic customs of a territory (Bellini et al., 2020), in a framework of sustainable wellness-driven tourism (Pencarelli, 2020). As argued by Volpe (2020, 195), the alliance between cultural heritage and tourism would allow tourism to become more cultured: not more elitist, but better quality, more respectful of landscape and heritage and also slower and offering an authentic, rich, pleasant experience of life and knowledge of the many cultural idiosyncrasies of even the more remote regions of Europe.

Together with the broad concept of cultural heritage outlined in the Faro Convention, this proactive participatory approach opens up untapped opportunities for "minor" cultural heritage. After the COVID-19 pandemic, it may also unlock new

---

[2]The scientific literature provides many methods and tools already experimented in Europe and beyond (Chirikure et al., 2010; Perkin, 2010; Huibin & Marzuki, 2012; Dragouni & Fouseki, 2018; Dragouni et al., 2018).

[3]The term is becoming an obsession and a near-mandatory component to "*feel* good about the work we do" (Waterton & Smith, 2010, 8).

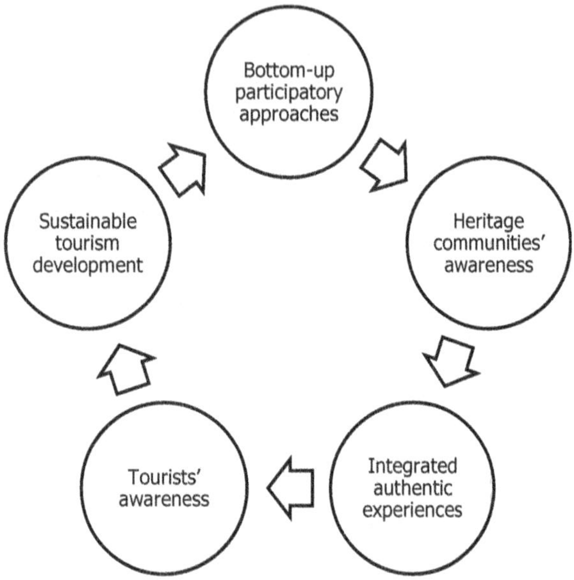

Fig. 3 Heritage-aware destinations for heritage-aware tourism (own elaboration)

paths for proximity tourism (Romagosa, 2020, 3), by uncovering new, as-yet unexplored destinations, and preventing overtourism and the touristification of cities (Franch & Peretta, 2020).

References

Albanesi, A. (2010). Le organizzazioni internazionali per la protezione del patrimonio culturale. In L. Casini (Ed.), *La globalizzazione dei beni culturali* (pp. 29–56). il Mulino.

Battilani, P. (2017). Si fa presto a dire patrimonio culturale. Problemi e prospettive di un secolo di patrimonializzazione della cultura. *Storia e Futuro. Rivista di storia e storiografia on line, 45*, 1–21.

Bellini, N., Campostrini, S., Pasquinelli, C., & Trunfio M. (2020). "Ripartenza" o "rinascimento"? Le politiche del turismo oltre l'emergenza. *Vision–The Post Covid Series*. Accessed June 13, 2022, from https://www.thinktank.vision/images/2020/Paper_Turismo.pdf

Bollwerk, E. (2015). Co-Creation's role in digital public archaeology. *Advances in Archaeological Practice, 3*, 223–234.

Bortolotto, C. (2007). From the "monumental" to the "living" heritage: A shift in perspective. In R. White, & J. Carman (Eds.), *World heritage: Global challenges, local solution*. Proceedings of a conference at Coalbrookdale, 4–7 May 2006 hosted by the Ironbridge Institute (pp. 39–45). Archeopress.

Bottesi, A. (2018). *Patrimonio: incertezza di un concetto in mutazione* (pp. 112–128). Archivio per l'Antropologia e la Etnologia.

Carmosino, C. (2013). La Convenzione quadro del Consiglio d'Europa sul valore del patrimonio culturale per la società. *Aedon*, 1. Accessed June 13, 2022, from http://www.aedon.mulino.it/archivio/2013/1/carmosino.htm

Cerquetti, M. (2015). Dal materiale all'immateriale. Verso un approccio sostenibile alla gestione nel contesto glocale. *Il capitale culturale. Studies on the Value of Cultural Heritage, 2*, 247–269.

Cerquetti, M., & Ferrara, C. (2018). Marketing research for cultural heritage conservation and sustainability: Lessons from the field. *Sustainability, 10*(3), 1–16.

Cerquetti, M., Nanni, C., & Vitale, C. (2019). Managing the landscape as a common good? Evidence from the case of "Mutonia" (Italy). *Land Use Policy, 87*, 1–11.

Cerreta, M., & Giovene di Girasole, E. (2020). Towards heritage community assessment: Indicators proposal for the self-evaluation in Faro convention network process. *Sustainability, 12*, 1–22.

Chirikure, S., Manyanga, M., Ndoro, W., & Pwiti, G. (2010). Unfulfilled promises? Heritage management and community participation at some of Africa's cultural heritage sites. *International Journal of Heritage Studies, 16*(1-2: Heritage and Community Engagement: Collaboration or Contestation?), 30–44. https://doi.org/10.1080/13527250903441739

Consiglio, S., & Riitano, A. (2015). *Sud innovation. Patrimonio culturale, innovazione sociale e nuova cittadinanza*. Franco Angeli.

Consiglio, S., & Zabatino, A. (2015). L'innovazione sociale per la rinascita del patrimonio dimenticato. In S. Consiglio & A. Riitano (Eds.), *Sud innovation. Patrimonio Culturale, Innovazione Sociale e Nuova Cittadinanza* (pp. 69–102). FrancoAngeli.

Council of Europe. (1969). *Explanatory report to the European convention on the protection of the archeological heritage*. Council of Europe Report.

Council of Europe. (1985). *Explanatory Report to the Convention for the Protection of the Architectural Heritage of Europe*. Council of Europe Report.

Council of Europe. (1992). *Explanatory Report to the European Convention on the Protection of the Archaeological Heritage (Revised)*. Council of Europe Report.

Council of Europe. (2005a). *50 years of the European Cultural Convention*. Council of Europe Report.

Council of Europe. (2005b). *Explanatory Report to the Council of Europe Framework Convention on the Value of Cultural Heritage for Society*. Council of Europe Report.

Covatta, L., & Cammelli, M. (2013). I beni e le attività culturali. In *Per il governo del Paese. Proposte di politiche pubbliche* (pp. 295–300). Passigli.

Cunsolo, F. (2019). La Convenzione UNESCO del 2005 sulla protezione e promozione della diversità delle espressioni culturali. In E. Baroncini (Ed.), *Il diritto internazionale e la protezione del patrimonio culturale mondiale* (pp. 223–244). Dipartimento di scienze giuridiche dell'Università di Bologna.

D'Alessandro, A. (2015). La Convenzione di Faro e il nuovo Action Plan del Consiglio d'Europa per la promozione di processi partecipativi. I casi di Marsiglia e Venezia. In L. Zagato & M. Vecco (Eds.), *Citizens of Europe. Culture e diritti* (pp. 77–92). Venezia. https://doi.org/10.14277/6969-052-5/SE-3-4

Della Torre, S., Montella, M., & Sciullo, G. (2016). s.v. Paesaggio. In M. Montella (Ed.), *Economia e gestione dell'eredità culturale. Dizionario metodico essenziale* (pp. 60–63). Cedam.

Directorate of Culture, Cultural and Natural Heritage. (2001). *Report of the Secretary General on the 5th Session of the European Conference of Ministers responsible for Cultural Heritage* (Portoroz, Slovenia, 6–7 April 2001). Accessed June 13, 2022, from https://rm.coe.int/16804e2402

Dolff-Bonekämper, G. (2009). The social and spatial frameworks of heritage–what is new in the Faro convention? In *Heritage and beyond* (pp. 69–83). Council of Europe Publishing.

Dragouni, M., Fouseki, K., & Georgantzis, N. (2018). Community participation in heritage tourism planning: Is it too much to ask? *Journal of Sustainable Tourism, 26*(5), 759–781. https://doi.org/10.1080/09669582.2017.1404606

Dragouni, M., & Fouseki, K. (2018). Drivers of community participation in heritage tourism planning: An empirical investigation. *Journal of Heritage Tourism, 13*(3), 237–256. https://doi.org/10.1080/1743873X.2017.1310214

Duvelle, C. (2017). Aventuras y desventuras de una hermosa convención internacional. *Revista Andaluza de Antropología, 12*, 37–47. https://doi.org/10.12795/RAA.2017.12.02

Fairclough, G. (2009). New heritage frontiers. In *Heritage and beyond* (pp. 29–41). Council of Europe Publishing.

Fairclough, G., Harrison, R., Jameson, J. H., & Schofield, J. (2008). *The heritage reader*. Routledge.
Faircoulgh, G., Dragićević Šešić, M., Rogač Mijatović, L., Auclair, E., & Soini, K. (2015). The Faro convention, a new paradigm for socially–and culturally–sustainable heritage action? Култура/Culture, 8, 9–19, Accessed June 13, 2022, from https://journals.cultcenter.net/index.php/culture/article/view/111
Fojut, N. (2009). The philosophical, political and pragmatic roots of the convention. In *Heritage and beyond* (pp. 13–21). Council of Europe Publishing.
Franch, M., & Peretta, R. (2020). *Turismo, fragilità, emergenze*. McGraw-Hill.
Francioni, F., & Lenzerini, F. (2008). *The 1972 world heritage convention. A commentary*. Oxford University Press.
Galasso, G. (1991). Definire i beni culturali. *Spunti e appunti. Economia della cultura, 1*(1), 8–23.
Golinelli, G. M. (Ed.). (2015). *Cultural heritage and value creation. Towards new pathways*. Cham.
Greffe, X. (2009). Heritage conservation as a driving force for development. In *Heritage and beyond* (pp. 101–112). Council of Europe Publishing.
Guermandi, M. P. (2018). La Convenzione di Faro: il difficile cammino di un altro concetto di patrimonio. *IBC*, 36(4). Accessed June 13, 2022, from http://rivista.ibc.regione.emilia-romagna.it/xw-201804/xw-201804-a0003
Harrison, R. (2013). *Heritage. Critical approaches*. Routledge.
Hartog, F. (2020). Il patrimonio, una nozione per i tempi di crisi. *il Mulino, 4*, 571–585.
Heinich, N. (2009). *La fabrique du patrimoine. De la cathédrale à la petite cuillère*. Maison des Science de l'Homme.
Huibin, X., & Marzuki, A. (2012). Community participation of cultural heritage tourism from innovation system perspective. *International Journal of Services Technology and Management, 18*(3/4), 105–126.
Lattanzi, V. (2013). Towards a Museum of Possible Worlds. In S. Ferracuti, E. Frasca, & V. Lattanzi (Eds.), *Beyond modernity: Do ethnographic museums need ethnography?* (pp. 3–20). Espera.
Lázaro Ortiz, S., & Jiménez de Madariaga, C. (2022). The UNESCO convention for the safeguarding of the intangible cultural heritage: A critical analysis. *International Journal of Cultural Policy, 28*(3), 327–341. https://doi.org/10.1080/10286632.2021.1941914
Loulanski, T. (2006). Revising the concept for cultural heritage: The argument for a functional approach. *International Journal of Cultural Property, 13*, 207–233. https://doi.org/10.1017/S0940739106060085
Montella, M. (2009). *Valore e valorizzazione del patrimonio culturale storico*. Electa-Mondadori.
Montella, M. (2016). La Convenzione di Faro e la tradizione culturale italiana. In P. Feliciati (Ed.), *La valorizzazione dell'eredità culturale in Italia, Atti del convegno di studi in occasione del 5° anno della rivista (Macerata, 5–6 novembre 2015)* (pp. 13–17). Eum.
Montella, M., & Petraroia, P. (2016). s.v. Valorizzazione. In M. Montella (Ed.), *Economia e gestione dell'eredità culturale. dizionario metodico essenziale* (pp. 112–117). Cedam.
Möller, K. (2019). Will they or won't they? German heritage laws, public participation and the Faro convention. *European Journal of Post-classical Archaeologies, 9*, 199–217.
O'Brien, D. (2014). *Cultural policy. Management, value and modernity in the creative industries*. Routledge.
Pavan-Woolfe, L. (2019). L'azione del Consiglio d'Europa in materia culturale dal 1954 alla Convenzione di Faro. In L. Pavan-Woolfe & S. Pinton (Eds.), *Il valore del patrimonio culturale per la società e la comunità. La Convenzione del Consiglio d'Europa tra teoria e prassi* (pp. 35–71). Linea edizioni.
Pencarelli, T. (2020). Cultura e turismo: sfide per una nuova dialettica virtuosa nella prospettiva del wellness tourism. *Il capitale culturale. Studies on the Value of Cultural Heritage. Supplementi, 11*, 307–331. https://doi.org/10.13138/2039-2362/2537

Perkin, C. (2010). Beyond the rhetoric: Negotiating the politics and realising the potential of community-driven heritage engagement. *International Journal of Heritage Studies*, *6*(1-2: Heritage and Community Engagement: Collaboration or Contestation?), 107–112. https://doi.org/10.1080/13527250903441812

Petraroia, P. (2010). Tutela e valorizzazione. In M. Montella & P. Dragoni (Eds.), *Musei e valorizzazione dei beni culturali. Atti della Commissione per la definizione dei livelli minimi di qualità delle attività di valorizzazione* (pp. 43–54). Clueb.

Pinton, S. (2019). La Convenzione di Faro: alcuni profili di diritto internazionale. In L. Pavan-Woolfe & S. Pinton (Eds.), *Il valore del patrimonio culturale per la società e la comunità. La Convenzione del Consiglio d'Europa tra teoria e prassi* (pp. 73–98). Linea edizioni.

Poirrier, P. (2004). *Les enjeux de l'histoire culturelle*. Seuis.

Romagosa, F. (2020). The COVID-19 crisis: Opportunities for sustainable and proximity tourism. *Tourism Geographies, 22*(3), 690–694. https://doi.org/10.1080/14616688.2020.1763447

Rosenstein, C. (2016). *Understanding cultural policy*. Routledge.

Saviano, M., Di Nauta, P., Montella, M. M., & Sciarelli, F. (2018). Managing protected areas as cultural landscapes: The case of the Alta Murgia National Park in Italy. *Land Use Policy, 76*, 290–299. https://doi.org/10.1016/j.landusepol.2018.03.052

Schofield, J. (2015). Forget about 'heritage': Place, ethics and the Faro convention. In T. Ireland & J. Schofield (Eds.), *The ethics of cultural heritage* (Ethical archaeologies: The politics of social justice) (Vol. 4, pp. 197–209). Springer. https://doi.org/10.1007/978-1-4939-1649-8_12

Schofield, J. (2016). Heritage expertise and the everyday: Citizens and Authority in the Twenty-first Century. In *Who needs experts? Counter-mapping cultural heritage* (pp. 1–11). Routledge.

Senatore, G. (2013). *Storia della sostenibilità. Dai limiti della crescita alla genesi dello sviluppo*. Franco Angeli.

Smargiassi, G. (2012). s.v. Carta europea. In *Dizionario di Economia e Finanza Treccani*. Accessed June 13, 2022, from https://www.treccani.it/enciclopedia/carta-europea_%28Dizionario-di-Economia-e-Finanza%29/

Smith, L. (2006). *Uses of heritage*. Routledge.

Sokka, S., Badia, F., Kangas, A., & Donato, F. (2021). Governance of cultural heritage: Towards participatory approaches. *European Journal of Cultural Management & Policy, 11*(1), 4–19.

Tasso, M. (2017). Participated planning of a heritage walk: A conscious involvement of the community. In S. Pinton & L. Zagato (Eds.), *Cultural heritage. Scenarios 2015-2017 (pp. 745–757)*. Edizioni Ca' Foscari. https://doi.org/10.14277/6969-052-5/SE-4-46

Thérond, D. (2009). Benefits and innovations of the Council of Europe Framework Convention on the value of cultural heritage for society. In *Heritage and beyond* (pp. 9–11). Council of Europe Publishing.

Throsby, D. (2001). *Economics and culture*. Cambridge University Press.

Thurley, S. (2005). Into the future. Our strategy for 2005-2010. *Conservation Bulletin*, 49, 26–27. Accessed June 13, 2022, from https://historicengland.org.uk/images-books/publications/conservation-bulletin-49/cb-49/

UNESCO (2007). *10 clés pour la Convention sur la protection et la promotion de la diversité des expressions culturelles*. Accessed June 13, 2022, from https://unesdoc.unesco.org/ark:/48223/pf0000149502_fre/PDF/149502fre.pdf.multi

Van der Auwera, S., & Schramme, A. (2011). Civil society action in the field of cultural heritage. *A European Perspective. Heritage & Society, 4*(1), 59–82.

Van Velthem, L. H., Kukawka, K., & Joanny, L. (2017). Museus, coleções etnográficas e a busca do diálogo intercultural. *Boletim do Museu Paraense Emílio Goeldi. Ciências Humanas, 12*(3), 735–748.

Vecco, M. (2007). *L'evoluzione del concetto di patrimonio culturale*. FrancoAngeli.

Vecco, M. (2010). A definition of cultural heritage: From the tangible to the intangible. *Journal of Cultural Heritage, 11*, 321–324.

Vícha, O. (2014). The concept of the right to cultural heritage within the Faro convention. *International and Comparative Law Review, 14*(2), 25–40. https://doi.org/10.1515/iclr-2016-0049

Volpe, G. (2019). *Il bene nostro. Un impegno per il patrimonio culturale*. Edipuglia.

Volpe, G. (2020). Un Faro per il patrimonio culturale nel post-Covid-19. *Scienze del Territorio. Rivista di Studi Territorialisti*, 194–202. https://doi.org/10.13128/sdt-12355

Waterton, E., & Smith, L. (2010). The recognition and misrecognition of community heritage. *International Journal of Heritage Studies*, *6*(1-2: Heritage and Community Engagement: Collaboration or Contestation?), 4–15. https://doi.org/10.1080/13527250903441671

Zabatino, A. (2015). Storie di innovazione spontanea e necessaria. In S. Consiglio & A. Riitano (Eds.), *Sud innovation. Patrimonio Culturale, Innovazione Sociale e Nuova Cittadinanza* (pp. 25–68). FrancoAngeli.

Zagato, L. (2008). La Convenzione sulla protezione del patrimonio intangibile. In L. Zagato (Ed.), *Le identità culturali nei recenti strumenti UNESCO. Un approccio nuovo alla costruzione della pace?* (pp. 27–70). Cedam.

Zagato, L. (2015). The notion of "heritage community" in the Council of Europe's Faro convention. Its impact on the European legal framework. In *Between imagined communities of practices: Participation, territory and the making of heritage* (pp. 141–168). Göttingen University Press.

Cultural and Creative Industries as Activators and Attractors for Contemporary Culture-Driven Nomadism

Paola Bertola, Valeria Iannilli, Alessandra Spagnoli, and Angelica Vandi

Abstract The aim of this chapter is to analyse why cultural and creative industries are becoming particularly significant for the development and activation of local territories (Lazzeretti, L., Capone, F., & Casadei, P., *Tourism in the City. Towards an Integrative Agenda on Urban Tourism.* Springer, 2017). More specifically, it focuses on innovative Italian realities characterized by creative production, cultural tourism and territorial regeneration and on their relative impacts on creating collective imaginaries with regard to site-specific sociocultural identities (Iannilli, V., & Linfante, V., *ZoneModa Journal* 9:141-165, 2019). Culture-intensive goods and services (Hesmondhalgh, D., *The Cultural Industries.* Sage, 2002; see also Bovone, L., & Mora, E., *Saperi e mestieri dell'industria culturale.* Franco Angeli, 2003) are a critical component of cultural and creative industries. As such, they are spurring the activation of local communities and attracting global, culture-driven travellers within our contemporary, nomadic society (Bertola, P., Colombi, C., & Vacca, F., *Strategic Design Research Journal,* 2022). In light of this scenario, two main trajectories concerning the type of community involved and the actions within the territory have been identified and applied as variables for the development of four scenarios with the aim to guide future design strategies. Therefore, four exemplifying cases are provided which represent respectively four possible design strategies and innovation trajectories for an impactful, positive, and sustainable cultural and creative consumption.

The aim of the present study is not to define an immediately generalizable model, but to provide a first synthesis, identifying general features and scenarios related to cultural regeneration processes and multidisciplinary actors. By so doing, future research paths can be opened capable of practically engaging with the theoretical implications provided in this chapter.

P. Bertola · V. Iannilli · A. Spagnoli · A. Vandi (✉)
Department of Design, Politecnico di Milano, Milan, Italy
e-mail: paola.bertola@polimi.it; valeria.iannilli@polimi.it; alessandra.spagnoli@polimi.it; angelica.vandi@polimi.it

E. Borin et al. (eds.), *Cultural Leadership in Transition Tourism*, Contributions to Management Science, https://doi.org/10.1007/978-3-031-14121-8_4

1 Introduction: Cultural and Creative Industries and Tourism: Links and Opportunities for Meaningful and Valuable Synergies

The link between tourism and cultural and creative industries (CCIs) has been progressively strengthened in light of the creative turn (Florida, 2002; see also Richards & Wilson, 2007) that has affected the economic, productive and distributive models in the recent decades. This has led to an increasing focus on activities and experiences with high added value, which are symbolically significant and based on contents and services rather than products. The core asset of the cultural and creative sector—which includes the visual arts, but also design, fashion, craft, audiovisual and performing arts—lies in the intangible dimension of production and creative practices based on strong knowledge and skills and in their nature firmly oriented towards experimentation and innovation (Bertola et al. 2022). Moreover, the sector's intrinsic potential to activate collaborative and co-participative value production processes is manifested both in its ability to offer hybrid and highly experiential models of cultural fruition and consumption and in the development of platforms capable of making consumers active participants in the creation of the actual contents (Hesmondhalgh, 2002; Bovone & Mora, 2003).

In light of the foregoing and of the increasing relevance and growth of the creative economy, essential synergies between the cultural, creative and tourism sectors are consolidating (OECD, 2014). The traditional paradigm of cultural consumption mainly revolves around the enjoyment of built heritage and its preservation. However, said paradigm is evolving into a model of creative consumption which is shifting the focus from heritage goods to content and distribution platforms (Richards, 2011). This evolution allows focusing on how CCIs can affect tourism positively and promisingly by delivering considerable added value, increasing tourism demand and diversifying tourism supply (OECD, 2020), even and above all at a crucial phase of the crisis in the sector.

Within the context of market globalization and spatio-temporal compression of exchange of goods, services and experiences, the COVID-19 crisis has led to a temporary and unpredictable de-globalization of the economic, social and cultural models that have characterized the recent decades (Niewiadomski, 2020). The traditional cultural and creative consumption—understood as structured and experienced before the pandemic crisis—has also been strongly affected. In fact, reduced international and, at times, even local mobility,[1] temporary inaccessibility to physical places of experience, social distancing and related containment measures have forced a reset and relative rethinking of the paradigms related to the access,

[1]Data from smartphone localization collected by Apple show that due to the first COVID-19 containment measures put in place by EU national governments (March–May 2020), mobility has dropped dramatically in major European cities, with the Mobility Index range dropping from more than 1.2 points to 0.2 (JRC data processed on Apple Mobility Trends Report available at www.apple.com/covid19/mobility in Montalto et al., 2020: 10–11).

distribution and fruition of cultural places and content on a global scale. The impact on the entire creative sector and, in conjunction, on the tourism sector, has been significant. Indeed, from a socio-economic view, there has been a drastic reduction in incomes and related risks for the sector's employees[2]; while from a sociocultural view, there has been a fallout in terms of social disintegration and impoverishment of cultural exchanges.

However, the cultural and creative sector comprises a diversified and fluid ecosystem of actors and relies on a highly intangible and value-creating production. Therefore, it has been able to react to the crisis in a prompt and dynamic manner by proposing and experimenting new cultural consumption models and formats, in terms of content, products and services.

The interconnectedness and interdependence of the broad cultural and creative network—made up of public cultural institutions, major private players, non-profit organizations and several small micro-enterprises and professionals—have undoubtedly been a weakness for the sector (mainly in terms of structural support policies). Still, as confirmed by Borin and Donato (2015), cultural networks effectively promote a strategic cooperation between public and private institutions and are also a strong driver for innovation and experimentation. For instance, the sudden inaccessibility to physical venues has led to breaking down the cultural sector's prejudices towards the adoption of the digital in its production, distribution and engagement processes (NEMO, 2021). New virtual and immaterial models have been promoted for the audience's active participation in the cultural and creative experience. Moreover, unedited formats and languages have been experimented capable of reaching and reactivating new audiences, directing the gaze towards unprecedented needs for culture and shedding new light on contexts and places that are alternative to the circuits of mass cultural tourism.

From this perspective, the pandemic crisis can be considered an opportunity to speed up the integration process between tourism and CCIs. In this acceleration, the latter has positively contributed to the development of alternative, sustainable and participatory models (de Bruin & Jelinčić, 2016) with spin-offs characterized by new and creative-based audiences, original ways of considering the local viewed as a source for high-potential cultural and creative ecosystems and unedited and more systematic partnerships based on the sharing of creative skills and expertise. In particular, it is possible to establish cultural and creative networks within the sector

[2]Compared to 2019, 2020 EU cultural employment decreased by 2.6% (about 195,000 people), twice as much as the 1.3% decrease seen in total employment. The most affected sector was the "creative, arts and entertainment activities" sector (NACE 90) (source: Eurostat, Culture Statistics—Cultural Employment, https://ec.europa.eu/eurostat/statistics-explained/index.php/Culture_statistics_-_cultural_employment#Self-employment, accessed 15 February. 2022). In addition, as stated in NEMO's survey, 75% of the museums report an income loss between 1000 and 30,000 Euros per week even in the 2020 summer months resulting in the deletion of programs and projects (55%), the reduction in visitor capacity (49%) and the reallocation of staff tasks (42%) (source: NEMO's survey involving 600 museums, mostly European, carried out at the end of 2020—NEMO, 2021).

among cultural institutions, or inter-sectoral collaboration models can be developed between cultural and creative organizations and other companies in territorial or transnational frameworks (Littoz-Monnet, 2013). All this can contribute to the territory's development, both from an economic and from a social perspective, promoting inclusiveness and participation (Barker, 2019).

In light of this scenario, the present chapter starts by analysing two dimensions of ongoing transformations: the changing dynamics of contemporary tourism, where new geographies and landscapes are emerging beyond the traditional tourist landmarks, and the consequent increasing role of CCIs in revamping hidden cultural tangible and intangible reservoirs, transforming them into new cultural attractors. Within this context, an interpretative framework is provided, supported by a few exemplifying cases highlighting four possible strategies for territorial re-signification and reimagination. The aims of said strategies are to activate and attract larger and unprecedented audiences and for the cultural and creative industries to act as promoters and enablers. The exemplifying cases, based in Italy, were selected for their ability to attract creative communities operating in the sociocultural-environmental scope of action of the identified territories in a transformative and not purely contemplative way, to then generate a positive impact on the identity and symbolic values of the regions and their communities. Therefore, the interpretative framework allows identifying innovation trajectories and suggesting an initial conceptualization of intervention models and practices that can have a positive impact on tourism and CCIs.

2 Landscapes and Territories Within the Contemporary Culture-Driven Mobility

Landscapes and territories have undergone a great transformation in their interaction with human activities and flows (Appadurai, 1996) since postmodernity. Currently, they are configuring themselves as a fluid context (Bauman, 2000; see also Bauman, 2007), where global flows of people, technologies, knowledge and cultures interweave. They are becoming an ecosystem of territorial mobility which no longer finds expression in the mechanical principle of the functional interdependence of the industrial city, but in new dynamic connections in which the territory becomes a node of networks and relationships (Castells, 2004), based on the real distribution of cultural heritage and the characteristics of the region (Borin & Donato, 2015). Therefore, the emerging context is characterized by an ongoing tension between the geographically localized physical dimension and the non-spatialized, timeless virtual dimension. Moreover, it can be observed that new valorization processes are increasingly driven by the link between knowledge economy (Rullani, 2004), experience economy (Pine & Gilmore, 1999) and digital economy (Rifkin, 2011).

In the global city, the intensified movement of people, objects, images, narratives and meanings contributes to creating new economic and social phenomena (Sheller

& Urry, 2006), including new expressions of tourism. These new phenomena are having an impact on local spaces, interstitial areas and global networks of local spaces, growing as systems capable of engaging virtuous territorial processes starting from unprecedented innovations of meaning (Bertola & Manzini, 2006). Moreover, the depicted scenario is facilitated by an economic context that is enhancing the links between learning processes, innovation and competitiveness within new forms of sharing economy (Frenken & Schor, 2017; see also Bernardi & Diamantini, 2018). This is leading to an ecosystem characterized by deterritorialization issues, in which territorial mobility flows are aimed at displacing and relocating individuals, also to increase the latter's cultural and social options. The processes of un-embedding and re-embedding—formerly identified by Anne Giddens (1990) as results of modernity—are expressed in new sociocultural, political and economic interdependencies among territories. Consequently, alongside globalization phenomena, the concept of glocalization (Robertson, 1995) has been rising, giving birth to new complex cultural landscapes where flows of people, technologies, media, financial resources and narratives are interwoven (Appadurai, 1996), thus increasing the complexity of interpreting the socio-territorial space. While globalization tends to standardize behaviours, production and consumption patterns, glocalization allows for specific local and territorial characteristics to be placed in the spotlight and valued, thanks to dynamics of knowledge accessibility enabled by global digital networks.

Moreover, if the concepts of "standard quality" and cultural homologation lay at the very centre of globalization, glocalization is driven by an increasing search for "authentic quality" and cultural diversity (Pine & Gilmore, 2007; Iannilli & Linfante, 2019). Especially the Millennials' generation, having been highly exposed to homologating globalization effects, is re-embracing this search for cultural differentiation, trying to escape from mass market consumption trajectories. In addition, the so-called Generation Z is progressively embedding, in its consumption processes, the dimension of social and environmental impacts of goods and services as key components driving its choices (AA.VV. 2021).

In light of this scenario, travels represent an ideal response to the new communities' increasing demand for experience and culture-driven consumption. Travel is a cultural artefact precisely because of its ability to create meaningful connections between the individual, society and territories. It is undoubtedly one of the most enduring historical cultural elements (Toscano, 1996) enabling the passage from identity to otherness (ibidem) which, together with the dimension of freedom of movement, well represents the shared attitude of the contemporary generations. Indeed, the Millennials and Gen Z are driven by a pursuit for cultural experiences, which can be nourished by journeys where memory– seen as a layering of unique narratives and stories—can enrich the archive of the individual's biographical episodes. Moreover, they are spurred by a project-oriented approach to cultural exploration, which no longer conceives travel as a "standardized product" and package, but as a customized and self-designed experience.

Starting from the first pilgrimages up to travels as educational experiences and cultural enrichment—mainly referring to the Grand Tour (Bonadei, 2007) and mass tourism (Urry, 1995)—travel is now part of a broader interpretative context, recontextualized within new glocal and phygital geographies. These new trajectories are configured as complex networks of global and local places and territories, where their digital and virtual projections enhance their potential and accessibility, allowing project-oriented travellers to reach and experience unedited geographies.

From the "new paradigm of mobility" (Sheller & Urry, 2006), which starts from studies on migration (Clifford, 1997) and tourism sociology (Rojek & Urry, 1997) to anthropology (Appadurai, 1996; see also Appadurai, 2001) and geography (Cresswell, 2006), travel is now seen in its multidimensional and "imaginative" dimension. Indeed, "virtual mobility" (Urry, 2000) is becoming the nourishment for the social practice of imagination (Appadurai, 1996) which then drives the planning of the physical experience of travelling. In this context, "spatiality" is the product of people's mediations, mental constructions, social relationships and narratives with which specific social meanings are built and shared through physical experiences and digital representations. These multiple interactions and interdependencies underlying contemporary mobility can be taken as a key factor for interpreting the new relationships that individuals are establishing within physical and virtual territories.

The COVID-19 pandemic has only been accelerating some of the depicted phenomena by increasing access to the digital and virtual representation of territories and by developing new forms of mobility. In the attempt to escape from crowds and traditional landmarks, closed by tourism, small local regions and communities are being looked at with a refreshed interest. The effects and duration of this new attitude are still unpredictable, but a more distributed and locally based tourist mobility paradigm could have several beneficial impacts on the sociocultural and environmental sustainability. Within this potential scenario, culture and creativity can play a crucial role in supporting the transition, not only since they are key drivers of the new generations' mobility, as mentioned, but also due to their entrepreneurial dimension enabling territorial transformation and enhancement.

3 Cultural and Creative Industries as a Vector of Territorial Regeneration

According to the vision of the Territorialist School, the conceptions of territory as a "highly complex living organism" (Magnaghi, 2000, 25) and of its inhabitants as bearers of creative abilities who shape their life project within a rich and meaningful relational environment lie at the heart of all the reflections introduced above.

Within this scenario, cultural and creative industries have the ability to generate virtuous relationships among three constituent components of the territory: the natural environment, the built environment and the anthropic environment. Territorial regeneration strategies go beyond traditional recovery and redevelopment interventions. Therefore, CCIs favour resilient projects capable of relinking spaces and people within networks of new meanings. Said projects are based on strategies involving physical requalification, social inclusion and cultural enhancement through creative entrepreneurship, which feeds the new economic development.

In a context in which the territory implies a relationship between culture, nature and history, the concept of "sustainability" refers not only to environmental quality but also to a more comprehensive interpretation dealing with "virtuous connections between environmental, social, territorial, economic, and political sustainability, [making] primary needs consistent with self-reliance and eco-development, to finally reach self-sustainability" (Magnaghi, 2000, 72). This is underpinned by a form of culture that has rediscovered travel as a pursuit for traditional knowledge. Indeed, the latter "through oral memory not only conveyed models, techniques and unitary languages of construction and production of landscapes but also handed down a symbolic mental map that marked with shared meanings the whole territory" (Decandia, 2019, 129). The result is a scenario of contemporary travel which, instead of homogeneous models, has at its core a variety of paths that can be explored by individuals partaking in co-creating their unique experiences.

It is within this context that, through the lens of "weak thought" (Vattimo & Rovatti, 2010), CCIs emerge as promoters of proposals capable of reimagining experiential alternatives to mass tourism by "combining and sewing what seemed to us irreconcilable: tradition and modernity, past and future, localism and globalisation, social inequality and consumerism" (Fiorani, 2006, 11). In fact, CCIs are increasingly oriented to the cultural and social enhancement of "fragile areas" (Lombardini et al., 2019) within urban and territorial contexts and their communities. Their driving vision is often represented by the concept of "sustainable local development" (Magnaghi, 2000). Thanks to creativity and design-driven processes, the said development interprets innovation and technology as tools for enabling the creation of inclusive, shared and participatory projects, in synergy with local inhabitants (Cabiddu et al., 2013). According to this interpretation and within the context of contemporary collaborative economy, local tourist destination communities become producers of new values, engaged by actions and interactions promoted by creative entrepreneurs (Marques, 2013, see also Meijer et al., 2012).

Along with this trend, new evidence is emerging owing to projects focused on the creative development of territories. In particular, for example, Playable City Network (Bristol)[3] calls upon creative industries in tackling local regeneration challenges, or there are platforms, such as Likealocal,[4] set up to engage host communities that have been transformed into tourist guides capable of depicting their territory authentically through experiences escaping from traditional mass pathways. Similarly, the emergence of a strong tourist demand connected to the so-called slow tourism is giving rise to a whole series of initiatives aimed at stimulating the creation of local community networks. A specific example is CittàSlow,[5] a movement established in the administrative context of some small

---

[3] https://www.playablecity.com/, accessed 16 November 2021

[4] https://www.likealocalguide.com, accessed 20 November 2021

[5] https://www.cittaslow.it/, accessed 20 November 2021

Italian municipalities with the aim to integrate the slow food philosophy into the behavioural practices and hospitality habits of local communities and city governments.

All the foregoing clearly highlights a transition from a vision focused on preserving tangible and intangible heritage to a vision based on generating—through the CCIs' unique entrepreneurial approach—new flows of services, cultural infrastructures and tangible and intangible goods. Such approach empowers the peculiar "cultural capital" (Throsby, 1999) embedded in traditional heritage while enhancing, reproducing and re-signifying it (Daldanise et al., 2019).

Consequently, these revamped places rooted in local and peripheral contexts are becoming new cultural tourism destinations, especially in their "creative extensions", vehiculated through their digital representations (Richards & Raymond, 2000; see also Richards, 2011; Richards & Marques, 2012; OECD, 2014), which are also linking them to global networks and their material and immaterial nodes. Indeed, said destinations are characterized by a new and unprecedented spatiality and temporality. By creating new interactions between physical and virtual spaces, they are being transformed into glocal landscapes. Therefore, the aim is to consolidate a new trend where local systems are no longer isolated but interwoven through meaningful and continuous relationships with the global dimension and through CCI-led regeneration processes. Therefore, they are being prompted to reinvest their traditional cultural capital in new forms of cultural value-embedded assets and services (Throsby, 1999).

The COVID-19 pandemic has reinforced this trend by strengthening the search for non-traditional tourist destinations and especially by activating CCIs in new ways. Indeed, new entrepreneurial approaches are being based on specific dynamics such as the reshoring of productive activities, the relocation in hometowns through home-office practices and the investment in proximity economy. Moreover, the growing sensitivity towards environmental sustainability, social inclusivity and cultural diversity is leading to a recovering and reimaging of abandoned and peripheral territories. In light of these phenomena, the Italian territories—due to their highly diversified and historicized nature and their traditional know-how in the arts and culture-intensive industries—represent an ideal context for observing these ongoing dynamics.

4 Cultural and Creative Industries as New Attractors for Culture-Driven Mobility: Evidence from Italy

The theme of cultural tourism is increasingly connected to CCIs, which often operate in public/private partnership initiatives. In this direction, pioneering projects can be found primarily in Italy, possibly owing to its unique historical layering of heritage, as well as to its strongly diversified and distributed cultural and environmental reservoirs.

The Farm Cultural Park in Favara, analysed further on in the chapter, is among the several initiatives focused on regenerating the most marginal areas, representing about 60% of the Italian national territory. This line of projects falls within the National Strategy for Inland Areas (SNAI) promoted by the Agency for Territorial Cohesion, oriented towards developing and enhancing services in the cultural, agri-food and tourism sectors. 72 projects have been launched involving approximately 1060 municipalities and about 2 million inhabitants with a 17% coverage of the national territory. We can also list the Valore Paese project promoted by the State Property Agency within the same policy trajectory. Its aim is to make provincial paths and routes visible and reachable so as to encourage slow and experiential tourism projects capable of enhancing the national territory's tangible and intangible heritage.

Not by chance, Italy is one of the drivers of the so-called slow-food movement, which actually originated in the country much earlier with regard to the depicted contemporary tourist trends. As a matter of fact, Italy gave rise to the concept of slow tourism (Dickinson & Lumsdon, 2010; see also Dickinson et al., 2011). Indeed, the emergence of a demand for slow travelling (Fullagar et al., 2012)—an authentic travelling capable of offering immersive experiences (Cohen, 2005)—as a response to a standardized and mass tourism offer, well matches with the Italian hospitality industry, characterized by a unique cultural heritage, widespread in local communities. The search for experiences anchored to traditions, culture and local contexts acknowledges that "slow is embodied in the qualities of rhythm, pace, tempo and velocity that are produced in the sensory and affective relationship between the traveller and the world " (Fullagar et al., 2012, 18). Consistently with this trend, also the idea of "Albergo Diffuso" ("Diffused Hotel") emerged in the country (Dall'Ara, 2002; see also Dall'Ara, 2015) as a form of distributed hospitality, one of the first strategic experimentations aimed at enhancing local territories. The concept of "Albergo Diffuso" was initially launched in the 1980s and is characterized by a hospitality model capable of regenerating and enhancing the heritage of small villages. In particular, it promotes the territory's development by integrating public (natural, historical and cultural) and private resources (private homes, small businesses and local products) (Confalonieri, 2011).

Owing to the paradigm of slowness and the growing demand for culture-driven tourism, the *Made in Italy* territory—home to internationally valued cultural-intensive industries, such as agri-food, fashion and design—has nurtured new forms of experiential tourism. The contemporary travel horizons have expanded, including destinations such as craft and industrial districts connected to their traditional manufacturing system, to their industrial archaeology as well as to the legacy still embedded in contemporary Italian products and services (Lazzeretti et al., 2017). Although little research has been carried out, this type of tourism can contribute to revamping and restoring landscapes shaped throughout centuries by the interlink between the natural environment, human settlements and productive activities, which urbanization and deindustrialization have been erasing. An example along this direction is given by the recovery of the ancient medieval village of Solomeo on the hills around Perugia, in the Umbria region. Pivoted on a renewed "humanistic capitalism", Solomeo is acknowledged for its excellence in using culture-intensive industries, such as fashion, for promoting territorial regeneration. Here, apparel entrepreneur Brunello Cucinelli, through what he calls "the economy of beauty", recovered the village from abandonment by setting up his company's headquarters, surrounded by an ecosystem of cultural services, including a craft school and a theatre. This intervention was aimed at rebuilding the identity of the place with the primary object to recover and maintain social, cultural and productive practices already inherent the territory. As a result, the restored landscape and the cultural production systems were transformed into a tourist attraction.

Indeed, all the initiatives described above are deeply rooted in the local contexts. Therefore, they can be enhanced by the "pluralization of the territory" (Crosta, 2001), which is connected in real time with global networks, and thrives through the typical logic of digital communication infrastructures, which inherent the "network society" defined by Castells (1996). This specific shape of society is characterized by individualized bonds with several networks and the individuals' belonging to a growing number of communities, irrespective of the fact of sharing a common physical place. Hence, these communities have become protagonists of the new social phygital contexts, linked to forms of enhancement and regeneration of peripheral and localized territories, thus involving local inhabitants. Nonetheless, they also involve globally displaced audiences and actors such as creative professionals, research centres, universities and city-maker movements. Within these pluralized contexts, cultural capital represents the starting point for activating a process of territorial regeneration, often informed by the implementation of circular economy principles through a multidimensional production of value (Daldanise et al., 2019). Other aspects that are also becoming relevant in the cultural and creative sector (Cerreta & Savino, 2020) involve life cycle extension, reuse and repair in production cycles, business models (MacArthur, 2013) and creation through these practices of economic, social and environmental value (de Vries & Petersen, 2009). Strategies based on the "reuse" of cultural and landscape heritage for the regeneration of micro-communities can contribute to improving the quality of life, job creation as well as the preservation and recovery of natural resources, all of which can also revamp tourism dynamics. These actions aim to include a plurality of local actors in the decision-making process and build long-term development strategies that acknowledge the central role of social capital and the territory's tangible and intangible cultural capital in order to trigger new development trajectories. Within this process, CCIs play an increasingly important role "with the ability to create communication between management models, sustainability assessments, deliberative approaches and conservation strategies, regeneration and enhancement of cultural heritage" (Cerreta & Savino, 2020, 1016).

Only recently, the local cultural and territorial heritage has been identified in the European debate as an essential resource for sustainable development, acknowledging the value of some pioneering Italian experiences (European Commission, 2014a, b, 2015). It has also been acknowledged that CCIs can contribute to this change by tracing multiple and multidimensional trajectories that cross global, physical and virtual networks.

Collaborative and participatory projects can be developed through anticipatory actions—meaning production—thus generating new cultural phenomena. With regard to territorial interstices,[6] abandoned contexts are connected to a web of complex relationships. Owing to forms of creative hubs, new working contexts are welcoming contemporary digital nomads (Pisanu, 2019), among whom artists, designers, filmmakers and app developers. Indeed, start-ups often promote design approaches and projects open to the community and its cooperation, mixing cultural laboratories with business incubators, stable locations with dynamic and distributed ones and physical places with online networks (Daldanise et al., 2019).

Moreover, an extensive network of new material and immaterial meanings are being introduced in the social, cultural and territorial world[7]: unedited symbols, cultural codes and vocabularies, which are making it possible to overcome traditional practices and behaviours established in a territory and to develop a series of experiments based on co-creation. This is perceived as "going beyond", where creative actors are capable of drawing unprecedented relationships between the familiar and the new, between the continuous and the discontinuous.

Along these trajectories of experiences driven by CCIs, several urban regeneration projects support the development of the tourism industry, approaching circular economy models.[8] By feeding the recovery of the traditional know-how of culture-intensive industries, these projects are aimed not only at reducing the consumption of raw materials, extending the life cycle of products, repairing and reusing waste in the context of production cycles but also at creating new economic, social and environmental values, thus also becoming drivers for tourist flows (Van Buren et al., 2016).

Finally, not only can CCIs transform peripheral local contexts into new forms of tourist destinations, but they can also act at a larger urban level. Contemporary cities are anchored to their historical heritage, but at the same time, they are transformed by new transits, fragmentation and temporariness. Globalization, relocation of companies, the transformation of the population's social and cultural demand and the diffusion of information technologies have led to social marginalization, new migrations and deterritorialization, underlying the ongoing transformation from a geography of places to a geography of flows. This dynamic eliminates the idea of cities as we know them and transforms their nature, while new urban areas replace traditional landmarks becoming more capable of better welcoming contemporary flows and nomadic communities. Within these contexts, CCIs often promote projects,[9] highlighting how re-semantization and re-functionalization are stronger sustainable alternatives to demolition and are capable of restoring a more ecological and inclusive urban vision.

---

[6]Please refer to the exemplary case of Farm Cultural Park described in Sect. 5.2.1.

[7]Please refer to the exemplary case of Dolomiti Contemporanee described in Sect. 5.2.3.

[8]Please refer to the exemplary case of Lottozero in Prato described in Sect. 5.2.4.

[9]Please refer to the exemplary case of Lenz Fondazione described in Sect. 5.2.2.

All the evidence depicted above shows an emerging direction for tourism transformation where CCIs can act as new activators and attractors, redesigning travel geographies within a more equitable and sustainable paradigm.

5 Interpretation

5.1 Interpretative Framework Methodology: The Design-Orienting Scenario Approach

According to the directions and assumptions provided in the previous sections based on literature review and evidence observations, as well as on the informed perspective concerning CCIs as a catalyst for sociocultural development, the proposed interpretative model was based on a design-orienting scenario approach.

The scenarios were initially developed to imagine possible visions of the future capable of being used as a tool for decision-making and in order to map the related potential implications within uncertain contexts (Swartz, 1991). More recently, scenario design practices have been developed to be used in design-driven processes with the aim to present alternative and plural possible future scenarios, introducing a disruptive vision of design strategies, products or services (Manzini et al., 2008).

In this section, four scenarios are described, developed following the 2×2 matrix technique, a typical methodological approach in scenario development (Rhydderch, 2017). The said approach is based on a four-quadrant scheme defined by two pairs of variables describing possible polarities or tensions driving ongoing trends. Within each quadrant, one possible and preferable future scenario (Voros, 2003) is provided based on the analysis of potential interactions between the two variables through the lens of specific agencies which are identified as the main drivers for future transformation. These agencies specifically identify behavioural models and attitudes that help describe and qualify the different scenarios.

Starting from literature review, evidence observation and background knowledge in the field of CCIs, two pairs of variables are introduced in this study. Each variable ranges along two possible polarities in order to understand the nature of potential design directions in the field of regenerating cultural and territorial heritage. On the one axis, the variables represent the type of communities involved, which can range from inner to outer communities. On the other axis, the variables represent the possible actions within the territories, which can range from re-signifying heritage reservoirs by reconnecting them to specific cultural roots up to completely reimagining them as new identities and functions. Two possible behavioural models were identified as transformative agencies, ranging from purely preserving heritage to introducing broader environmental and social goals in the process.

More specifically, the two pairs of interpretative variables have the aim to better define the two main trajectories of the framework, which are as follows:

Types of Communities involved Variables: inner/outer.

The previous paragraphs demonstrated that current creative communities have a peculiar impact on the evolution of specific regions with quiescent potentials. Some people travel and contribute to global cultural flows, taking their expertise around the world while being willing to be contaminated by other cultures. Within this framework, they are defined as *outer communities* as they are the ones that participate in and support the project over a period of time and have an impact on the territory with their multidisciplinary competencies.

On the other hand, *inner communities* are local actors rooted in the territory, who know its traditions and know-how, closely feeding the involved reality with a profound vocation to preserve the heritage yet to spread culture and social inclusivity and participation. Such actors could be the ones who light up the project or who live nearby and nurture the current dialogue among various perspectives in and for their land.

Actions within the Territory Variables: reimagining/re-signifying.

The research on this trajectory involves the role of art and culture in the reuse of abandoned buildings and spaces, as well as the activation, involvement and self-empowerment of the inhabitants. The Italian regions are cultural deposits of traditions, know-how and peculiar cultural material that have occasionally become underexposed and forgotten. This also happens to local communities living in those territories (Centis & Micelli, 2021).

On the one hand, the cultural agency has the power to *re-signify* these places, both physically and symbolically, bringing back to life quiescent traditions and inner characteristics that could become meaningful for the local economy, switching the perspective in terms of approaches and values. On the other hand, the territories can be *reimagined*, meaning that culture, in this case, can give rebirth to a forgotten and abandoned place while having the ability to create the necessary conditions to encourage a new collaborative cultural regeneration of public spaces and areas (Cerreta et al., 2018).

As mentioned, the four quadrants are also interpreted through two transformative behavioural models, more specifically:

Heritage Preservation The aim of the case study is to reactivate ancient practices and traditions of the territory/district through the design of culture-led actions, making geographical identity a dynamic element to be reinterpreted.

Social Activism The reality becomes a driver for social change and inclusion, being open to dialogue with minorities, marginalized communities and new generations, with the primary aim to bring about a change in society.

The scenarios analysed in the present study are viewed as an "exploratory" tool (Colombi & Zindato, 2019). Therefore, four exemplifying cases are proposed, each showing a visionary initiative within the scope of the 2×2 scheme (Fig. 2), which already represents a possible trajectory and its related design strategy for positive and sustainable cultural regeneration. The analytical observation was focused on cultural centres with eradicated realities in the Italian territory, capable of attracting various communities and favour new typologies of global cultural flows (Appadurai, 1996). In this context, the cultural agency—referring to creative actions enabling social and environmental change (Sommer, 2006)—is the driver for economic growth and social development, capable of impacting already breached and conventional circumstances in light of the regeneration of the territory's values and its networking dynamics.

Indeed, the Italian landscape is characterized by vast and historical cultural heritage patrimonies as well as diverse, sociocultural regional identities. Therefore, it is a fertile ground for the development of innovative realities in which cultural production and consumption are at the core of their business models.

The four cases analysed are the following:

- Farm Cultural Park, a visionary project founded in Sicily with the mission to reimage the City of Favara and its community through the lenses of art and architecture
- Lenz Rifrazioni, a creative hub in Parma, which reactivated an abandoned industrial neighbourhood through dialogue with social minors and by performing art experimentations
- Dolomiti Contemporanee, a diffuse lab for contemporary art and culture in potentially high-value disused public spaces in the Dolomites alpine area
- Lottozero, a co-working space for textile design, art and culture based in the textile district of Prato, aimed at promoting circular economy practices involving interdisciplinary local and international actors (Fig. 1)

5.2 Exemplifying Cases Within the Interpretative Framework

5.2.1 Farm Cultural Park: Culture for Community Regeneration

In a context paralysed by fear and disenchantment, Farm Cultural Park (https://www.farmculturalpark.com/) changed the history of the Sicilian City of Favara in the province of Agrigento, a small rural town with a population of around 30,000 inhabitants, characterized by an extraordinary abandoned historic centre. In light of this scenario, Farm was established in 2010, between the abandoned walls of "Sette Cortili", as an independent cultural-led centre and laboratory of urban experimentation. Its aim was to profoundly reimagine the meaning of a southern city whose evidence of a glorious past was hidden by an inadequate administration. Decay and neglection were the regeneration engines that fed Farm's creation. Thanks to its cultural agency, it transformed Sette Cortili into an art gallery, hosting international artists' residences and reviving forgotten artisanal traditions through civic participation and strong inner community resilience. Therefore, it is not to be considered a traditional cultural centre because it completely regenerated Favara's central area, rebuilding a sense of belonging among citizens through culture and

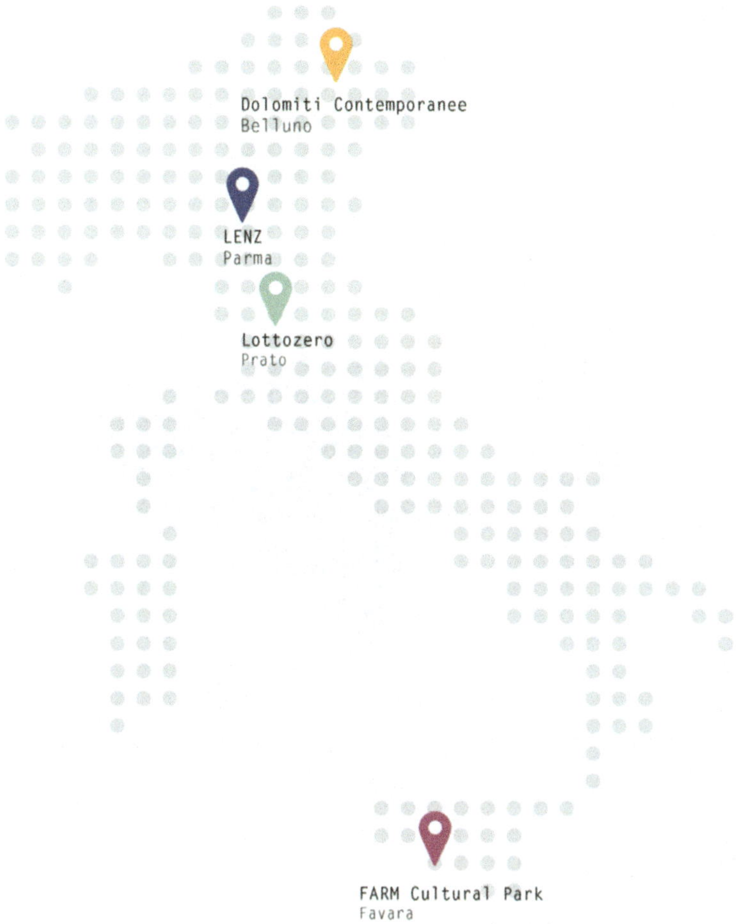

Fig. 1 The geographical location of the exemplifying cases described in Sect. 5.2

creativity. In the last 5 years, Farm has invested a lot of efforts in the future with many new initiatives, such as the "The School of Architecture for Children" or the "Società per azioni buone" (literally *joint-stock company for good*). During the pandemic, it strongly pushed its digital communication through different media, with the aim to make its reality known, especially by Italian citizens who were investing in micro and local tourism, unable to travel abroad.

The philosophy of "taking care" of the place one lives in and of the relationships within the community is Farm's driving element. Indeed, its cultural production and the profound social transformation obtained started to attract artists and curious tourists to Favara, which in these years has become Agrigento's second attraction right after the Valle dei Templi.

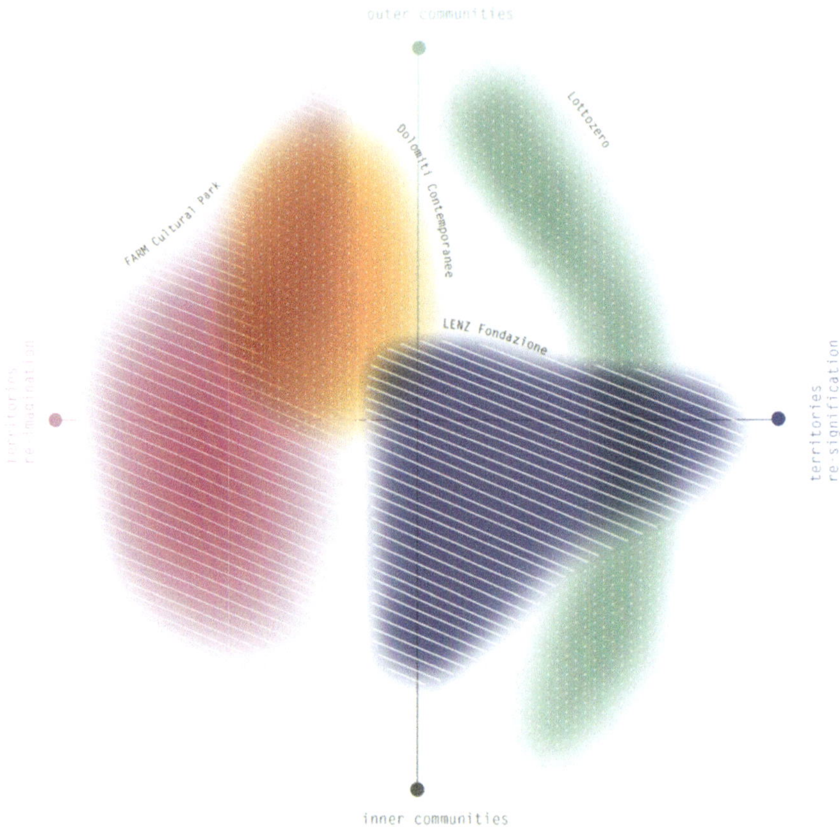

Fig. 2 Interpretative framework

Andrea Bartoli, Farm's cofounder, admits that 'We live in the era of "social innovations" that is not made for selfishness and betrayals but of collaboration, sharing and knowledge [. . .] In a moment in which resources and institutional support are extremely hard to find, the only way to make things possible is to share our knowledge and expertise. [. . .] Waiting for someone else to improve our territory is a utopia' (Lombardo, 2019).

Cultural production aimed at community regeneration is what leads to participative actions in abandoned territories, engaging local and outer communities and generating fruitful and successful tourist flows. All this, in turn, leads to new forms of business model innovation in the cultural and creative sector. For these very reasons—when considering the framework proposed in Fig. 2—Farm Cultural Park is positioned close to the "territory reimagination" variable. With regard to the vertical axes of "community", Farm involves and has more influence on local citizens but also embraces cultural diversity from foreign actors. Its pillar is "social activism", as it was founded to regenerate a city through local community inclusivity, becoming a living museum of people in which civic participation in culture-led actions is what made the community resilient.

5.2.2 LENZ Fondazione: Culture for Social and Spatial Inclusion

The second reality analysed—Lenz Fondazione (https://lenzfondazione.it)—is located in Parma, one of the largest cities in the Emilia Romagna region (about 130,000 inhabitants) and home of one of the most renowned traditional theatres in Italy, Teatro Regio. Its extensive musical tradition and its strong roots in the performing arts are just two of the reasons why the City of Parma was designated Capital of Culture in 2021 by the Italian Ministry of Culture.

Set up in a peripheral area of the city in 2015 by joining two existing cultural associations, LENZ is a creative factory that aims to investigate the contemporary theatrical language and create a bridge between the artistic tradition rooted in the territory and a profound vocation for international performing culture with an inclusive and participatory attitude. Lenz is constantly open to dialogue with minorities and multidisciplinary artistic residencies. It promotes art to marginal communities and is open to the new generations' perspectives while organizing social theatre practices and awareness-raising activities for disabled people, people with addiction disorders and autism. Its collaborations with the AUSL laboratory and the University of Parma have led to the creation of installations, live performances, creative workshops, seminars and dialogues with the public, social theatre practices and even kid labs. It is particularly involved in searching for an expressive and linguistic bridge between irrational and impulsive visions, theatrical practices and stage design.

A further critical aspect concerns Lenz's location. In fact, it advocated for the reactivation of a potential space, the San Leonardo District, which had become unproductive after the 1970s, following the closing of glass factories in the area. After years of degradation and loss of the majority of social, economic and urban functions, LENZ took over the production spaces, leveraging on the preservation of their identity but converting the previous industrial production into cultural production. Moving the artistic activities from the centre to the peripheries and away from the traditional places of culture is evidence of the fact that one of the primary vocations of culture is to reappropriate forgotten places in the city. The revitalization of territories through new structures and meanings reveals the need to reinterpret spaces in an unconventional context and with contemporary tools, encoding their deeper meanings and reconfiguring them in new stories. From a business innovation perspective, the case of Lenz is emblematic: the cultural agency implements social inclusion actions within urban dismissed areas, revamping traditions and identities, supporting new communities in reappropriating spaces and cultural practices to seize opportunities arising from a constantly evolving economy.

This is why Lenz occupies the bottom right-hand corner of the framework (Fig. 2). Indeed, it has re-signified the former industrial neighbourhood by being socially active and by mainly involving city realities with occasional external influences from international artists.

5.2.3 Dolomiti Contemporanee: Culture for Territory Activation

Dolomiti Contemporanee (http://www.dolomiticontemporanee.net/) was founded in 2011 with the aim to regenerate the territory of the Belluno Dolomites (an alpine area of 32 km^2 among the regions of Veneto, Trentino Alto Adige and Friuli Venezia Giulia), declared a UNESCO World Natural Heritage Site in 2009.

Through re-functionalization practices and by redefining the identity of places thanks to contemporary art and culture, the Dolomiti Contemporanee project falls within this specific environmental, social and cultural context. In particular, it proposes "spatial and conceptual reconfiguration" actions, consciously distant from the traditional alpine naturalistic imagery.

The project is mainly characterized as an interlinking of networks: networks of places, mainly abandoned; networks of actions, permanent and temporary; and networks of communities, local and extended. Since the beginning, the places of interest have been closed and disused spaces that used to play an active role in the life and economy of the territory. In particular, said places are industrial archaeology structures, schools, villages, military or civil architectures which, after decades of neglect and silence, have gone back to expressing their potential as sources of cultural and artistic production. Alongside several projects involving the sites' temporary occupation with artistic purposes, there are two long-term projects of wider relevance that represent both the nodes and engines of the cultural reactivation actions: the Nuovo Spazio di Casso, an experimental centre for contemporary mountain culture opened in the former school of Casso, renovated and reactivated by Dolomiti Contemporanee more than 50 years after the Vajont disaster, and Progettoborca, a project for the re-functionalization of the former Eni Village of Borca di Cadore conceived by Enrico Mattei and realized by architect Edoardo Gellner. The locations of both projects still represent an unhealed social wound. However, they have no commemorative value but rather a generative aim, owing to the creation of new spaces, functions and meanings with their own specificity (Trevisani, 2020). This process was activated thanks to the involvement of creative communities, artists, architects, designers and philosophers and often through the formula of residence: these spaces are inhabited by creative communities that have the support of public and private mountain communities. Said communities, in fact, provide assistance and support for the realization of artworks and inhabit these spaces activating transformative and regenerative actions revealing the related hidden potential, also from a commercial point of view (Bagnara, 2021).

In light of this scenario, Dolomiti Contemporanee is a pivotal example of the importance of cultural agencies, as it has completely reframed the traditional imaginary connected to winter sports destinations and activated new culture-driven, business opportunities.

Dolomiti Contemporanee is a multidisciplinary laboratory and a strategic programme that sees the involvement of an extensive network of local actors and "temporarily settled" creative communities. This has allowed developing and implementing reimagination and regeneration actions (Fig. 2) addressed to abandoned places, making the latter again available to the communities and the territories by imagining new shared values and identities.

5.2.4 Lottozero: Culture for Sustainable Expertise Protection

Prato—a Tuscan city with a population of about 192.000 inhabitants—is widely recognized as an exemplifying case of the typical Italian industrial district (Becattini, 2000). In fact, it reflects its main features in terms of organizational approach and high-street fashion textile and knitwear productivity.

Following the third industrial revolution, Prato became Europe's most important textile and fashion centre (Espon, 2020). This was also due to an impactful Chinese migration wave which established an extensive network of small firms operating in an informal and flexible economy. This led to the development of a "parallel district" (Espon, 2020) in areas vacated by Italian textile firms which were abandoning the business due to the 2010's crisis. Specifically, the area is Macrolotto Zero which, from an urban point of view, was a degraded neighbourhood in the city, characterized by social mistrust towards the Chinese people even if they heavily contributed to the supplier and manufacturing firms.

For these many reasons, Prato's municipality—especially the Department of Culture—is involved in some of the main regenerative initiatives of the area, and Lottozero (https://www.lottozero.org/) is one of the most successful projects promoted with the aim to revitalize one of the oldest textile districts in Europe.

Conceived as a centre for textile design, art and culture, Lottozero started its business in 2016 as an open lab for textile production, co-working, experimentation and research. Thanks to its strong relation to the city's industrial environment, its space represents a renovated initiative that follows collaborative and circular practices by means of classes, workshops and events. In this context, multidisciplinarity becomes a design tool (Bertola et al., 2016) promoting participation and social innovation. Indeed, many of the creative initiatives developed are addressed to citizens and tourists of all nationalities, thanks to talent scouting and the many residencies offered to leverage on different background expertise. Moreover, the open-access modality favours the establishment of an international network of actors in the textile sector, having Lottozero as the central hub.

Despite the attractive international power, the citizens of Prato remain the primary beneficiaries of the initiatives taking place in and for the territory (Espon, 2020). Relevant is the work carried out thanks to the partnership with the "China Association". In fact, it led to the activation of specific activities aimed at merging the local and Chinese communities, trying to bring together know-how and economic power in order to contribute collaboratively to the district's livelihood and innovation.

Lottozero's exemplifying case highlights how cultural agencies can become the core in the renewal of traditional expertise with regard to culture-intensive products and services. This favours the creation of circular and sustainable practices which become attractors for new audiences and economic opportunities.

In light of this scenario, the trajectories traced by Lottozero's representation in the framework are transversal when looking at the "community" polarities, whereas its activities for the territory fall more likely under the "re-signification" variable. "Heritage preservation" is Lottozero's guiding value. In fact, it was primarily established to safeguard the most rooted European textile district's know-how and to make Prato a leading city as regards environmental, economic and social sustainability, as well as for its cultural aspects.

6 Conclusions

The present study shows the potential role of cultural and creative industries within the scope of tourism development. The main hypothesis introduced is that cultural consumption can be the driver for cultural and environmental heritage preservation as well as for attracting new tourist flows. In fact, as highlighted by literature review and evidence observation, CCIs are emerging in their recent evolution for their capacity to promote new entrepreneurial actions connected both with local ecosystems of private and public actors and global networks. These initiatives are focused on implementing innovative business models characterized by a rich offer of cultural products and services that are becoming new attractors for tourist flows. Moreover, they often incorporate a dimension of participative design that can engage the local community and be appealing to global networks of creative professionals and users, driven by cultural production and consumption interests. These dynamics have a positive impact on peripheral and unusual destinations within local territories aside from traditional tourist pathways, as well as on minor urban areas not usually considered among cities' landmarks.

To support this hypothesis, a design-orienting scenario methodology was applied to elaborate four potential exploratory scenarios with the aim to guide future design strategies:

- Scenario 1: Cultural production aimed at regenerating the community, bringing participative actions into abandoned territories (i.e. Favara village), thus allowing the engagement of the local community and the attraction of new ones not commonly part of the traditional tourist flows
- Scenario 2: Cultural production aimed at promoting social inclusion actions within urban dismissed areas, renewing cultural identity (i.e. theatre tradition in Parma) and supporting new communities in reappropriating spaces and cultural practices
- Scenario 3: Cultural production aimed at activating the territory through the development of culture-intensive activities completely reframing the traditional

imaginary connected to a specific area (i.e. the Dolomites purely as sports destination)

- Scenario 4: Cultural production aimed at renewing traditional expertise with regard to culture-intensive products and services (i.e. textile in Prato) developing circular and sustainable practices which become attractors for new audiences and for culture-driven travellers

Limitations to the present study are represented by the lack of quantitative validation for the four scenarios. Therefore, future studies are highly recommended in order to collect and cluster other relevant cases.

Finally, the four scenarios are intended as exclusive. Hence, future strategies should consider combining actions belonging to different scenarios, depending on the specific context of the proposed design strategy.

References

Appadurai, A. (2001). Grassroots globalization and the research imagination. *Public Culture, 12*(1), 1–19. https://doi.org/10.1215/08992363-12-1-1

Appadurai, A. (1996). *Modernity at large*. University of Minnesota Press.

Bagnara, I. (2021, July 24). *Conversazioni sulla montagna*. Intervista a Gianluca D'Incà Levis di Dolomiti Contemporanee. Artribune. Retrieved June 8, 2022, from https://www.artribune.com/professioni-e-professionisti/who-is-who/2021/07/montagna-intervista-gianluca-dinca-levis-dolomiti-contemporanee/

Barker, V. (2019). The democratic development potential of a cultural ecosystem approach. *Journal of Law, Social Justice and Global Development, 24*, 86–99. https://doi.org/10.31273/LGD.2019.2405

Bauman, Z. (2000). *Liquid Modernity*. Polity Press.

Bauman, Z. (2007). *Liquid Times: Living in an Age of Uncertainty*. Polity Press.

Becattini, G. (2000). *Il bruco e la farfalla. Prato: una storia esemplare dell'Italia dei distretti*. Le Monnier.

Bernardi, M., & Diamantini, D. (2018). Shaping the sharing city: An exploratory study on Seoul and Milan. *Journal of Cleaner Production, 203*, 30–42. https://doi.org/10.1016/j.jclepro.2018.08.132

Bertola, P., Colombi, C., & Vacca, F., (2022). Designing culture-intensive Artefacts. How the design process interprets craft reiteration to build future diversities. *Strategic Design Research Journal*.

Bertola, P., & Manzini, E. (2006). Design Multiverso. Ed. Polidesign.

Bertola, P., Vacca, F., Colombi, C., Iannilli, V. M., & Augello, M. (2016). The cultural dimension of design-driven innovation. A perspective from the fashion industry. *The Design Journal, 19*(2), 237–251. https://doi.org/10.1080/14606925.2016.1129174

Bonadei, R. (2007). I sensi del viaggio. Franco Angeli.

Borin, E., & Donato, F. (2015). Unlocking the potential of IC in Italian cultural ecosystems. *Journal of Intellectual Capital., 16*(2), 267–284. https://doi.org/10.1108/JIC-12-2014-0131

Bovone, L., & Mora, E. (2003). Saperi e mestieri dell'industria culturale. Franco Angeli.

Cabiddu, F., Lui, T. W., & Piccoli, G. (2013). Managing value co-creation in the tourism industry. *Annals of Tourism Research, 42*, 86–107. https://doi.org/10.1016/j.annals.2013.01.001

Castells, M. (2004). *The network society: A cross-cultural perspective*. Edward Elgar Publishing.

Castells, M., (1996). *The rise of the network society: The information age: Economy, society and culture*. Blackwell Publishers, Inc.

Centis, L., & Micelli, E. (2021). Regenerating places outside the Metropolis. A Reading of three global art-related processes and development trajectories. *Sustainability, 13*(22), 12359. https://doi.org/10.3390/su132212359

Cerreta, M., Daldanise, G., & Sposito, S. (2018). Culture-led regeneration for urban spaces: Monitoring complex values networks in action. *Urbani izziv, 29*(supplement), 9–28. https://doi.org/10.5379/urbani-izziv-en-2018-29-supplement-001

Cerreta, M., & Savino, V. (2020). Circular enhancement of the cultural heritage: An adaptive reuse strategy for Ercolano Heritagescape. *Computational Science and Its Applications–ICCSA, 2020*, 1016–1033. https://doi.org/10.1007/978-3-030-58808-3_72

Clifford, J. (1997). *Routes: Travel and translation in the late twentieth century*. Harvard Business Press.

Cohen, E. (2005). Principales tendencias en el turismo contemporáneo. *Política y Sociedad, 42*(1), 11–24.

Colombi, C., & Zindato, D. (2019). Design scenarios and anticipation. In R. Poli (Ed.), *Handbook of anticipation* (pp. 821–842). Springer. https://doi.org/10.1007/978-3-319-91554-8_52

Confalonieri, M. (2011). A typical Italian phenomenon: The "albergo diffuso". *Tourism Management, 32*(3), 685–687. https://doi.org/10.1016/j.tourman.2010.05.022

Cresswell, T. (2006). *On the move mobility in the modern Western Worl*. Routledge.

Crosta, P. L. (2001, November 15-16). Società delle differenze, pluralizzazione del territorio e il ruolo dell'interazione sociale nella produzione di "pubblico", al plurale. *Convegno Internazionale "Dalla città diffusa alla città diramata"*, Politecnico di Torino, Dip. Scienze e tecniche per i processi di insediamento.

Daldanise, G., Gravagnuolo, A., Oppido, S., Ragozino, S., Cerreta, M., & De Vita, E. G. (2019). Economie circolari per il patrimonio culturale: processi sinergici di riuso adattivo per la rigenerazione urbana. *XXI Conferenza Nazionale SIU_ Società Italiana degli urbanisti*, Firenze, 6–8 giugno 2018. Planum Publisher.

Dall'Ara, G. (2002). Albergo diffuso: un'idea che piace. *La Rivista del Turismo, 4*(1), 36–40.

Dall'Ara, G. (2015). *Manuale dell'albergo diffuso. L'idea, la gestione, il marketing dell'ospitalità diffusa*. FrancoAngeli.

de Bruin, A., & Jelinčić, D. A. (2016). Toward extending creative tourism: Participatory experience tourism. *Tourism Review, 71*(1), 57–66. https://doi.org/10.1108/TR-05-2015-0018

de Vries, B. J. M., & Petersen, A. C. (2009). Conceptualizing sustainable development: An assessment methodology connecting values, knowledge, worldviews and scenarios. *Ecological Economics, 68*(4), 1006–1019. https://doi.org/10.1016/j.ecolecon.2008.11.015

Decandia, L. (2019). Orme di pietra nel paesaggio: un museo multimediale e interattivo per riannodare il rapporto tra una comunità e il suo territorio. *Scienze Del Territorio, 7*, 128–136. https://doi.org/10.13128/sdt-10958

Dickinson, J. E., & Lumsdon, L. M. (2010). Slow travel and tourism. Earthscan.

Dickinson, J. E., Lumsdon, L. M., & Robbins, D. (2011). Slow travel: Issues for tourism and climate change. *Journal of Sustainable Tourism, 19*(3), 281–300. https://doi.org/10.1080/09669582.2010.524704

Ellen MacArthur Foundation. (2013). *Towards the circular economy. Opportunities for the consumers good sectors*. Retrieved June 8, 2022, from https://www.mckinsey.com/~/media/mckinsey/dotcom/client_service/sustainability/pdfs/towards_the_circular_economy.ashx

ESPON. (2020). *SHARING stocktaking and assessment of typologies of urban circular collaborative economy initiatives. Case study Prato–regeneration of urban spaces and social cohesion*. Retrieved June 8, 2022, from https://www.espon.eu/sites/default/files/attachments/SHARING_Annex2_CS_Prato.pdf

European Commission. (2014a, July 7). Towards a circular economy: A zero waste programme for Europe. *Communication from the commission to the European Parliament,* The Council, The European Economic And Social Committee And The Committee Of The Regions. Retrieved June 8, 2022, from https://eur-lex.europa.eu/resource.html?uri=cellar:50edd1fd-01ec-11e4-831f-01aa75ed71a1.0001.01/DOC_1&format=PDF

European Commission. (2014b, July 22). Towards an integrated approach to cultural heritage for Europe. *Communication from the commission to the European Parliament,* The Council, The European Economic And Social Committee And The Committee Of The Regions. Retrieved June 8, 2022, from https://resources.riches-project.eu/wp-content/uploads/2015/10/CELEX-52014DC0477-EN-TXT.pdf

European Commission, Directorate-General for Research and Innovation. (2015). *Getting cultural heritage to work for Europe: Report of the horizon 2020 expert group on cultural heritage*. Publications Office. Retrieved June 8, 2022, from https://data.europa.eu/doi/10.2777/745666

Fiorani, E. (2006). Moda, corpo e immaginario. Il divenire del mondo fra tradizione e innovazione. Ed. .

Florida, R. (2002). *The rise of the creative class: And how It's transforming work, leisure*. Basic Books.

Frenken, F., & Schor, J. (2017). Putting the sharing economy into perspective. *Environmental Innovation and Societal Transitions, 23*, 3–10. https://doi.org/10.1016/j.eist.2017.01.003

Fullagar, S., Wilson, E., & Markwell, K. (2012). slow: Thinking through slow mobilities and experiences. In S. Fullagar, K. Markwell, & E. Wilson (Eds.), *Slow tourism: Experiences and Mobilities*. Channel View Publications.

Giddens, A. (1990). *The consequences of modernity*. Stanford University Press.

Hesmondhalgh, D. (2002). *The cultural industries*. Sage.

Iannilli, V., & Linfante, V. (2019). Nuovi percorsi della moda tra globale e locale. Dai grandi centri alla disseminazione culturale del fashion system. *ZoneModa Journal, 9*(2), 141–165. https://doi.org/10.6092/issn.2611-0563/9966

Lazzeretti, L., Capone, F., & Casadei, P. (2017). The role of fashion for tourism: An analysis of Florence as a manufacturing fashion city and beyond. In N. Bellini & C. Pasquinelli (Eds.), *Tourism in the City. Towards an integrative agenda on urban tourism* (pp. 207–220). Springer.

Littoz-Monnet, A. (2013). *The European Union and culture: Between economic regulation and European cultural policy*. University Press.

Lombardini, G., Butelli, E., & Rossi, M. (2019). *Dai territori della resistenza alle comunità di patrimonio: percorsi di autorganizzazione e autogoverno per le aree fragili*. Ricerche E Studi Territorialisti–SdT Edizioni.

Lombardo, E. (2019, November 19). #peripheries: Culture-led urban regeneration in Sicily: The story of farm Cultural Park and Favara | Italy | ASEF culture360. *ASEF culture360*. https://culture360.asef.org/magazine/peripheries-culture-led-urban-regeneration-sicily-story-farm-cultural-park-and-favara-italy/

Magnaghi, A. (2000). Il Progetto locale. *Bollati Boringhieri.*

Manzini, E., Jégou, F., & Penin, L. (2008, February). Creative communities for sustainable lifestyles. *Proceedings of the 2nd Conference on Sustainable Consumption Research Exchange* (pp. 259–276).

Marques, L. (2013). Constructing social landscape through events: The glocal project of 's Hertogenbosch. In G. Richards, M. de Brito, & L. Wilks (Eds.), *Exploring the social impact of events* (pp. 84–94). Routledge.

Meijer, A., Grimmelikhuijsen, S., & Brandsma, G. J. (2012). Communities of public service support: Citizens engage in social learning in peer-to-peer networks. *Government Information Quarterly, 29*(1), 21–29. https://doi.org/10.1016/j.giq.2011.06.004

Montalto, V., Sacco, P. L., Alberti, V., Panella, F., & Saisana, M. (2020). *European cultural and creative cities in COVID-19 times: Jobs at risk and the policy response. EUR 30249 EN*. Publication office of the European Union. https://doi.org/10.2760/624051

NEMO (2021, January). Follow-up survey on the impact of the COVID-19 pandemic on museums in Europe. *Final Report*. Retrieved June 8, 2022, from https://www.ne-mo.org/fileadmin/Dateien/public/NEMO_documents/NEMO_COVID19_FollowUpReport_11.1.2021.pdf

Niewiadomski, P. (2020). COVID-19: From temporary de-globalisation to a re-discovery of tourism? *In Tourism Geographies, 22*(3), 651–656.

OECD. (2014). Tourism and the creative economy. *OECD studies on tourism*. OECD Publishing. https://doi.org/10.1787/9789264207875-en

OECD. (2020). Culture shock: COVID-19 and the cultural and creative sectors. *OECD policy responses to coronavirus (COVID-19)*. OECD Publishing. https://doi.org/10.1787/5b0fd8cd-en

Pine, B. J., & Gilmore, J. H. (1999). *The experience economy*. Harvard Business Review Press.

Pine, B. J., & Gilmore, J. H. (2007). *Authenticity: What consumers really want*. Harvard Business Review Press.

Pisanu, V. (2019). Il nomade digitale tra radicamento e sguardo cosmopolita. *Geography Notebooks, 4*(1), 79–95. https://doi.org/10.7358/gn-2019-001-pisa

Rhydderch, A. (2017). *Scenario building: The 2x2 matrix technique. Futuribles international*. Retrieved June 8, 2022, from https://www.futuribles.com/en/group/prospective-and-strategic-foresight-toolbox/document/scenario-building-the-2x2-matrix-technique/

Richards, G. (2011). Creativity and tourism: The state of the art. *Annals of Tourism Research, 38*(4), 1225–1253. https://doi.org/10.1016/j.annals.2011.07.008

Richards, G., & Raymond, C. (2000). Creative tourism. *ATLAS news*, 23, 16–20.

Richards, G., & Wilson, J. (2007). *Tourism, creativity and development*. Routledge.

Richards, G., & Marques, L. (2012). Exploring creative tourism: Editor's introduction. *Journal of Tourism Consumption and Practice, 4*(2), 1–11.

Rifkin, G. (2011). *The third industrial revolution. How lateral power is transforming energy, the economy, and the World*. Palgrave MacMillan.

Robertson, R. (1995). Glocalization: Time-space and homogeneity-heterogeneity. In M. Featherstone, S. Lash, & R. Robertson (Eds.), *Global Modernities* (pp. 25–44). Sage Publications.

Rojek, C., & Urry, J. (1997). *Touring cultures*. Routledge.

Rullani, E. (2004). *Economia della conoscenza*. Creatività e valore nel capitalismo delle reti.

Sheller, M., & Urry, J. (2006). The new mobilities paradigm. *Environment and Planning A: Economy and Space, 38*(2), 207–226.

Sommer, D. (2006). *Cultural Agency in the Americas*. Duke University Press. https://doi.org/10.2307/j.ctv11qdz2k

Swartz, P. (1991). *The art of the long view*. Doubleday.

Throsby, D. (1999). Cultural capital. *Journal of Cultural Economics, 23*, 3–12. https://doi.org/10.1023/A:1007543313370

Toscano, M. A. (1996). Per una sociologia del viaggio. In AA.VV. *Il viaggio. Dal Grand Tour al turismo post-industriale*. Atti del Convegno Internazionale–Roma 5-6 dicembre 1996. Edizioni Magma.

Trevisani, M. (2020). Allevare il cervo a sei zampe. In F. Andreoni, P. Bizzini, M. De Vidi, G. Gregnanin, A. Monaci, & M. Trevisani (Eds.), *Bagliore*. Sei scrittori raccontano i nuovi centri culturali. Il Saggiatore.

Urry, J. (1995). *Lo sguardo del turista*. Il tempo libero e il viaggio nelle società contemporanee.

Urry, J. (2000). *Sociology beyond societies*. Mobilities for the Twenty-First Century. Routledge.

Van Buren, N., Demmers, M., Van der Heijden, R., & Witlox, F. (2016). Towards a circular economy: The role of Dutch logistics industries and governments. *Sustainability, 8*(7), 647. https://doi.org/10.3390/su8070647

Vattimo, G., & Rovatti, P. A. (2010). *Il pensiero debole*. Feltrinelli.

Voros, J. (2003). A generic foresight process framework. *Foresight, 5*(3), 10–21. https://doi.org/10.1108/14636680310698379

Cultural and Creative Industries in the Brussels Region: Challenges and Opportunities of the Ecosystem

Ruba Saleh and Christian Ost

Abstract Brussels Region has a very ambitious circular economy program which is starting to show its fruits in the sectors of interest of the region, namely, construction, resources and waste, logistics, and commerce. Although the region is also very rich of diversified cultural and creative industries (CCIs), the contribution of CCIs to this regional green transition has not yet been really taken into consideration neither in the regional circular economy agenda nor by the lifelong learning programs. The 2019 edition of the Cultural and Creative Cities Monitor places Brussels 92nd out of 190 European cities in terms of overall cultural performance. This research paper aims at mapping the CCI ecosystem in the Brussels Region and exploring what reskilling/upskilling programs are needed to enhance their role as levers for local and sustainable tourism development.

For this preliminary mapping, desk research was carried out between April and October 2021, while for mapping the reskilling/upskilling programs, a survey was sent to the mapped ecosystem. Throsby's concentric circles model theoretical framework was adopted, and the CCIs were classified accordingly. This preliminary mapping of 1170 CCIs took into consideration the 19 municipalities of the Brussels Region and consisted of desk research. It is worth mentioning that this mapping is neither conclusive nor exhaustive. It is instead a first step toward creating a common understanding of the richness and diversity of existing actors and activities. A nugget of information aimed at triggering reflection on the needs and how the ecosystem could be enhanced in the future.

1 Introduction

The cultural and creative industries (CCIs) are now part of a context of economic transition and sustainable development (EY, 2021; KEA, 2020a; KEA & PPMI, 2019; IDEA Consult et al., 2021; OECD and ICOM, 2019; Potts, 2021; Throsby,

R. Saleh (✉) · C. Ost
ICHEC Brussels Management School, Brussels, Belgium
e-mail: ruba.saleh@ichec.be; Christian.ost@ichec.be

E. Borin et al. (eds.), *Cultural Leadership in Transition Tourism*, Contributions to Management Science, https://doi.org/10.1007/978-3-031-14121-8_5

2018; UNESCO and the World Bank, 2021; Voices of culture, 2021). The economic transition is based on the theories of long economic movements that have gone through history and in particular ensured the industrial revolution and its expansion through long cycles of growth separated by ebbs and major crises. The succession over time of basic inventions (steam engines, electricity, iron and steel, transport technology, digital technology and biotechnology) has anchored new industries in a dynamic economic and social fabric. The principle of innovation, a phenomenon derived from creativity itself, made it possible to transform inventions into new products, services, and organizations (Delbeke, 1984). The example of the automobile industry is characteristic of it, since the invention of the thermic engine was accompanied by innovations of cars, innovations of associated services, and organizational innovation (assembly lines).

Even if creativity is an artistic concept at its origin, the creative industries have only accompanied these long economic movements in recent years. The marginal economic nature of cultural activities in the global production network (GPN), together with the difficulty of applying economic and managerial theory to these nonmarket sectors, explains why the cultural and creative industries have taken longer to develop than in other productive sectors (Blaug, 1980).

It should be noted that the processes of growth, of cyclical crisis, and of transition toward new forms of production, consumption, and organization are also cultural and social phenomena which can explain a wide disparity in the forms of development of CCIs in different countries. Culture and artistic activities are not managed or financed in the same way everywhere, and the innovative character of the CCIs sometimes comes from a vision of culture as a public or common good or a vision of culture based on entrepreneurship and individual risk-taking.

It has also been argued that creativity was a dominant urban feature. Richard Florida's creative class (2002) used to be broken down into super creative class (like engineering, research, arts, design, media workers), creative class (like healthcare, business, finance, administration, education), and bohemians (small and marginal groups). Florida (2002) found that companies were keen to locate their facilities where creative workers were and that the urban realm would become attractive for more of them. Florida also argued that urban authorities could invest in cultural and artistic activities, sport facilities, accommodations for creative workers, and quality urban design, instead of keeping ancient industrial areas which are destined to collapse in the long run. Here again, the timing is crucial, because urban development taken as a long-term process would accompany new innovations, rather than facing radical changes that could jeopardize the whole city and spur gentrification. It is generally accepted that a robust link does exist between culture and urban areas' appeal. In 2009, INEUM Consulting conducted a study for the Forum d'Avignon. The study "Culture—A symbolic or economic success factor for urban development planning?" documents 32 cities across the world and posits a barometer of the cultural appeal of territories. It concludes that urban area's appealing features were made of the combination of cultural policies, development of knowledge-based economy, and living environment. Although it is not mentioned explicitly, creativity

Type of feature	Features
Features linked to the level of cultural activity	• Action-packed towns • Infrastructures and cultural services • Showcasing architecture • Diversity and multiculturalism
Features linked to the knowledge-based economy	• Quality of education • Knowledge of languages • Qualification of human capital • Research and innovation
Features linked to living environment and the general environment	• Quality of health care • Nightlife • Living environment • Quality of life • Quality of telecommunications • Internal transport
Features linked to business development	• Competitiveness over wages • Economic diversity • Competitiveness of office space prices • Market access • External connectivity

Fig. 1 Urban area's appealing features. Source: INEUM Consulting (2009: 4)

is part of the knowledge-based economy and hence contributes to one of the four features for success (Fig. 1).

This research paper aims at mapping the CCI ecosystem in the Brussels Region and exploring what reskilling/upskilling programs are needed to enhance their role as levers for local and sustainable tourism development.

2 Cultural and Creative Industries (CCIs) in Brussels Region

In 2010, INEUM Consulting documented the cultural appeal of the Brussels Region. The conclusion was displayed as a SWOT grid that provides strengths, weaknesses, opportunities, and threats of the possible role of culture in the territorial development (Fig. 2). The analysis highlights creativity as part of historic Brussels brand: "A creative city, which was the cradle of many artistic movements (Art nouveau, Cobra, Art Nouveau, Surrealism ...) and has been the refuge of many thinkers (Victor Hugo, Brontë, Marx...)" (INEUM Consulting, 2010: 3).

Strengths	Weaknesses
▪ *First-rate artistic and cultural offer: contemporary dance, opera, jazz scene, plastic arts, film, rich museum heritage, electronic music, etc.* ▪ *European crossroads between Amsterdam, London, Cologne and Paris (by Thalys, Brussels is 1hour 20 minutes from Paris, under 2 hours from London and 2 hours 38 minutes from Amsterdam).* ▪ *Officially bilingual and multicultural city (French and Dutch).* ▪ *Very good quality of life: Brussels ranks 14th in the world.* ▪ *Other distinction: Brussels ranks 18th on the 2007 Anholt City Brand Index, mainly due to the presence of European institutions.*	▪ *Impacts, strengths and dynamics of the economy of culture in Brussels inadequately understood* ▪ *Metropolis lacks a strategic and comprehenensive cultural vision* ▪ *"Cultural" know-how and funding fragmented among various actors whose policies and actions are not streamlined* ▪ *No space for culture not strictly French- or Dutch-speaking* ▪ *Few incentives for making financial contributions to culture (tax exemptions, for example)* ▪ *Lack of clarity regarding higher learning opportunities in culture and the arts in Brussels* ▪ *Economic growth gauged primarily by productivity with little regard for job creation* ▪ *Funding for territorial development of the Region in a context where taxes benefit the hinterland instead of the Capital-Region.*
Opportunities	**Threats**
▪ *Good coordination between cultural actors working at ground level: Cultural Plan (RAB-BKO), "Metropolitan Brussels" (Chamber of Commerce), etc.* ▪ *Growing awareness of the links between cultural economy, creativity, innovation and growth (since 2009-2010)*	▪ *Growing urban social fracture and social unrest* ▪ *Increasing competition from the other urban regions of Europe*

Fig. 2 SWOT of the possible role of culture in the territorial development of the Brussels Region. Source: INEUM Consulting (2010: 1)

The Cultural and Creative Cities Monitor 2019 provides a tool for assessing and benchmarking 190 European cities from 30 countries. It is composed of three major facets of the cultural, social, and economic vitality of a city that break out into nine dimensions (Fig. 3). 29 indicators correspond to the 9 dimensions and provide quantitative data measuring cultural performance at urban level (Montalto et al., 2017; 2019a, b) (Van Puyenbroeck et al., 2021).

According to the Cultural and Creative Cities Monitor, in 2017, Brussels was a leading city in culture and creativity within its population group.[1] Indeed, it was ranked fifth out of 21 cities under this category (Fig. 4). For the first subindex, cultural vibrancy, Brussels is ranked 10th; for the second subindex, creative economy, it is ranked 3rd; while for the third subindex, enabling environment, it is ranked 13th.

In 2017, under cultural vibrancy, Brussels exceled in the number of museums and art galleries and theatres which reaffirms the findings of the INEUM Consulting study in 2010. Under creative economy, Brussels outshined particularly under the dimension D2.3 New Jobs in Creative Sectors where it is positioned one of the top five cities in its population group especially for the following indicators: jobs in new arts, culture, and entertainment enterprises, jobs in new media and communication enterprises, and jobs in new enterprises in other creative sectors. This is an important improvement in comparison with the conclusions of the INEUM Consulting study

[1]Twenty-one cities with a population greater than one million inhabitants: in alphabetic order, Barcelona, Berlin, Birmingham, Brussels, Bucharest, Budapest, Cologne, Hamburg, Lille, London, Lyon, Madrid, Marseille, Milan, Munich, Paris, Prague, Rome, Sofia, Vienna, Warsaw

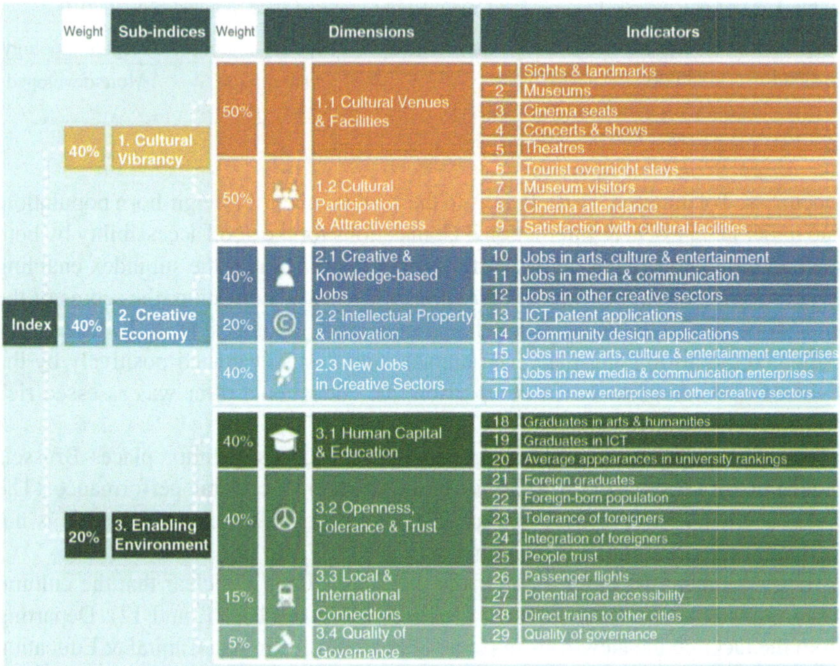

Fig. 3 The Cultural and Creative Cities Monitor's conceptual framework, weighting scheme, and indicators (Montalto et al., 2017)

Index Rank		Sub-index Ranks		
#	City-country [confidence intervals]	1. Cultural Vibrancy	2. Creative Economy	3. Enabling Environment
[XXL group] >1,000,000 inhabitants (21 cities)				
1	Paris-FR [1, 1]	1	1	2
2	Munich-DE [2, 2]	9	2	3
3	Prague-CZ [3, 6]	2	6	16
4	Milan-IT [3, 4]	3	11	8
5	Brussels-BE [5, 9]	10	3	13
6	Vienna-AT [4, 8]	4	17	5

Fig. 4 Top five cities in the C3 index per population group (source: Montalto et al., 2017)

which found at the time that although the number of businesses was rising, the number of jobs was decreasing (2010: 4). However, it is worthwhile mentioning that the scores of these indicators have been estimated due to missing data (Montalto et al., 2017: 70). Under enabling environment, Brussels scored highly under D3.2

Table 1 Dataset Brussels. Source: The Cultural and Creative Cities Monitor dataset 2019

City	GDP group	Employment rate group	Population group	Region typology
Brussels	>45,000	63–70%	>One million XXL	More developed

Openness, Tolerance & Trust, thanks to the high number of foreign-born population, and under D3.3 Local & International Connections for the good accessibility by both road and rail. Nonetheless, it is worth mentioning that under the subindex enabling environment particularly under D3.1 Human Capital & Education, the scores of the indicators, graduates in arts and humanities and graduates in ICT, have been estimated due to missing data. The human capital was evaluated positively by the INEUM Consulting study. In comparison, the educational offer was assessed rich but fragmented (2010: 3).

The 2019 edition of the Cultural and Creative Cities Monitor places Brussels 92nd out of 190 European cities in terms of overall cultural performance (The Cultural and Creative Cities Monitor 2019) (Annex 9). Moreover, Brussels is not anymore one of the top five per population group (Table 1).[2]

If we compare the index's data from 2017 and 2019, it is clear that the cultural performance has not much evolved (Table 2) (Annexes 10, 11, and 12). Departing from the fact that the New Jobs in Creative Sectors and Human Capital & Education data were estimated due to missing data, we assume that working toward enhancing these two dimensions will boost the cultural performance of the Brussels Region. According to the UNCTAD[3] "the quantity and quality of a country's 'human capital' set the parameters for success" (2004: 3).

In 2016, the Regional Circular Economy Programme was adopted by Brussels regional government. This ambitious program is starting to show its fruits in the sectors of interest of the region, namely, construction, resources and waste, logistics, and commerce. Although the region is a hotbed of diversified cultural and creative industries (CCIs), the contribution of CCIs to this regional green transition has not yet been really taken into consideration neither in the regional circular economy agenda nor by the lifelong learning programs.

The cultural and creative industries are more concerned with continuing education and learning programs that can accelerate the implementation of innovation by giving its workforce a mix of creative, digital, managerial, network, and entrepreneurial skills (EC, 2018; OECD, 2018). A new wave of innovation is always correlated with an adaptive curricula and training programs (Ost & Saleh, 2021). As new needs emerge in CCIs, a new education market should offer tailored teaching and learning opportunities for CCI stakeholders. Although the movement is global

[2]Twenty cities with a population greater than one million inhabitants: in alphabetic order, Barcelona, Berlin, Birmingham, Brussels, Bucharest, Budapest, Cologne, Hamburg, London, Lyon, Madrid, Milan, Munich, Paris, Prague, Rome, Rotterdam, Sofia, Vienna, Warsaw.

[3]UNCTAD: United Nations Conference on Trade and Development

Table 2 Comparison between Brussels 2017 and 2019 index. Source: The Cultural and Creative Cities Monitor dataset 2019

C3 index 2017	C3 index 2019	Weight	Subindex	2017 score	2019 score	Weight	Dimensions	2017 score	2019 score
24,28	24,64	0,4	1. Cultural vibrancy	20,22	20,19	0,5	D1.1 Cultural Venues & Facilities	16,35	16,47 ↑
						0,5	D1.2 Cultural Participation & Attractiveness	24,10	23,91 ↓
		0,4	2. Creative economy	24,38	25,05	0,4	D2.1 Creative & Knowledge-Based Jobs	33,77	35,97 ↑
						0,2	D2.2 Intellectual Property & Innovation	17,85	10,84 ↓
						0,4	D2.3 New Jobs in Creative Sectors	18,24	21,24 ↑
		0,2	3. Enabling environment	32,18	32,73	0,4	D3.1 Human Capital & Education	27,07	28,77 ↑
						0,4	D3.2 Openness, Tolerance & Trust	33,18	33,23 ↑
						0,15	D3.3 Local & International Connections	37,18	37,18 =
						0,05	D3.4 Quality of Governance	50,08	47,07 ↓

and indicates the common goal of the education sector to face these challenges, CCIs draw on a very wide range of artistic and cultural actors. Multidisciplinary programs are needed, as well as international and intercultural programs that reflect the global reach of the new wave of innovation (VVA et al., 2021).

In November 2020, the European Commission launched the Pact for Skills, a model of shared engagement for skills development in Europe. Five months later, the European Commission organized a high-level roundtable on skills for the ecosystem of cultural and creative industries. The objectives of this roundtable were to discuss upskilling and reskilling challenges of the cultural and creative industries workforce and identify the necessary support for skills development at European level and, more specifically, how the new European skills strategy, including the Pact for Skills, could support the different CCIs and secure commitments from participants engaged in the discussion regarding their contribution to the Pact for Skills. The Pact for Skills for the ecosystem of cultural and creative industries was formally launched on 28 April 2022 by the EC, Directorate General of Employment, Social Affairs, and Inclusion. In that occasion, the large-scale skills partnership for the cultural and creative industries ecosystem endorsed the Creative Pact for Skills (C-P4S) Manifesto which highlights the challenge, the ambition, the priority needed skills, the proposal and objectives, the road map, and the key performance indicators (KPIs) of this large-scale skills partnership.

To date, there is no agreed-upon definition of the cultural and creative industries, and discrepancies still exist on how to measure their social and economic impacts. A number of international organizations collect data and publish statistics with no harmonized categories or definitions (Eurostat, 2021, KEA, 2020a, b, UNCTAD, 2004, 2018, UNESCO Institute for Statistics 2009, UNCTAD and UNDP, 2008, OECD, 2020a, b; EY, 2021, UNESCO and The WORLD BANK, 2021). This lack of internationally comparable statistics hinders the design of tailored policies aimed at boosting CCIs (Montalto et al., 2017; Kloosterman et al., 2019; OECD, 2021). Scholars from H2020 project CICERONE conducted a thorough literature review of the existing different taxonomies and came to the conclusion that the definition of empirical measures is problematic because there are "three ways in which economic activities are classified for statistical purposes: occupation, industry and trade. There are two more significant categories of problem. First, the debate about 'creative and non-creative' activities... Second, the problem of material and immaterial/invisible goods and services" (Kloosterman et al., 2019: 13).

This research paper embraces the type of value embodied and generated by the cultural goods and services. According to Throsby, cultural goods and services are experience goods that can be defined according to the value they embody and/or generate. The duality of value is crucial for distinguishing cultural goods and services from commercial commodities. As highlighted by Throsby, "Cultural value...is multifaceted concept reflecting qualities such as the aesthetic, symbolic, spiritual or historical values attaching to a particular item" (Throsby, 2010: 20). Departing from the fact that cultural goods and services embody cultural value and yield cultural and commercial values, they are therefore valued by the person who makes them but also by those who consume them because of a cultural value attributed to the good/service. In addition to the duality of value, Throsby argues that cultural goods and services encompass three distinctive qualities: "they require some input of human creativity in their production; they are vehicles for symbolic messages for those who consume them; ...they contain, at least potentially, some intellectual property" (Throsby, 2010: 16).

Throsby elaborates on the relationship between the cultural goods and services and the creative goods and services and suggests that the former is a subset of the latter which he defines as "products that require some reasonably significant level of creativity in their manufacture, without necessarily satisfying other criteria that would enable them to be labelled 'cultural'" (Throsby, 2010: 16–17). The main difference is the value that these products embody and the one(s) they produce.

For the sake of this research paper, we define cultural and creative industries (CCIs) as the sectors and activities that rely on the input of human creativity for creating symbolic, economic, and social values through intellectual property and frame of mind that prospect an integrated value chain (creation, production, dissemination, distribution, and engagement).

3 Mapping Methodology

Under the framework of C-SHIP[4] project, in April 2021 and with the help of an intern,[5] a preliminary desk mapping of the cultural and creative industries in the Brussels Region kicked off. Throsby's concentric circles model (Fig. 5) theoretical framework was adopted, and the CCIs were classified accordingly (Throsby, 2001, 2008). It should be noted that this tentative rendering of CCI classification is open to discussion. Cultural significance may vary from places to places, and there is no global consensus about "what makes culture and arts unique." Although the core creative arts may be commonly considered as central, other disciplines are not. Let us just mention the example of cultural heritage made of both tangible and intangible assets, whose specific significance may change across different cultural context. It is also important to highlight that external circles gather activities directly or indirectly related to core creative arts, with the methodological constraint of dealing with very large industries. Advertising, for example, has a component of artistic creativity but relies in a substantial way on marketing and management fields. This preliminary mapping of 1170 CCIs (Fig. 6) took into consideration the 19 municipalities of the Brussels Region and consisted of desk research (Table 3). It is worth mentioning that this mapping is neither conclusive nor exhaustive. It is instead a first step toward creating a common understanding of the richness and diversity of existing actors and

---

[4]C-SHIP is a professional training program in cultural entrepreneurship aimed at making culture sustainable, impactful and viable (https://www.ichec.be/en/c-ship-project).

[5]In total, three interns worked on this mapping exercise. The first intern worked on building the database and creating the map. The second and third intern worked on updating the dataset.

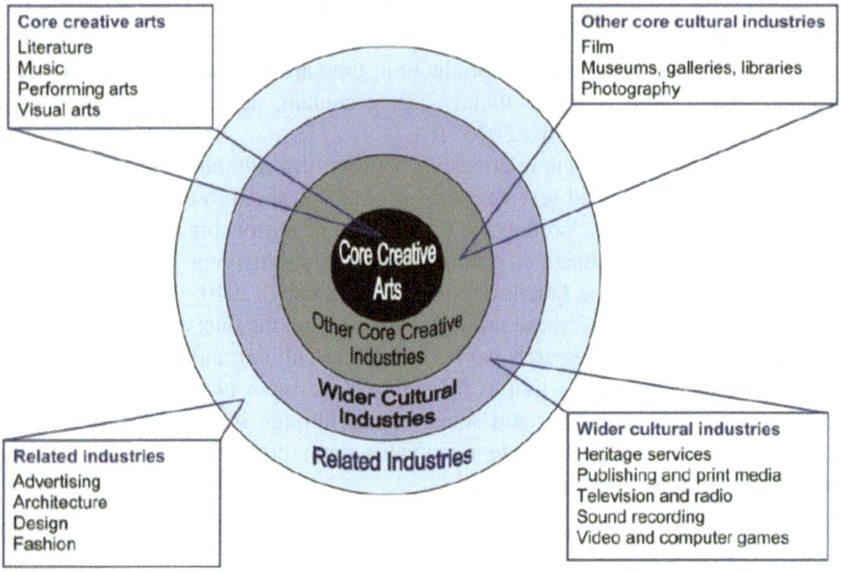

Fig. 5 The concentric circles model of the cultural industries (Source: Throsby, 2008)

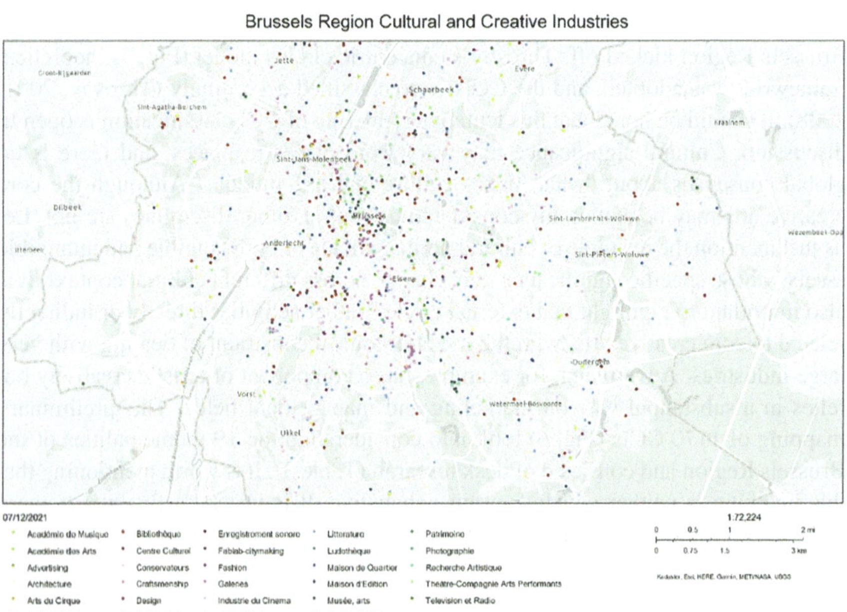

Fig. 6 Brussels Region cultural and creative industries, preliminary mapping

Table 3 Mapped municipalities and number of mapped CCIs in the Brussels Region

Mapped municipalities in the Brussels Region	
Municipality	Number of mapped CCIs
Anderlecht	50
Auderghem	28
Berchem-Sainte-Agathe	9
Bruxelles-Ville	307
Etterbeek	46
Evere	21
Forest	62
Ganshoren	10
Ixelles	179
Jette	15
Koekelberg	19
Molenbeek-Saint-Jean	51
Saint-Gilles	90
Saint-Josse-ten-Noode	26
Schaerbeek	97
Uccle	71
Watermael-Boitsfort	33
Woluwe-Saint-Lambert	24
Woluwe-Saint-Pierre	32
Total	1170

activities. A nugget of information aimed at triggering reflection on the needs and how the ecosystem could be enhanced in the future (Table 4).

4 Upskilling/Reskilling Needs

In order to better understand the needs, a short bilingual questionnaire (FR/ENG) was launched. It was available from 22 September until 30 November 2021. The questionnaire was disseminated via a mailing list composed of 1176 email addresses[6] collected online, to a list of directors of the key cultural players, and via LinkedIn. Although the English version was viewed by 435 persons and the French version by 237 persons on LinkedIn, only 29 persons completed the French questionnaire, and 6 persons completed the English version, out of which 1 person was not from the Brussels Region. Therefore, for the English versions, only 5 questionnaires were taken into consideration. In total, 34 questionnaires were taken into consideration. Table 5 summarizes the social and institutional framework of the

[6] In compliance with the privacy regulation, the emails of the recipients were not visible.

Table 4 Cultural and creative industries in the Brussels Region classification

Circle	Activity	Typology
Core cultural industries	Literature	Authors, scriptwriters, poets and illustrators, novelist, screenwriter, dramatists, comics cartoonists, and caricaturist
	Music	Academy of music, music house, opera houses, concert halls, associations
	Performing arts and visual arts	Theatre/performing companies, cirques art, art academies, artist atelier, curators, artistic research, artistic production, multidisciplinary artistic groups/programs
	Craftsmanship	Craftsmen, academy of arts and crafts
Other core cultural industries	Film	Creation, production, postproduction, and distribution houses in addition to polyvalent centers renting equipped halls and facilities
	Museums	Arts museum and other types of museums, cultural and arts centers, foreign cultural institutes, community centers, youth centers, cultural associations, and foundations
	Galleries	Galleries
	Libraries	Libraries and game libraries (ludothèque)
	Photography	Photography studios, foundations, exhibition, production, and distribution spaces focused on creative photography
Wider cultural industries	Heritage services	Associations for the promotion of heritage, heritage conservators/restorers, archaeological sites
	Publishing and print media	Editing and publishing houses on comic books, visual and performing arts. Publishing houses all domains, diffusion-distribution of French-, Dutch-, and English-language press and books, distributors and publisher of digital books
	Television and radio	Radio stations and television
	Sound recording	Sound recording studios
	Video and computer games	Design and implementation of interactive experiences in the fields of education, culture, tourism, communication, and leisure. Phygital technologies at the service of heritage/creator of immersive experiences for heritage
Related industries	Advertising	Advertising companies
	Architecture	Architectural studios
	Design	Design studios
	Fashion	Fashion design studios
	Fablab-citymaking	Fablabs

sample. Table 6 outlines the sector distribution. Table 7 condenses the professional representation. Finally, Table 8 encapsulates the expressed training preference.

As seen in the structural precarious nature of the CCIs, the sample confirmed the diversity of sector typologies and the multiplicity of jobs/activities performed by

Table 5 Sample social/institutional framework

Gender	#	Age range	#	Experience in the cultural field	#	In which type of organization do you work?	#	What is your work status	#
Female	16	20–24	3	1–5 years	8	Public sector (institution, organization, agency, etc.)	10	Full time	16
Male	17	25–34	5	5–10 years	5	Private for profit (institution, foundation, organization, etc.)	9	Part time	4
Other	1	35–44	10	10–15 years	4	NGO (nongovernmental organization)	0	Freelancer	8
		45–54	8	15–20 years	4	IGO (intergovernmental organization)	0	Volunteer	3
		55–64	6	More than 20 years	13	Self-employed	8	Student	3
		65+	2			Other	7		
Total	34	Total	34	Total	34	Total	34	Total	34

person. In terms of training needs, business model innovations, technology, ecosystem and partnership building, sustainability and positive impacts, and internal and external governance were considered pertinent. The relevance of these topics to CCIs was explored by recent research and publications (Polivtseva, 2020; OECD, 2020a, b; KEA, 2020b; VVA et al., 2021).

It appears that CCI actors want to learn how to develop sustainable business models because, by its very nature, the sector is content-driven and is eager to develop its management skills. It can be argued that such skills are now crucial for accessing funding mechanisms (Kern, 2020).

In addition, there is an increasing difficulty in project management, partnership building, and strategy (Ost & Saleh, 2021). This is probably due to the uncertainties of the pandemic but also to the new historical issues related to the digital environment and how to exploit it in an ethical and fair way and how to copyright in the digital age (VVA et al., 2021).

2021 kicked off as the International Year of Creative Economy for Sustainable Development. In this regard, the UN General Assembly highlighted that "the creative economy encourages creativity and innovation in attaining inclusive, equitable and sustainable growth and development, while facilitating life transitions and supporting women, youth, migrants and older persons, as well as empowering people in vulnerable situations" (UN, 2019: 3). In a post-pandemic scenario, resilience is a very critical issue, and it should be read primarily in terms of professional status and how to make work become more sustainable (Comunian & Lauren, 2020), as one-fourth of the respondents to the abovementioned questionnaire are self-employed. The recent report of the European Network of Culture and Audiovisual Experts, which investigated the status and working conditions of artists and creative

Table 6 Sample sector distribution

Circle	Activity	Typology
Core cultural industries	Literature	0
	Music	Lyrics/concerts (1), music (1), music teacher, concert performer (1)
	Performing arts and visual arts	Plastic arts (1), theatre (direction, comedian), cultural education (1), visual arts (1), dance (1), performance (1), dance, performing and visual arts (1)
	Craftsmanship	Craftsmen, academy of arts and crafts
Other core cultural industries	Film	Film and cinema production (3), audiovisual production (2)
	Museums	Museum (7), cultural center (1)
	Galleries	0
	Libraries	0
	Photography	0
Wider cultural industries	Heritage services	Built heritage (1), higher education (1), cultural service of a municipality (1), urban development, neighborhood committee, history, education (1),
	Publishing and print media	History and literature, editing, and publishing (1)
	Television and radio	Radio (1), journalism, and audiovisual production (1)
	Sound recording	0
	Video and computer games	Web design (1), media/XR for heritage/culture (1)
Related industries	Advertising	0
	Architecture	0
	Design	0
	Fashion	0
	Fablab-citymaking	0

professionals, showed that a consistent number of freelancers in the sector have not been able to take advantage of any of the public rescue packages for various technical/bureaucratic reasons (EENCA, 2020: 51).

5 CCIs and Cultural Entrepreneurs Upskilling/Reskilling as Drivers for Local, Inclusive, and Sustainable Tourism Development

At the beginning of this research paper, reference was made to the long-wave economic theory which constitutes the current context of the cultural and creative industries. It was mentioned that creation is a basis for innovation as bringing new

Table 7 Sample profession representation

What is your profession?	
Independent art consultant	Producer of cultural events
Producer, editor, graphic designer, ASBL coordinator, colorist, etc.	Concertist, music teacher
Author and lyricist/running a theatre café and writing workshops	Dancer
Journalist	Production coordinator
Museum administrative and financial officer	Actress
Cultural center director	Dancer, costume designer
Comedian, theatre director	Architect
Website project support	Lecturer
Program coordinator	Audiovisual producer
General coordinator	Film producer
Producer/screenwriter	Cinema producer
Editor	Museum scientific attaché
Artist	Museum communication (press service)
Head of culture and heritage department	Museum head of audiences and activities
Retired	Museum sales manager
Trainee journalist/production assistant	Museum educator
Director	Museum curator
Total	34

products and services into the market. Tourism is no exception to such process, first, because it is one of the most emblematic industries of the postwar growth phase leading up to the Golden Sixties. This phase saw the opening of borders, the development of trade, and major innovations in transportation. The resulting growth in tourism has created not only tens of millions of jobs around the world but also new skills and occupations, likewise in any long movement of economic growth.

The economic downturn of the 1970s, due to the recognition of the limits of growth and the extensive waste of non-exhaustible resources, made the rapid development of tourism being reconsidered. It took several decades for the tourism sector to reach diminishing marginal returns which forced players to reposition themselves in an increasingly segmented market. It will be recalled that low-cost travel is one of the examples of repositioning of players for whom wage and cost pressures no longer made it possible to generate enough growth margins. Major players have disappeared by more sustainable tourism activities. This has been manifested with innovative formulas among others: agritourism, ecotourism, participatory cultural tourism, smart (low and high tech) and inclusive tourism, etc. (OECD, 2020b, 2021).

CCI's own upskilling/reskilling corresponds exactly to that moment of transition when product and service innovations will be implemented through organizational innovations. The arrival of the Internet and the social networks has enabled the shift

Table 8 Sample expressed training preference

Is business model innovation relevant to your career?	#	Is technology relevant to your career?	#	Is ecosystem and partnership building relevant to your career?	#	Is sustainability and positive impact relevant to your career?	#	Is internal and external governance relevant to your career?	#
5	20	5	12	5	13	5	12	5	11
4	5	4	12	4	13	4	9	4	8
3	5	3	7	3	2	3	9	3	11
2	3	2	2	2	5	2	2	2	1
1	1	1	1	1	1	1	2	1	3
Total	34	Total	34	Total	34	Total	34	Total	34

from a growth of mass tourism to more sustainable, individual, and local forms of tourism. The current trend is not over because economic transition has not yet affected all tourism sectors, in particular transportation (think about cruise ships) and the hospitality sector (think of large infrastructures) (OECD, 2021).

According to the United Nations World Tourism Organization (UNWTO), 2019 was the tenth year of sustained growth which witnessed the arrival of 1.5 billion international tourists. COVID-19 wreaked havoc and caused the loss of 74% of international tourists registering, thus, 381 million international tourists bringing tourism back again to the levels of 30 years ago.[7]

This is when the pandemic highlighted the growing need to contribute to local growth and well-being while cherishing and strengthening solidarity and bonds with peers around the world. Keeping an eye on what is happening internationally goes hand in hand with the enhancement of local resources and their reinterpretation according to contemporary local needs. In fact, the pandemic completes this cultural transition by bringing a new challenge to tourism-related sectors. Obstacles to travel, social distancing measures, or travel restrictions have sparked a resurgence of local, inclusive, and sustainable activities. It is in such transitional economic context that cultural creativity must be revisited to find sustainable and local tourism development niches.

The concern for implementing upskilled capacities in the field of sustainable tourism in Brussels can be observed in some initiatives, for example, the training initiatives of the Brussels Quality Academy[8] which is capitalizing on post-COVID experience and new practices in order to offer practitioners comprehensive training on CCI's contribution to sustainable tourism. In particular, the Academy aims to help the sector provide quality work by offering training modules on ongoing activities in Brussels. In addition to training courses to improve knowledge about the region and customer experience know-how, the academy will soon be offering additional support through brand-new training courses in the following areas: crisis management, digital marketing, public relations, social networks, organization of events, and eco-responsible businesses. The aim of such trainings is not just an awareness-raising concept for actors and practitioners in the field of sustainable tourism but also to create a network of multidisciplinary (co)-creation initiatives in developing strategies, products, and services through active collaboration. Moving forward together in sustainable tourism requires coordinated individual and collective insights of both supply and demand sides. Sustainable tourism cannot succeed only by bringing new digital technologies on supply side that make tourists' experience more satisfactory. It also requires commitment from demand-side consumers, users of heritage, visitors of museums and historic places, and engaged tourism intermediaries. Similarly in 2021, ICHEC Brussels Management School launched C-SHIP (Cultural Entrepreneurship), an action learning training program aimed at enabling professionals in the cultural and creative industries to ensure

[7] https://www.unwto.org/covid-19-and-tourism-2020 (accessed 21 September 2021)

[8] https://visit.brussels/en/article/brussels-quality-academy (accessed 20 February 2022)

financial viability and develop and/or strengthen their management capacities for a sustainable, flourishing, and resilient activity while maximizing their societal contribution and their positive impacts.

Recently, a growing number of CCIs are embracing the circular economy philosophy in Brussels. We will explore some of the emerging innovative added value activities and experiences. In terms of local know-how and craftsmanship, les Ateliers de la rue Voot[9] is a cooperative association which offers weekly artistic workshops on ceramics, sculpture, drawing, digital creation, photography, and film and original training courses focused on sustainability. It raises local awareness about cultural and creative activity environmental impacts and promotes responsible attitudes. Recovery, repair, and recycling are at the heart of their repair bicycle workshop, while self-building and maintenance are at the basis of the solar techniques' workshop. The Ateliers de la rue Voot is a place where curious tourists could meet locals and learn about sustainable artistic innovations but also rent second-hand refurbished bicycles. The Ateliers de la rue Voot's project "a bicycle for 10 years" is a member of circlemade[10] network.

Another interesting initiative is "Fais-le-toi-même" (Do it yourself) which defines as a "network, a place, to unite, share and consolidate the sustainable creation sector."[11] It facilitates and promotes the use of recycled materials in innovative ways in order to reduce the environmental impact of design and the creative sectors at large. This network built a collaborative ecosystem which allows each artistic project to evolve independently while being part of a supportive community sharing a common vision. It also encourages the participation of neighbors and consults with local businesses as essential components of the ecosystem's development. A yearly Christmas market of handmade sustainable artistic products is organized by the network in two municipalities of the Brussels Region as well as creative thematic artisanal markets.

The use of virtual technologies introduces innovations that allow mass consumption to avoid creating bottlenecks and makes accessible remote or invisible places. This is the philosophy behind establishing Mixua ecotourism and XR4heritage. Mixua[12] offers ecotours, online workshops, and a webshop of zero-waste, handmade ecological products. XR4heritage[13] brings together cultural mediators, scriptwriters, creatives, tourism and education professionals, social innovators, and professionals of emerging technologies with the objective of putting XR[14] at the service of heritage. Its motto is XR for inclusion, education, and preservation. XR4heritage

[9]https://voot.be/ (accessed 21 September 2021)

[10]Circlemade is Brussels Region network for circular economy pioneers. https://www.circlemade.brussels/en/ (accessed 21 September 2021)

[11]https://faisletoimeme.be/ (accessed 21 September 2021)

[12]https://en.mixua.be/ (accessed 21 September 2021)

[13]https://xr4heritage.com/ (accessed 21 September 2021)

[14]XR includes mixed reality and interactivity, virtual experience, sensorial activations, video mapping, 360° exhibition, augmented reality.

offers three services: the lab, a B2B incubator for innovative territories; the factory, a sustainable attractions factory; and the desk, a tailor-made service to boost projects.

In terms of circular design and fashion design, there is a rich variety of initiatives among others, Isatio, Wear a Story, Konligo, and Design with Sense. Isatio[15] is a Belgian brand which creates 100% circular textile. All the products are handmade and produced in Brussels in collaboration with local suppliers and artisans in a closed loop. It is based on a collection of unique pieces made from new coupons and samples recovered from the textile industry. Isatio, thus, transforms a negative externality (waste) into a positive externality (high-value clothing) and environmentally friendly. Wear a Story[16] offers unique women garments named after a woman who could have been the old owner. The upcycled old garments into modern cuts are made in Brussels. Konligo[17] offers sustainable and locally produced innovative technology for outdoor and indoor event structures. Design with Sense[18] is a cooperative project for the creation of interior spaces and manufacturing in wood. It is led by architects, craftsmen, and designers who are convinced that the creative process must be centered on people and materials. The reuse of materials is embraced in the entire value chain. The cooperative gives priority to rational use and extension of the life cycle of materials. Tourists have the pleasure of enjoying eco-design in shops, coffee shops, and restaurants built with local sourcing and reused sustainable materials.

The common denominator of these innovative projects/initiatives is the critical entrepreneurial skills ignition. Thanks to the vast network of subsidized business incubators in Brussels Region, most of these projects were either incubated or followed closely for the entirety of the project or for some specific aspects/initiatives. Different entities provide diverse packages, but in general the incubation boosts the business model and business plan, improves the marketing strategy, accompanies entrepreneurial initiative toward sustainability, and/or enhances the partnership and networking capacities of the project/initiative. In 2020, hub.brussels[19] commissioned Dr. Marlen Komorowski to analyze the economic impact of CCIs in Brussels-Capital Region. The study demonstrated that CCIs play an important role in the region's economy. It was illustrated that CCIs generated around €3213.3 million of net value added in 2018 which represents a contribution of 3.8% to the Brussels region economy. According to the study, in 2018, CCIs generated a turnover of more than 12,511.6 million euros (Komorowski, 2020).

[15] https://en.isatio.com/a-propos (accessed 21 September 2021)

[16] https://www.wearastory.be/ (accessed 21 September 2021)

[17] https://www.konligo.com/ (accessed 21 September 2021)

[18] http://designwithsense.be/ (accessed 21 September 2021)

[19] hub.brussels is Brussels Agency for Business Support which offers enterprises (including CCIs) from the region a wide range of free advice, services, and tools to help them be successful and sustainable. https://hub.brussels/en/ (accessed 28 February 2022)

6 Conclusions

The preliminary mapping presented in this research paper is neither conclusive nor exhaustive. It aims to open the door to fieldwork, enriching the mapped activities, and to further explore CCI training needs in the Brussels Region. It would be ideal to combine desk research with field research in the future and to enhance the core creative arts by mapping also individual artists/creatives.

The new qualifications and skills of CCIs must be grafted onto the innovative factors of day/proximity local, inclusive, and sustainable tourism. Similar developments can allow regions to re-appropriating local benefits of tourism, in closed loops, without the risk of resource exploitation while enhancing local know-how and skills. This is also aligned with the New European Bauhaus which aims to make the Green Deal a cultural, human-centered, and positive, tangible experience based on its three principles: sustainability (including circularity), quality of experience (including aesthetics), and inclusion (including accessibility and affordability).

It emerges from this research that innovative cultural projects/initiatives need critical entrepreneurial skills ignition. It is clear from both the questionnaire sample and the earmarking Brussels-based cultural entrepreneurs that business and operation model innovation is a critical training requirement. In addition, sustainability in terms of sustainable working conditions, viable projects, and positive sustainable impacts is also a key training component. Finally, ethical and cooperative governance, solid partnerships, and disruptive technologies are subsidiary success ingredients for a holistic upskilling training of CCIs.

Annexes

Annex 9 Cities index ranking

#	Cities	C3 index score	#	Cities	C3 index score	#	Cities	C3 index score
1	Paris (FR)	65,95	64	Turku (FI)	28,12	127	Zagreb (HR)	21,03
2	Zurich (CH)	49,25	65	Athens (EL)	28,11	128	Saint-Étienne (FR)	20,94
3	Bern (CH)	46,87	66	Linz (AT)	28,08	129	Salamanca (ES)	20,73
4	Copenhagen (DK)	46,78	67	Madrid (ES)	27,9	130	Genoa (IT)	20,55
5	Lisbon (PT)	46,34	68	Norwich (UK)	27,8	131	Cluj-Napoca (RO)	20,52
6	Basel (CH)	44,42	69	Poznan (PL)	27,77	132	Bradford (UK)	20,48

(continued)

Annex 9 (continued)

#	Cities	C3 index score	#	Cities	C3 index score	#	Cities	C3 index score
7	Stockholm (SE)	43,78	70	Hannover (DE)	27,58	133	Leiden (NL)	20,47
8	Luxembourg (LU)	42,21	71	Mannheim (DE)	27,4	134	Szeged (HU)	20,47
9	Munich (DE)	41.22	72	Ljubljana (SI)	27,33	135	Birmingham (UK)	20,44
10	Dublin (IE)	41,03	73	Warsaw (PL)	26,99	136	Brescia (IT)	20,28
11	Geneva (CH)	40,13	74	Antwerp (BE)	26,85	137	Essen (BE)	20,09
12	Stuttgart (DE)	40,12	75	Limerick (IE)	26,69	138	Brno (CZ)	19,96
13	Amsterdam (NL)	39,21	76	Cologne (DE)	26,62	139	Seville (ES)	19,92
14	Lund (SE)	38,88	77	Montpellier (FR)	26,58	140	Namur (BE)	19,91
15	Weimar (DE)	37,66	78	Nantes (FR)	26,58	141	Győr (HU)	19,81
16	Florence (IT)	37,56	79	Granada (ES)	26,57	142	Sofia (BG)	19,7
17	Heidelberg (DE)	36,99	80	Rome (IT)	26,34	143	Maribor (SI)	19,56
18	Glasgow (UK)	36,23	81	Amersfoort (NL)	26,14	144	Matera (IT)	19,56
19	London (UK)	36,11	82	Kortrijk (BE)	25,96	145	Lecce (IT)	19,52
20	Dresden (DE)	35,47	83	Lyon (FR)	25,89	146	Bilbao (ES)	19,15
21	Milan (IT)	35,31	84	Wrocław (PL)	25,89	147	Veszprém (HU)	19,08
22	Karlsruhe (DE)	35,08	85	Aarhus (DK)	25,85	148	Osijek (HR)	19,01
23	Frankfurt (DE)	34,64	86	Tampere (FI)	25,84	149	Katowice (PO)	18,92
24	Helsinki (FI)	34,63	87	Pula (HR)	25,58	150	Nitra (SK)	18,68
25	Venice (IT)	34,48	88	s-Hertogenbosch (NL)	25,56	151	Ostend (BE)	18,61
26	Cork (IE)	34,62	89	Santiago (ES)	25,16	152	Guimarães (PT)	18,57
27	Bristol (UK)	34,19	90	Faro (PT)	24,92	153	Ravenna (IT)	18,47
28	Tallinn (EE)	34,06	91	Trieste (IT)	24,70	154	Liège (BE)	18,26
29	Berlin (DE)	33,98	92	Brussels (BE)	24,64	155	Sibiu (RO)	18,24
30	Tartu (EE)	33,76	93	Maastricht (NL)	24,34	156	Lille (FR)	18,22
31	Brighton & Hove (UK)	33,51	94	Riga (LT)	24,23	157	Liepāja (LV)	18,12
32	Vienna (AT)	33,39	95	The Hague (NL)	24,23	158	Timişoara (RO)	18,1
33	Oslo (NO)	33,25	96	Uppsala (SE)	24,12	159	Limassol (CY)	18,06

(continued)

Annex 9 (continued)

#	Cities	C3 index score	#	Cities	C3 index score	#	Cities	C3 index score
34	Eindhoven (NL)	33,19	97	Bergen (NO)	24,09	160	Pécs (HU)	17,91
35	Budapest (HU)	32,78	98	Coimbra (PT)	24,04	161	Split (HR)	17,68
36	Prague (CZ)	32,66	99	Waterford (IE)	23,93	162	Thessaloniki (EL)	17,47
37	Manchester (UK)	32,4	100	Odense (DK)	23,87	163	Marseille (FR)	17,39
38	Graz (AT)	32,27	101	Bruges (BE)	23,59	164	Lleida (ES)	17,26
39	Edinburgh (UK)	31,97	102	Liverpool (UK)	23,33	165	Naples (IT)	17,23
40	Nuremberg (DE)	31,93	103	Stavanger (NO)	22,83	166	Zaragoza (ES)	17,2
41	Bratislava (SK)	31,9	104	Umea (SE)	22,82	167	Bochum (DE)	17,12
42	Avignon (FR)	31,7	105	Bucharest (RO)	22,81	168	Łódź (PO)	17,1
43	Galway (IE)	31,36	106	Leeuwarden (NL)	22,75	169	Terrassa (ES)	16,56
44	Malmo (SE)	31,23	107	Rotterdam (NL)	22,67	170	Karlovy vary (CZ)	16,18
45	Vilnius (LT)	31,16	108	Turin (IT)	22,58	171	Iasi (RO)	16,12
46	Barcelona (ES)	31,15	109	Bordeaux (FR)	22,5	172	Veliko Tarnovo (BG)	15,91
47	Krakow (PL)	31,1	110	Espoo (FI)	22,43	173	Rijeka (HR)	15,89
48	Mainz (DE)	30,4	111	Norrköping (SE)	22,41	174	Burgos (ES)	15,86
49	Utrecht (NL)	30,24	112	Gdansk (PO)	22,23	175	Braga (PT)	15,64
50	Porto (PT)	30,17	113	Dundee (UK)	22,23	176	Olomouc (CZ)	15,63
51	Gothenburg (SE)	29,81	114	Valletta (MT)	22,22	177	Pilsen (CZ)	15,4
52	Bologna (IT)	29,69	115	Perugia (IT)	22,12	178	Debrecen (HU)	15,18
53	Leuven (BE)	29,59	116	Cagliari (IT)	21,95	179	Las Palmas (ES)	15,14
54	York (UK)	29,59	117	Nicosia (CY)	21,95	180	Torun (PO)	14,64
55	Leipzig (DE)	29,58	118	Sintra (PT)	21,78	181	Cordova (ES)	14,64
56	Hamburg (DE)	29,12	119	Kaunas (LT)	21,72	182	Baia Mare (RO)	14,22
57	Leeds (UK)	29,03	120	Klaipeda (LT)	21,66	183	Varna (BG)	13,77
58	Toulouse (FR)	28,98	121	San Sebastián-Donostia (ES)	21,66	184	Košice (SK)	13,23

(continued)

Annex 9 (continued)

#	Cities	C3 index score	#	Cities	C3 index score	#	Cities	C3 index score
59	Bremen (DE)	28,76	122	Parma (IT)	21,51	185	Lublin (PO)	12,83
60	Groningen (NL)	28,59	123	Valencia (ES)	21,34	186	Presov (SK)	12,42
61	Trento (IT)	28,58	124	Mons (BE)	21,32	187	Ostrava (CZ)	12,26
62	Ghent (BE)	28,26	125	Pesaro (IT)	21,26	188	Plovdiv (BG)	12,08
63	Nottingham (UK)	28,24	126	Limoges (FR)	21,25	189	Kalamata (EL)	11,75
						190	Patras (EL)	10,05

Source: Cultural and Creative Cities Monitor dataset 2019

Annex 10 Final data Brussels 2017 and 2019

Final data Brussels 2019		Unit 0–100			
Subindex	Dimensions	Weight	Indicators	Brussels score 2017	Brussels score 2019
1.Cultural vibrancy	D1.1 Cultural Venues & Facilities	50%	Sights and landmarks	13,58	13,40
		50%	Museums and art galleries	24,01	26,54
		100%	Cinemas	9,42	9,25
		100%	Concert and music halls	13,60	13,42
		100%	Theatres	23,58	23,24
	D1.2 Cultural Participation & Attractiveness	50%	Tourist overnight stays	21,29	21,29
		100%	Museum visitors	22,98	22,69
		100%	Cinema attendance	26,46	26,06
		100%	Satisfaction with cultural facilities	24,29	24,29
2.Creative economy	D2.1 Creative & Knowledge-Based Jobs	100%	Jobs in arts, culture, and entertainment	48,58	50,91
		100%	Jobs in media and communication	29,16	30,14
		100%	Jobs in other creative sectors	33,58	26,86
	D2.2 Intellectual Property & Innovation	100%	ICT patent applications	14,77	6,09
		100%	Community design applications	20,94	15,60
	D2.3 New Jobs in Creative Sectors	100%	Jobs in new arts, culture, and entertainment enterprises	13,78	17,22

(continued)

Annex 10 (continued)

Final data Brussels 2019		Unit 0–100			
		100%	Jobs in new media and communication enterprises	18,54	26,42
		100%	Jobs in new enterprises in other creative sectors	22,41	20,07
3.Enabling environment	D3.1 Human Capital & Education	100%	Graduates in arts and humanities	23,46	27,66
		100%	Graduates in ICT	9,37	10,25
		100%	Average appearances in university rankings	48,39	48,39
	D3.2 Openness, Tolerance & Trust	100%	Foreign graduates	23,16	31,34
		100%	Foreign-born population	76,69	77,76
		100%	Tolerance of foreigners	20,37	20,37
		100%	Integration of foreigners	20,00	20,00
		100%	People's trust	16,67	16,67
	D3.3 Local & International Connections	100%	Accessibility to passenger flights	34,74	34,74
		100%	Accessibility by road	40,98	40,98
		100%	Accessibility by rail	35,84	35,84
	D3.4 Quality of Governance	100%	Quality of governance	50,08	47,07

Source: Cultural and Creative Cities Monitor dataset 2019

Annex 11 Index for the 19 cities with a population greater than 1 million inhabitants 2017

Cities			Subindex ranks					
(XXL group) >1,000,000 inhabitants (21 cities)								
#	C3 index	city	1. Cultural vibrancy score	City	2. Creative economy score	City	3. Enabling environment score	City
1	64,62	Paris	56,27	Paris	81,36	Paris	56,05	London
2	42,68	Munich	41,18	Prague	55,05	Munich	47,85	Paris
3	35,88	London	39,11	Vienna	44,65	Bucharest	41,62	Barcelona
4	35,01	Prague	33,98	Milan	37,40	Hamburg	41,12	Munich
5	34,72	Milan	32,31	Berlin	39,25	London	41,02	Birmingham
6	34,46	Vienna	31,08	Munich	38,31	Budapest	40,70	Madrid
7	33,97	Berlin	27,92	Rome	37,29	Milan	39,03	Vienna
8	29,92	Hamburg	27,71	Barcelona	37,14	Warsaw	33,85	Berlin
9	29,74	Barcelona	26,66	Lyon	35,69	Berlin	32,18	Brussels
10	28,36	Budapest	23,85	Cologne	34,71	Prague	31,05	Milan
11	27,02	Madrid	22,66	Hamburg	33,16	Sofia	29,45	Hamburg
12	27,00	Bucharest	22,44	London	27,91	Cologne	29,38	Cologne
13	26,58	Cologne	20,22	Brussels	27,52	Vienna	26,14	Lyon
14	26,56	Warsaw	20,71	Budapest	27,15	Madrid	23,78	Budapest
15	25,54	Rome	20,05	Madrid	25,83	Barcelona	23,28	Prague
16	24,84	Lyon	19,67	Warsaw	25,59	Rome	22,91	Bucharest
17	24,28	Brussels	16,01	Birmingham	24,38	Brussels	20,65	Rome
18	19,59	Birmingham	11,40	Bucharest	22,37	Lyon	19,18	Warsaw
19	18,53	Sofia	7,97	Sofia	12,45	Birmingham	10,37	Sofia

Source: Cultural and Creative Cities Monitor dataset 2019

Annex 12 Index for the 20 cities with a population greater than 1 million inhabitants 2019

Cities			Subindex ranks					
(XXL group) >1,000,000 inhabitants (20 cities)								
#	C3 index	city	1. Cultural vibrancy score	City	2. Creative economy score	City	3. Enabling environment score	City
1	65,95	Paris	56,45	Paris	83,23	Paris	56,01	London
2	41,22	Munich	40,62	Prague	51,87	Munich	50,42	Paris
3	36,11	London	39,18	Vienna	47,32	Budapest	41,61	Munich
4	35,31	Milan	33,17	Milan	40,07	London	41,45	Birmingham
5	33,98	Berlin	31,58	Berlin	39,91	Milan	40,61	Barcelona
6	33,39	Vienna	30,38	Munich	37,46	Warsaw	39,88	Madrid
7	32,78	Budapest	29,48	Rome	36,59	Berlin	37,46	Vienna
8	32,66	Prague	28,19	Barcelona	35,90	Sofia	33,56	Berlin
9	31,15	Barcelona	26,55	Lyon	35,86	Hamburg	32,73	Brussels
10	29,12	Hamburg	23,24	Cologne	34,13	Bucharest	31,35	Rotterdam
11	27,90	Madrid	22,44	Hamburg	29,38	Barcelona	30,50	Cologne
12	26,99	Warsaw	22,20	Rotterdam	29,33	Madrid	30,40	Milan
13	26,62	Cologne	22,19	London	28,06	Prague	30,13	Lyon
14	26,34	Rome	22,13	Budapest	28,06	Cologne	29,02	Hamburg
15	25,89	Lyon	20,49	Madrid	25,82	Rome	25,96	Prague
16	24,64	Brussels	20,19	Brussels	25,56	Vienna	25,01	Budapest
17	22,81	Bucharest	19,85	Warsaw	25,05	Brussels	21,96	Bucharest
18	22,67	Rotterdam	16,73	Birmingham	23,11	Lyon	21,09	Rome
19	20,44	Birmingham	11,91	Bucharest	18,79	Rotterdam	20,36	Warsaw
20	19,79	Sofia	8,14	Sofia	13,66	Birmingham	10,87	Sofia

Source: Cultural and Creative Cities Monitor dataset 2019

References

Blaug, M. (1980). *The methodology of economics or how economists explain*. Cambridge University Press.

Comunian, R., & Lauren, E. (2020). Creative and cultural work without filters: Covid-19 and exposed Precarity in the creative economy. *Cultural Trends, 29*(2), 112–128. https://doi.org/10.1080/09548963.2020.1770577

IDEA Consult, Goethe-Institut, Amann, S., & Heinsius, J. (2021). *Research for CULT committee–cultural and creative sectors in post-Covid-19 Europe: Crisis effects and policy recommendations, European Parliament, policy department for structural and Cohesion Policie*. Retrieved February 10, 2022 from https://www.europarl.europa.eu/RegData/etudes/STUD/2021/652242/IPOL_STU(2021)652242_EN.pdf

Delbeke, J. (1984). Recent long wave theories: A critical survey. In C. Freeman (Ed.), *Long Waves in the World Economy* (pp. 1–10). Frances Pinter.

EENCA. (2020). *The status and working conditions of artists and cultural and creative professionals*. Retrieved February 10, 2022 from http://www.eenca.com/index.cfm/publications/the-status-and-working-conditions-of-artists-and-cultural-and-creative-professionals/

European Commission. (2018). *The role of public policies in developing entrepreneurial and innovation potential of the cultural and creative sectors. Report of the OMC (Open method of coordination) working group of Member States' experts–Study*. Retrieved February 10, 2022 from https://op.europa.eu/en/publication-detail/-/publication/5d33c8a7-2e56-11e8-b5fe-01aa75ed71a1

Eurostat. (2021). *Culture statistics–cultural employment*. Retrieved February 10, 2022 from https://ec.europa.eu/eurostat/statistics-explained/index.php?title=Culture_statistics_-_cultural_employment#Self-employment

EY. (2021). *Rebuilding Europe: The cultural economy before and after the COVID-19 crisis, GESAC–The European Grouping of Societies of Authors and Composers*. Retrieved May 13, 2021 from https://1761b814-bfb6-43fc-9f9a-775d1abca7ab.filesusr.com/ugd/4b2ba2_1ca8a0803d8b4ced9d2b683db60c18ae.pdf

Florida, R. (2002). *The rise of the creative class and how it's transforming work, leisure, community and everyday life*. Basic Books.

Ineum Consulting. (2009). *Culture–A symbolic or economic success factor for urban development planning? Study*. Retrieved May 13, 2021 from http://www.forum-avignon.org/en/surveys

INEUM Consulting. (2010). What strategies for sustainable employment and urban development planning? Case-study: Brussels. Retrieved May 13, 2021 from http://www.forum-avignon.org/en/surveys

KEA, and PPMI. (2019). Research for CULT committee–culture and creative sectors in the European Union–Key future developments, challenges and opportunities. In *European Parliament. Policy department for structural and cohesion policies*. Retrieved May 13, 2021 from https://www.europarl.europa.eu/RegData/etudes/STUD/2019/629203/IPOL_STU(2019)629203_EN.pdf

KEA European Affairs. (2020a). The impact of the COVID-19 pandemic on the cultural and creative sector. *Report for the Council of Europe*. Retrieved May 13, 2021 from https://keanet.eu/wp-content/uploads/Impact-of-COVID-19-pandemic-on-CCS_COE-KEA_26062020.pdf.pdf

KEA European Affairs. (2020b). New market analysis of the cultural and creative sectors in Europe. *European Investment Fund*. Retrieved May 13, 2021 from https://keanet.eu/wp-content/uploads/ccs-market-analysis-europe-012021_EIF-KEA.pdf.

Kern, P. (2020). *The future of cultural policies* (1st ed.). Les Editions Kea.

Kloosterman, R. C, Pratt, A. C, d'Ovidio, M, Greco, L, Borén, T (2019). *Creative and cultural industries and global production networks approaches so far*. A brief review of GPN literature and its relevance for the creative and cultural industries. *(CICERONE report D1.1)*. Retrieved May 13, 2021 from https://doi.org/10.5281/zenodo.4535064.

Komorowski, M., (2020). L'impact économique des industries culturelles et créatives dans la Région de Bruxelles-Capitale. *Hub Brussels*. Retrieved February 10, 2022 from https://hub.brussels/fr/etude-les-industries-culturelles-et-creatives-a-bruxelles/

Montalto V, Jorge Tacao Moura C, Langedijk S, & Saisana M (2017). *The cultural and creative cities monitor*. https://doi.org/10.2760/58643.

Montalto V, Tacao Moura C. J, Alberti V, Panella F, & Saisana M (2019a). *The cultural and creative cities monitor. 2019 edition, EUR 29797 EN*, Publications Office of the European Union.

Montalto, V., Moura, C. J. T., Langedijk, S., & Saisana, M. (2019b). Culture counts: An empirical approach to measure the cultural and creative vitality of European cities. *Cities, 89*, 167–185.

OECD. (2018). *Conference on culture and local development: Creating creative Jobs: Parallel session B1, discussion note*. Retrieved May 13, 2021 from http://www.oecd.org/cfe/leed/venice-2018-conference-culture/documents/B1-DiscussionNote.pdf

OECD. (2020a). *Culture shock: COVID-19 and the cultural and creative sectors*. OECD Publishing. Retrieved May 13, 2021 from http://www.oecd.org/coronavirus/policy-responses/culture-shock-covid-19-and-the-cultural-and-creative-sectors-08da9e0e/.

OECD. (2020b). *OECD tourism trends and policies 2020, OECD Publishing*. Retrieved May 13, 2021 from https://doi.org/10.1787/6b47b985-en.

OECD. (2021). *G20 Rome guidelines for the future of tourism: OECD Report to G20 Tourism Working Group, OECD Tourism Papers, 2021/03, OECD Publishing*. Retrieved February 10, 2022 from https://doi.org/10.1787/d11080db-en

OECD and ICOM. (2019). *Culture and local development: Maximising the impact: A guide for local governments, communities and museums*, OECD Local Economic and Employment Development (LEED) Papers, No. 2019/07, OECD Publishing. Retrieved May 13, 2021 from https://doi.org/10.1787/9a855be5-en.

Ost, C., & Saleh, R. (2021). *Cultural and creative sectors at a crossroad: From a mainstream process towards an active engagement*. Built Heritage. *5*, 14 Retrieved May 13, 2021 from https://doi.org/10.1186/s43238-021-00032-y.

Polivtseva, E (2020). *The moment for change is now COVID-19 learning points for the performing arts sector and policy-makers*. Retrieved May 13, 2021 from https://www.ietm.org/en/publications/the-moment-for-change-is-now-covid-19-learning-points-for-the-performing-arts-sector

Potts, A. (2021). *The role of culture in climate resilient development, UCLG committee on culture reports, n°10, and climate heritage network* (Working Group 5).

Throsby, D. (2001). *Economics and culture*. Cambridge University Press.

Throsby, D. (2008). The concentric circles model of the cultural industries. *Cultural Trends, 17*(3), 147–164. https://doi.org/10.1080/09548960802361951

Throsby, D. (2010). *The economics of cultural policy*. Cambridge University Press. https://doi.org/10.1017/CBO9780511845253

Throsby, D. (2018). Culturally sustainable development: theoretical concept or practical policy instrument? In A. Kangas, N. Duxbury, & C. De Beukelaer (Eds.), *Cultural policies for sustainable development* (pp. 5–19). Routledge.

UN. (2019). *International year of creative economy for sustainable development, 2021. General Assembly, Seventy-fourth session Second Committee*, A/C.2/74/L.16/Rev.1. Retrieved May 13, 2021 from https://undocs.org/A/C.2/74/L.16/Rev.1

UNESCO Institute for Statistics. (2009). *2009 UNESCO framework for cultural statistics (FCS)*. https://unesdoc.unesco.org/ark:/48223/pf0000191061

UNESCO and The WORLD BANK. (2021). *Cities, culture, creativity: Leveraging culture and creativity for sustainable urban development and inclusive growth*. Retrieved February 10, 2022 from https://documents1.worldbank.org/curated/en/104121621556036559/pdf/Cities-Culture-Creativity-Leveraging-Culture-and-Creativity-for-Sustainable-Urban-Development-and-Inclusive-Growth.pdf

UNCTAD. (2004). *UNCTAD XI High-level Panel on Creative Industries and Development Agenda*. Retrieved May 13, 2021 from http://www.unctad.org/en/docs/tdximisc1_en.pdf

UNCTAD and UNDP. (2008). The creative economy report. *The challenge of assessing the creative economy: towards informed policy-making*. UNCTAD/UNDP. Retrieved May 13, 2021 from https://unctad.org/system/files/official-document/ditc20082cer_en.pdf

UNCTAD (2018). *Creative economy outlook: Trends in international trade in creative industries 2002-2015.* Country profiles 2005–2014. United Nations Publications. Retrieved May 13, 2021 from https://unctad.org/system/files/official-document/ditcted2018d3_en.pdf

Van Puyenbroeck, T., Montalto, V., & Saisana, M. (2021). Benchmarking culture in Europe: a data envelopment analysis approach to identify city-specific strengths. *European Journal of Operational Research, 288*(2), 584–597.

Voices of culture. (2021). *Culture and the UN sustainable development goals: Challenges and opportunities'*. Report. Retrieved May 13, 2021 from https://voicesofculture.eu/2020/09/21/culture-and-the-sustainable-development-goals-challenges-and-opportunities/

VVA. (2021). *Creative FLIP final report, work package 2 on learning*. Retrieved May 13, 2021 from http://creativeflip.creativehubs.net/wp-content/uploads/2021/07/FINAL-WP2_Final-Report-on-Skills-mismatch-2.pdf

Part II
New Heritage(s) in the Spotlight

Industrial Tourism as a New Opportunity for Cultural Tourism Development in the Post-pandemic Era

Antonella Garofano, Angelo Riviezzo, and Maria Rosaria Napolitano

Abstract The main aim of the present chapter is to provide an exhaustive theoretical framework of the industrial tourism phenomenon, deepening its various definitions and manifestations. At the same time, we discuss some best practices and experiences of industrial tourism in Italy, providing data on its real impact in terms of visitors. The final aim is to shed light on the real potentialities of industrial tourism for promoting cultural tourism sustainable development. A broad definition of industrial tourism is adopted in this chapter, including three different forms and typologies of attractors: industrial archaeology, corporate museums, and factory tours. The study focuses on Italy, where, compared to the pioneering European countries, only more recently this phenomenon emerged.

1 Introduction

Before the pandemic crisis stopped the world, the limitless and uncontrolled growth of tourism was determining direct and indirect environmental and social damages to several places, especially to those attracting unsustainable flows of mass tourism. In this sense, even if the current COVID-19 pandemic crisis represents one of the biggest challenges the tourism sector has ever faced, it also offers to decision-makers an unmissable opportunity to exploit heritage resources for the development of sustainable tourism (Beretić et al., 2021). Effectively, the need for a sustainable

A. Garofano (✉)
Department of Economics, University of Campania 'Luigi Vanvitelli', Capua, Italy
e-mail: antonella.garofano@unicampania.it

A. Riviezzo
Department of Law, Economics, Management and Quantitative Methods, University of Sannio, Benevento, Italy
e-mail: angelo.riviezzo@unisannio.it

M. R. Napolitano
Department of Business and Quantitative Studies, Parthenope University of Naples, Naples, Italy
e-mail: mariarosaria.napolitano@uniparthenope.it

E. Borin et al. (eds.), *Cultural Leadership in Transition Tourism*, Contributions to Management Science, https://doi.org/10.1007/978-3-031-14121-8_6

paradigm for the tourism sector had been widely recognized by scholars and policymakers well before COVID-19 (Hunter, 1997; Hardy et al., 2002; Saarinen, 2006; Higgins-Desbiolles, 2018). Tourism, in fact, is one of the main sectors interested by the 2030 Agenda (UNWTO, 2017), and 2 of the 17 declared sustainable development objectives make explicit reference to this sector (Rosato et al., 2021). With the numerous travel restrictions due to the pandemic, more people have returned to nature and to slow tourism, engaging more with domestic travel (Benjamin et al., 2020). Alongside the huge economic losses, the forced suspension of over-tourism has brought unimaginable benefits to environmentally fragile destinations, including some of the most popular destinations of world cultural tourism, such as Venice or Florence in Italy (Istat, 2020). In the last 2 years, in fact, the number of international tourist arrivals has dramatically decreased, and, despite the high recovery capacity of tourism, it will presumably take a few years to return to 2019 numbers. Actually, it is still not possible to know how the "new normal" will be, but now is the time to reflect and redesign the tourism industry, in order to make it more sustainable and more equitable (Benjamin et al., 2020). Mainstream tourism practices have been severely disrupted by COVID-19, while new opportunities have arisen from the very same disruptions for emergent niche tourism activities (Ioannides & Gyimóthy, 2020). Specifically, the continuous growth of domestic tourism opens up new possibilities of a region's sustainable development, favoring the redistribution of visitors' flows to larger areas and less-known destinations. This scenario represents a great opportunity for inner areas rich in heritage, which are far from the most consolidated tourist circuits and generally represent a marginal choice, compared to the most famous tourist destinations (Mantegazzi et al., 2021). In this new landscape, significant opportunities arise for industrial tourism, "a type of tourism which involves visits to operational companies and industrial heritage" (Otgaar et al., 2010, p. 1). This particular niche of cultural tourism, combining pleasure with education, includes a broad spectrum of destinations and places, which are in many cases the stages of an alternative path of territorial discovery (Garofano et al., 2021). In this sense, industrial tourism experiences may lead the way to sustainable tourism development, since they give both firms and territories the opportunity to exploit a historical and cultural heritage often concentrated in places unknown to mass tourism. Therefore, not only firms that open their doors for visitors secure their reputation and promote their brands and products, but they also build sustainable relations with society (Otgaar, 2012). Particularly suitable for intercepting the new trends of the demand, industrial tourism represents a relatively unexplored research theme in the broader field of cultural tourism (Richards, 2018) and is still in need of clearer concepts and definitions.

The main aim of the present chapter is to provide an exhaustive theoretical framework of the industrial tourism phenomenon, deepening its various definitions and manifestations. At the same time, we discuss some best practices and experiences of industrial tourism in Italy, providing data on its real impact in terms of visitors.

The final aim is to shed light on the real potentialities of industrial tourism for promoting cultural tourism sustainable development.

2 Defining Industrial Tourism as a Multifaceted Niche of Cultural Tourism

In the last decades, cultural tourism has been recognized as a major component of international tourism consumption (Richards, 2018). The parallel and fast growth of research on cultural tourism has revealed fragmentation into a number of emerging niches, including heritage tourism, arts tourism, creative tourism, gastronomic tourism, and film tourism (UNWTO, 2018). This fragmentation of the concept is reflected in a broader operational definition from the UNWTO General Assembly, held in 2017(b) in China: "Cultural tourism is a type of tourism activity in which the visitor's essential motivation is to learn, discover, experience and consume the tangible and intangible cultural attractions/products in a tourism destination" (p. 18). As highlighted by Richards (2018), the shift from tangible to intangible heritage is a major trend of cultural tourism research and practice. The much broader nature of contemporary cultural tourism highlights the necessity to include not just sites and monuments but also a wide range of cultural practices. In this scenario, industrial tourism emerges as a relevant niche, offering unique opportunities for firms, but also for cities and territories (Otgaar et al., 2010). Industrial tourism is a complex concept encompassing many different tourist activities, and, despite the growing interest in the last years, a univocal definition is still missing. The most established definition refers to industrial heritage tourism, whose development is linked to the interest in artifacts and symbols of the manufacturing past (Hospers, 2002). In the case of industrial heritage tourism, former industrial sites are recovered and organized to facilitate their reuse by visitors of various kinds (Cole, 2004). This form of tourism is directly attributable to the area of cultural tourism, as it is based on the conversion of industrial sites into cultural tourist attractions and on the enhancement of industrial heritage as a component of a country's culture.

It must be noted, however, that industrial heritage tourism represents a narrow view of the phenomenon, which has gradually expanded its scope to include corporate museums and factory tours aimed at witnessing production processes in real time (Xie, 2015). In its broader definition, therefore, industrial tourism includes a set of heterogeneous activities and experiences, which go far beyond the mentioned industrial archaeology, consisting of disused production sites, old factories, and workers' villages, places that testify to the industrialization process of the territory (Edwards & Llurdés, 1996). A central role in the development of industrial tourism is played by corporate museums and all those places that preserve evidence of the past of companies as well as the territories in which they find hospitality (Napolitano et al., 2018; Riviezzo et al., 2021a). Compared to corporate museums, business archives play a limited role in the diffusion of industrial tourism, while they are entrusted with the crucial function of preserving and safeguarding the company's documentary heritage.[1] Another form to be included in the wider framework of industrial tourism is living industry tourism (Savoja, 2011), focused on active companies, which decide to open the doors of their factories and/or laboratories to allow visitors to discover the secrets of the processes and the most ancient traditions kept within the company walls (Garofano et al., 2017). In this perspective, living companies may attract different types of tourists interested for several reasons to understand "how things work" (Mitchell & Mitchell, 2001) and offer a particular view on production processes and traditions that are at the core center of local culture.

Thus, despite the growing interest that emerged in the last decades toward industrial tourism, it still remains a relatively unexplored research topic (Otgaar et al., 2010), and the meanings given to the term may vary significantly depending on the local context. In Germany and France, for instance, this form of cultural tourism has been generally associated with industrial heritage, intended as the complex of testimonies of the industrial past hosted in nonoperational firms (Frew, 2000). In the same countries, indeed, visits to operational and living firms are identified with specific terms, such as *Werktourismus* in Germany or *visit d'enterprises* in France. In the recent literature, however, a broader definition of industrial tourism is generally accepted, including three different forms and typologies of attractors (Garofano et al., 2021; Riviezzo et al., 2021a): industrial archaeology (Edwards & Llurdés, 1996), corporate museums (Nissley & Casey, 2002; Montella, 2018), and factory tours (Savoja, 2011; Garofano et al., 2017).

The main aim of this chapter is to try to shed light on the differences between these three typologies and manifestations of industrial tourism, providing some empirical evidence on the real impact they can have to enhance the cultural tourism attractiveness of specific places, by linking the creativity and quality of production processes with new kinds of tourist experience.

To this aim, several secondary data sources have been integrated, such as institutional reports on tourism, field studies, official websites of corporate museums and industrial heritage sites, and specialized press.

In particular, our study focuses on Italy. Compared to the pioneering European countries, only more recently Italy has paid adequate attention to industrial tourism both in terms of theoretical reflection and concrete development programs. A more comprehensive view of industrial tourism in the country is still lacking, and an investigation into possible forms of integration between industrial heritage and living industries seems to be interesting and challenging.

[1] For more information about Italian legislation on business archives and their management, see *Codice dei beni culturali e del paesaggio*, Legislative Decree 22/01/2004, n. 42, and subsequent amendments and additions.

3 Industrial Archaeology: On the Traces of Local Production Traditions of the Past

Industrial heritage represents an important component of a country's patrimony, as it serves as a witness of industrial processes which have changed society over time (Edwards & Llurdés, 1996). Industrial archaeology deals with the interpretation of the surviving physical evidence to understand past human industrial and manufacturing activities. Even if the terms industrial archaeology and industrial heritage are often taken to be synonymous, the distinction between the two meanings has gradually become crucial to the acceptance of industrial archaeology as an academic discipline (Palmer, 2005). In the present study, however, the term industrial archaeology is used to identify nonoperational industrial sites (industrial heritage) and distinguish them from operational firms, considering both the categories as possible attractions of industrial tourism (Otgaar et al., 2010).

According to Edwards and Llurdés (1996), industrial heritage tourism "is concerned with the development of touristic activities and industries on man-made sites, buildings and landscapes that originated with industrial processes from earlier periods" (p. 342). Industrial archaeology itineraries, therefore, allow the visitor to discover the identity and peculiarities of a territory starting from its industrial history. The broad definition of industrial archaeology includes past production plants, industrial villages, railway stations, and mines, along with service infrastructures, such as aqueducts, bridges, and dams (Cole, 2004). In the last decades, there has been an increasing emphasis on the display and interpretation of industrial archaeology, and many sites representing evidence of former industrial and manufacturing activities have undergone restoration and conservation. In the last years, a growing number of contributions highlighted the importance of reorienting the tourism industry to sustainable tourism by exploiting the postindustrial heritage, especially in declining industrial areas (Kuzior et al., 2021). Indeed, the development of postindustrial tourism not only reduces the negative impact of industrial archaeology on the environment but also provides appropriate conditions for regional restructuring (Hospers, 2002; Yang et al., 2019).

Despite the growing interest toward the tourism potential of industrial heritage, there is still a lack of systematic knowledge about many aspects of industrial archaeology, and former industrial sites usually present a high level of fragmentation.

Italy presents a rich heritage of industrial archaeology, even if most former industrial areas and sites do not converge in a structured tourist offer yet. Thus, industrial heritage tourism is difficult to quantify, both in terms of attractions and in terms of visitors. In the latest survey on "museums and similar institutions," Istat (2021)—the National Institute of Statistics—included 33 industrial archaeology artifacts, i.e., old factories, mines, and so on. The surveyed industrial sites are mainly located in northern Italy and in Sardinia and in 2019 recorded over 700,000 visitors. According to Istat (2021), the most visited industrial archaeology site in Italy is the *Risiera di San Sabba* in Trieste, a rice processing plant built at the end of the

Table 1 The most visited Italian industrial heritage sites in 2019

Attractions	Region	Number of visitors
Risiera di San Sabba	Friuli Venezia Giulia	137,164
Mulino di Bienno	Lombardy	65,000
Porto Flavia	Sardinia	40,567
Complesso Monumentale Belvedere di San Leucio	Campania	35,983
Campo di Concentramento e Transito di Fossoli	Emilia Romagna	28,200
Manifattura dei Marinati	Emilia Romagna	20,000
Sacrario dei Caduti D'oltremare – Sacrario Militare di Bari	Puglia	19,000
MuseumHinterPasseier – Bunker Mooseum	South Tyrol	10,008
Il Filatoio Rosso di Caraglio	Piedmont	9855
Complesso Minerario Montevecchio – Percorso S. Antonio	Sardinia	8200
Complesso Minerario Montevecchio – Percorso Officine	Sardinia	8200
Complesso Minerario Montevecchio – Percorso Piccalinna	Sardinia	8200

Source: Istat (2021)

nineteenth century,[2] followed by the *Mulino di Bienno*, an ancient and characteristic mill located in a small village in the province of Brescia (Table 1). *Porto Flavia*, the service infrastructure of the mining area of Masua, in southwestern Sardinia, appears as a spectacular "suspended" port in the middle of a cliff, and it is an increasingly popular tourist destination. Among the few sites surveyed in the south of Italy, the monumental complex of *Belvedere di San Leucio* (Caserta, Campania region) represents one of the most important examples of silk manufacturing.[3] The site, which includes the factory, the workers' village, and the Silk Museum, has been declared World Heritage Site by UNESCO.

Among other important industrial archaeology sites included in the World Heritage List, it is worth mentioning the following: *Ivrea*, an industrial city of the twentieth century;[4] the workers' village *Crespi d'Adda*, dating back to the end of the nineteenth century;[5] and the *Rhaetian Railway*, an engineering and architectural masterpiece that connects Italy to Switzerland through a long and breathtaking labyrinth of valleys and mountain passes.[6]

[2] https://risierasansabba.it/

[3] https://www.belvederedisanleucio.it/

[4] https://www.ivreacittaindustriale.it/

[5] https://villaggiocrespi.it/

[6] https://www.rhb.ch/it/home

Among the most visited industrial archaeology sites in Italy, there are also several mines, such as the *Great Mine of Serbariu*, in Sardinia,[7] anchor point of one of the itineraries promoted by ERIH (European Route of Industrial Heritage), the European network that brings together the most important industrial archaeology sites.[8] Italian industrial archaeology also includes numerous places and industrial buildings that have been at the core of redevelopment and conversion processes. Among the most significant experiences emerge the tower of Porto Marghera, known as the Venezia Heritage Tower, the only example in Europe of the conversion of an old cooling tower into an exhibition center;[9] the former Montemartini thermoelectric power plant in Rome, the first public plant for the production of electricity;[10] and the former Fiat Lingotto factory in Turin, today an important multifunctional center which also houses a shopping center, a multiplex cinema, and the Agnelli art gallery. Equally noteworthy is the Pirelli HangarBicocca, born from the recovery of two imposing former Breda buildings from the mid-twentieth century, now converted into one of the largest spaces for contemporary art in Italy.[11]

4 Corporate Museums: Capturing the Soul of Local Companies

Along with archives, to which they are often attached, corporate museums fulfill the dual role of preservation and enhancement of corporate heritage (Riviezzo et al., 2021a). In the last decades, they have emerged as precious sites of organizational memory (Nissley & Casey, 2002) and have been investigated as tools of heritage marketing, especially in the case of long-lived firms (Napolitano et al., 2018). Significant attention has been paid to corporate museums' role in the creation of a narrative organizational identity (Riviezzo et al., 2016) as well as on their importance for strengthening corporate image (Montella, 2010) and communicating the company's history and values to internal and external stakeholders (Montella, 2018). In recent years, corporate museums have turned into industrial tourism destinations, offering new opportunities of interaction between firms and society (Riviezzo et al., 2021b, 2022).

Corporate museums represent a consolidated phenomenon in industrialized countries, even if the lack of a definition univocally shared by management scholars and practitioners determines the difficulty of collecting data on their performance as tourist attractions.

[7] https://www.museodelcarbone.it/

[8] https://www.erih.net/

[9] https://www.veneziaheritagetower.com/progetto

[10] http://www.centralemontemartini.org/

[11] https://pirellihangarbicocca.org/#

In Italy, the phenomenon of corporate museums experienced considerable growth starting from the 1990s and continued to grow in recent years. In its latest analysis, Istat (2021) surveyed 136 corporate museums and archives. It should be noted that if the museum is the place traditionally dedicated to communicating the company's history and values to external and internal stakeholders, the archive is the main reservoir of sources and ideas through which the path taken by the company over time is reconstructed (Napolitano et al., 2018). Unlike the museum, which generally comes to life and nourishes itself through the historical research and reconstruction work promoted by the business archive, the latter, in most cases, is not open to the public, but only to scholars and experts.

Most of corporate museums surveyed by Istat are located in northern Italy (67%), with the highest numbers in Lombardy, Piedmont, and Emilia-Romagna, all regions with a deep-rooted industrial tradition. However, a growing number of corporate museums have been established in recent years in southern Italy, mainly by long-lived companies operating in food and wine sectors.

At a national level also, the highest number of Italian corporate museums surveyed is dedicated to food and beverage (Istat, 2021). Among the best known are "Giorgio Amarelli" Museum (liquorice),[12] Perugina Museum (chocolate),[13] Casa Martini (alcoholic drinks),[14] Lavazza Museum (coffee),[15] Campari Gallery (alcoholic drinks),[16] and Peroni Museum (beer).[17] The museum offer is wide also with regard to engines and transport systems, well represented by the Ducati Museum,[18] Ferrari Museum,[19] Lamborghini Museum,[20] Piaggio Museum,[21] and Fiat Historic Center.[22] Other well-known corporate museums include Salvatore Ferragamo Museum, Benetton Archive, Casa Zegna, and Rossimoda Shoe Museum in the fashion field and Alessi, Molteni, Poltrona Frau, and Kartell for design and furnishings.

With regard to visitors' numbers, according to Istat, in 2019, more than two million people visited Italian corporate museums, with flows of visits mainly polarized around the structures of greatest appeal. As shown in Table 2, the most visited museums represent the automotive sector, like the two Ferrari museums at the top of the ranking with more than 600,000 visitors in 2019. Remaining in the field of transport, the third most visited business museum in Italy is the National Railway

[12] https://www.museodellaliquirizia.it/

[13] https://www.perugina.com/it/casa-del-cioccolato/la-casa

[14] https://www.martini.com/casa-martini/

[15] https://www.lavazza.it/it/museo-lavazza.html

[16] https://www.campari.com/it/inside-campari/campari-gallery

[17] https://archiviostorico.birraperoni.it/

[18] https://www.ducati.com/it/it/borgo-panigale-experience

[19] https://www.ferrari.com/en-BA/museums/ferrari-maranello

[20] https://www.lamborghini.com/it-en/museo

[21] https://www.museopiaggio.it/

[22] https://www.fcaheritage.com/it-it/heritage/luoghi/centro-storico-fiat

Table 2 The most visited Italian corporate museums

Name	Region	Visitors
Museo Ferrari – Maranello	Emilia Romagna	401,000
Museo Enzo Ferrari – Modena	Emilia Romagna	205,000
Museo Nazionale Ferroviario di Pietrarsa	Campania	200,000
Mudetec, Museo delle Tecnologie Automobili Lamborghini	Emilia Romagna	112,014
La Macchina del Tempo – Museo Storico Alfa Romeo	Lombardy	90,000
Museo Piaggio	Tuscany	77,152
Museo Storico Perugina	Umbria	74,643
Museo Lavazza	Piedmont	65,000
Museo Ducati	Emilia Romagna	58,500
Museo – Ex Stabilimento Florio delle Tonnare di Favignana	Sicily	55,284
Museo della Liquirizia "Giorgio Amarelli"	Calabria	52,000
Museo Salvatore Ferragamo	Tuscany	41,802

Source: Istat (2021)

Museum of Pietrarsa in Naples, with as many as 200,000 visitors a year.[23] Set up in an industrial archaeology complex of great importance, the Royal Mechanical, Pyrotechnic and Locomotive Factory founded by Ferdinand II of Bourbon in 1840, the Pietrarsa Museum represents an excellent example of tourism enhancement of the industrial heritage.

Along with museums owned and managed by private companies, other categories of business museums should be considered as industrial tourism destinations, such as industrial museums, focused on the history and evolution of an industry, and district museums, which document the history of specific territorial production realities (Montella & Cerquetti, 2016). In this regard, some of the best-known experiences in Italy are the Textile Museum of Prato, once European capital of textiles; the International Museum of Ceramics, in Faenza, dedicated to the conservation and dissemination of the culture of ceramics; and the Pescia Paper Museum, which allows the visitor to discover how a sheet of paper could be produced by hand from a pile of rags.

The best practices mentioned above share some common elements, emerging from the analysis of multiple secondary sources. Specifically, in most cases, corporate museums not only contribute to strengthening corporate image but also represent valuable cultural resources of their territories, testifying productive traditions handed down over time (Riviezzo et al., 2022). Interestingly, almost all the main Italian corporate museums belong to *Museimpresa*, the national association of corporate museums and archives,[24] which offers its members several opportunities to share experiences and knowledge about corporate heritage enhancement, with positive outcomes both for firms and for territories. In line with many studies highlighting the importance of effective stakeholder engagement for tourism

[23] https://www.fondazionefs.it/content/fondazionefs/it/esplora-il-museo/visita-pietrarsa.html

[24] https://museimpresa.com/

development (Byrd, 2007; Waligo et al., 2013), the most successful experiences of industrial tourism driven by corporate museums seem to be based on the concrete collaboration between companies, policy-makers, and public institutions, which have a common interest in enhancing local cultural heritage preserved within museums' walls.

5 Factory Tours: Discovering How Things Work in Real Time

Even if industrial tourism is described as a relatively recent phenomenon, there is evidence from several countries in the world that factory tours have a long history (Otgaar et al., 2010).

Well-known examples include visits to French wineries or tours in some famous US distilleries, dating back to the nineteenth century. In recent years, visits to operational firms have become the core element of living industry tourism, an activity oriented toward the direct knowledge of production processes and products (Savoja, 2011). It should be noted that the term factory tour is not directly associated with tourism and leisure, since there are various reasons to visit a company, including educational and professional motives. However, factory tours meet the need for authentic and satisfying recreational experiences, based on the observation of what is "behind the scene" of manufacturing activities. As noted by Frew (2000), all firms can become tourist attractions if they want, including service companies, even if tangible goods are usually more attractive for visitors.

While in countries such as France and Spain factory tours are already widely established, in Italy, the visit to operational companies is a more recent and less structured phenomenon. Actually, the organization of factory tours is not new to many companies in the food and wine sectors, where living industry tourism presents some areas of overlap with other forms of experience tourism (Mitchell & Mitchell, 2001). However, factory tours are normally organized independently by companies, while there are only few cases in which industrial tourism is based on an organizational framework created with the collaboration of firms and other institutional actors. The most consolidated experience is represented by the industrial tourism program "Made in Turin. Tour the Excellent," established in 2005 on the proposal of the Chamber of Commerce and managed by Tourism Office of Turin Province. Still today, even considering some changes in the organizational framework, the project involves about 60 companies from various product sectors, which open their doors to individuals and groups. Food drives the project both in terms of the number of companies involved (including Lavazza, Guido Gobino, and Caffarel) and in attendance, since it represents just under half of admissions, but many other industries are involved, including automotive and aerospace.

Food is the leading industry of many other initiatives that allow firms to open their doors in a coordinated program of factory tours, organized throughout the year.

Among others, "Wefood" is an initiative that opens the doors of a hundred companies in the Northeast of Italy, by offering twice a year a weekend of factory tours and cooking show events. A similar initiative is "Visit Industry Marche," a project co-financed by the European Regional Development Fund for creative cultural enterprises, which has networked companies, services, and local authorities from this region of central Italy, building tourism itineraries that combine a visit to production realities with the discovery of natural and artistic resources. In the south of Italy, "Make it in Puglia" has been recently launched as an industrial tourism network that aims to support companies interested in opening their doors to visitors, offering services ranging from the definition of a standard for the provision of factory tours to their promotion and sale.

At a national level, "Open Factory" is one of the most important openings of industrial culture (Bonfanti et al., 2019). Now in its seventh edition, the project involves every year over 50 firms, which open their doors to strengthen the bond with the community and to make people perceive the social and economic value of factories.

Regarding visits carried out throughout the year, in many cases, companies that have a corporate museum also offer their visitors the opportunity to take a factory tour. This is the case, for example, of Birra Peroni in Rome, which has prepared a special walkway for visitors to the factory, a kind of historical journey that traces the technological evolution of the plant, or the case of Ferrari, which also offers to the corporate museum's visitors the opportunity to take a guided tour inside the gates of the historic Maranello factory. Remaining in the world of sport cars, for some years now, Maserati has been offering tourists and fans of the brand the opportunity to take a tour of the Modena plant to discover up closely how iconic cars, synonymous with luxury and elegance, are produced.

In some cases, factory tours are organized even in the absence of a corporate museum and bring visitors directly inside the company, to show them each phase of the productive process. An illustrative example is given by Callipo, historic canning in Calabria, well known above all for the production of tuna. The company has been using factory tours for many years as an important tool of relationship and dialogue with various target audiences, including not only tourists but also school groups and other stakeholders interested in the products and in the territory.

Remaining in the attractive segment of food and beverage industry, many distilleries—like the wineries—are increasingly welcoming tourists and visitors in a structured way, such as Poli Distillerie, based in Schiavon, in Veneto region. Strengthened also by the fascination exerted by the company museum, the well-known distillery has managed over time to prepare an effective factory tour system, welcoming up to 15,000 visitors in a month. Fundamental to achieving these results was the dedication of the owner, Jacopo Poli, who personally guides the visitors to discover the secrets of the entire grappa production process: from the distillation room you pass to the cellar where the grappa rests up to the end of the tour with an unmissable stop in the tasting room.

Tourism research has widely highlighted how the interdependence of multiple stakeholders makes the strategic planning of tourist destinations a particularly complex task (Jamal & Getz, 1995). In the most successful experiences of living industry tourism mentioned above, collaboration and networking brought together a range of companies, providing avenues to facilitate communication and knowledge sharing among them (Nguyen et al., 2019). Along with firms opening their doors to visitors, multiple private and public actors are called to take part in the creation of living industry tourism destinations and experiences. Since both territories and firms have an interest in a better image, they can find converging interests in industrial tourism based on factory tours. Obviously, the development of this form of tourism requires high organizing capacity, that is, the ability to create partnerships between actors from various sectors (including local governments, educational and knowledge institutions, and citizen groups) as well as the ability to create an environment in which such partnerships survive and develop (Otgaar et al., 2010).

6 Conclusions

Based on the observation of its main tools and manifestations, it seems clear that industrial tourism offers several opportunities for both firms and territories by enhancing the local cultural offer and satisfying the thirst for knowledge and the curiosity that characterize postmodern tourists. Nevertheless, these opportunities are often not really exploited, and the strategic contribution of industrial tourism to the competitiveness of local cultural tourism offer remains widely unexpressed.

Compared to other European countries that experienced the process of deindustrialization long before, Italy presents a clear delay, only partially made up with the efforts of public and private actors to remove several industrial heritage sites from deterioration. The data and the experiences we illustrated above clearly show that this form of tourism is attracting a rising group of visitors, even as a consequence of the recent pandemic crisis and the changes in touristic behaviors and preferences. In the post-lockdown scenario, a growing interest in the "behind the scenes" of factories pushed living industry tourism flows. Even if aggregated data on this form of tourism are not available, the positive trend is evidenced by the latest data available on food and wine tourism, which is obviously only part of a bigger and more complex phenomenon. According to the Nomisma-Unicredit Tourism Observatory, in 2020, the Italians spent 354 million euros by participating in themed tours or visiting farms, as a way to put aside the restrictions of the lockdown and rediscover moments of leisure and culture in nearby destinations, perceived as less risky. Therefore, industrial tourism experiences, in all the analyzed forms, may provide a pathway to connect people to firms, their heritage, and their host territories also in the post-pandemic era. Moreover, being typically short-break tourism, industrial tourism seems in line with both the needs of postcrisis tourists, particularly oriented toward destinations closer to their place of residence, and the promotion of more sustainable tourism, helping to reduce overall emissions (Romagosa, 2020).

To date, the promotion of industrial tourism is mainly entrusted to individual companies, which manage the visits independently, and to a few private agencies that include corporate tours in their offer. As shown above, there are some recent and interesting examples of pilot initiatives organized in Italy with the main aim to systematize the living industrial heritage of specific places, bringing together under a common organizational machine and a shared communication strategy wide and varied offers of corporate tours.

These last experiences clearly show the need of a strong leadership in the enhancement of this niche of cultural tourism. Entrepreneurs can often be leaders in their area and act as champions for promoting these activities, using their influence to mobilize others in their networks. These skills have been identified as important for tourism and community development leadership (Long & Nuckolls, 1994; Moscardo, 2005), since effective tourism leadership depends on strong community networks and the ability to include a wide range of people in tourism initiatives and decisions. Indeed, the real challenge in promoting industrial tourism seems to be the creation of strong community networks based on public-private partnerships. In fact, companies alone cannot guarantee a sustainable development of this form of tourism, for several reasons.

First, involving local firms in wide programs aimed at structuring the offer of industrial tourism is not an easy task. In most cases, as highlighted by empirical evidence (Otgaar et al., 2010; Garofano et al., 2017), the success of such initiatives depends on the proactiveness of promoters, which are often public institutions responsible for tourism development strategies. In these cases, one of the hardest phases of the project consists of engaging companies, whose interests may not converge with tourism development. In fact, as argued by Savoja (2011), in most cases, companies open their doors and enhance their heritage for purposes that have nothing to do with tourism, being primarily focused on company and brand image, product promotion and sale, corporate social responsibility, and education.

Second, it must be considered that transforming an industrial site into an attraction for tourists and visitors normally requires significant investments. Of course, the attractiveness of an industrial tourism product mainly depends on the company and its products. As revealed by our investigation, in fact, tourists have a special interest in visiting manufacturing firms that produce goods with a symbolic value for a specific region, based on old or fascinating production processes. Anyway, by investing in digital equipment, professional guides, communication, and marketing, firms can make more attractive their visits. And this is even more evident in the case of industrial archaeology that clearly needs relevant investments to become an interesting attractor for potential tourists.

Third, as noted in the literature (e.g., Riviezzo et al., 2021a; Garofano et al., 2017; Savoja, 2011), industrial archaeology, corporate museums, and living industries generally do not represent per se a push factor able to arouse the desire for travel in the potential tourists. Moreover, as mentioned above, industrial tourism is normally a short-break phenomenon, addressed to a niche of market that is often saturated by standard tourist attractions (Otgaar et al., 2010). In this line of thought, the development of industrial tourism is strongly influenced by other factors. Namely, the presence of "classic" tourist attractions such as beautiful landscapes, famous monuments, or events can make industrial tourism sites an interesting alternative for short-time visits or even create more appeal for second-time visitors (Garofano et al., 2017). Therefore, the best integration of industrial tourism attractors into the overall tourism offer of a specific place may represent a big opportunity. In fact, another way to make industrial tourism more attractive deals with its combination with other, more traditional, tourism attractions.

To sum up, industrial tourism can produce benefits for places and their industries, only with the support of public and private actors. To this aim, strong leadership and a high organizational capacity are required in order to create fundamental partnerships between stakeholders from various sectors and from different segments of tourism as well. Of course, different areas may use diverse definitions of industrial tourism, ranging from very narrow to very broad, and this makes it difficult to compare activities and results (Otgaar et al., 2010). However, policy-makers should take a broader view and pay particular attention to potential synergies between industrial heritage and living industry tourism, without neglecting the central role of corporate museums and science parks.

References

Benjamin, S., Dillette, A., & Alderman, D. H. (2020). "We can't return to normal": Committing to tourism equity in the post-pandemic age. *Tourism Geographies, 22*(3), 476–483.

Beretić, N., Talu, V., & Cecchini, A. (2021). Emergency-proof tourism: The heritage of industrial archaeology in internal areas as a potential for a sustainable tourism. *Sustainability, 13*, 3911. https://doi.org/10.3390/su13073911

Bonfanti, A., Castellani, P., Giaretta, E., & Brunetti, F. (2019). Developing entrepreneurial learning triggered by factory tours. *The Learning Organization, 26*(6), 574–587.

Byrd, E. T. (2007). Stakeholders in sustainable tourism development and their role: Applying stakeholder theory to sustainable development. *Tourism Review, 62*(2), 6–13.

Cole, D. (2004). Exploring the sustainability of mining heritage tourism. *Journal of Sustainable Tourism, 12*(6), 480–493.

Edwards, J. A., & Llurdés, J. C. (1996). Industrial heritage: Mines and quarries. *Annals of Tourism Research, 23*(2), 341–363.

Frew, E. A. (2000). *Industrial tourism: A conceptual and empirical analysis* [PhD thesis]. Victoria University.

Garofano, A., Riviezzo, A., & Napolitano, M. R. (2017). From industrial heritage to living industry tourism. An explorative study in Italy. In *II International 4 Conference "Tourism Dynamics and Trends"*, Universidad de Sevilla, Seville (Spain), 26-29 June.

Garofano, A., Mignone, G., Riviezzo, A., & Napolitano, M. R. (2021). Il Turismo industriale: una grande opportunità per il rilancio turistico e culturale del Bel Paese. In A. Morvillo & E. Becheri (Eds.), *Rapporto sul Turismo Italiano* (XXIV ed., 2019–2020, pp. 257–266). Cnr Edizioni.

Hardy, A., Beeton, R. J., & Pearson, L. (2002). Sustainable tourism: An overview of the concept and its position in relation to conceptualisations of tourism. *Journal of Sustainable Tourism, 10*(6), 475–496.

Higgins-Desbiolles, F. (2018). Sustainable tourism: Sustaining tourism or something more? *Tourism Management Perspectives, 25*, 157–160.

Hospers, G. (2002). Industrial heritage tourism and regional restructuring in the European Union. *European Planning Studies, 10*(3), 397–404.

Hunter, C. (1997). Sustainable tourism as an adaptive paradigm. *Annals of Tourism Research, 24*(4), 850–867.

Ioannides, D., & Gyimóthy, S. (2020). The COVID-19 crisis as an opportunity for escaping the unsustainable global tourism path. *Tourism Geographies, 22*(3), 624–632. https://doi.org/10.1080/14616688.2020.1763445

Istat. (2020). *Report Movimento Turistico in Italia* (Gennaio-Settembre 2020). Retrieved from https://www.istat.it/it/files/2020/12/REPORT_TURISMO_2020.pdf

Istat. (2021). *Indagine sui musei e le istituzioni similari*, Roma.

Jamal, T. B., & Getz, D. (1995). Collaboration theory and community tourism planning. *Annals of Tourism Research, 22*(1), 186–204.

Kuzior, A., Lyulyov, O., Pimonenko, T., Kwilinski, A., & Krawczyk, D. (2021). Post-industrial tourism as a driver of sustainable development. *Sustainability, 13*(15), 8145. https://doi.org/10.3390/su13158145

Long, P. T., & Nuckolls, J. S. (1994). Organising resources for rural tourism development: The importance of leadership, planning and technical assistance. *Tourism Recreation Research, 19* (2), 19–34.

Mantegazzi, D., Pezzi, M. G., & Punziano, G. (2021). Tourism planning and tourism development in the Italian inner areas: Assessing coherence in policy-making strategies. In M. Ferrante, O. Fritz, & Ö. Öner (Eds.), *Regional science perspectives on tourism and hospitality* (Advances in spatial science) (pp. 447–475). Springer.

Mitchell, M. A., & Mitchell, S. J. (2001). Consumer experience tourism: A powerful tool for food and beverage producers. *Journal of Food Products Marketing, 6*(3), 1–16.

Montella, M. M. (2010). Musei d'impresa come strumento di comunicazione. Possibili innovazioni di prodotto, processo, organizzazione. *Esperienze d'Impresa, 18*(2), 147–163.

Montella, M. M. (2018). *I musei d'impresa. Heritage and total relationship marketing*. Editrice Minerva Bancaria.

Montella, M. M., & Cerquetti, M. (2016). Il ruolo dei musei aziendali nel marketing del Made in Italy: tipologie e posizionamento strategico. In M. R. Napolitano & V. Marino (Eds.), *Cultural Heritage e Made in Italy. Casi ed esperienze di marketing internazionale* (pp. 407–437). Napoli.

Moscardo, G. (2005). Peripheral tourism development: Challenges, issues and success factors. *Tourism Recreation Research, 30*(1), 27–43.

Napolitano, M. R., Riviezzo, A., & Garofano, A. (2018). *Heritage Marketing. Come aprire lo scrigno e trovare un tesoro*. Editoriale scientifica.

Nguyen, T. Q. T., Young, T., Johnson, P., & Wearing, S. (2019). Conceptualising networks in sustainable tourism development. *Tourism Management Perspectives, 32*, art. 100575. https://doi.org/10.1016/j.tmp.2019.100575

Nissley, N., & Casey, A. (2002). The politics of the exhibition: Viewing corporate museums through the paradigmatic lens of organizational memory. *British Journal of Management, 13* (S2), S35–S45.

Otgaar, A. (2012). Towards a common agenda for the development of industrial tourism. *Tourism Management Perspectives, 4*, 86–91.

Otgaar, A. H. J., Van Den Berg, L., Berger, C., & Xiang Feng, R. (2010). *Industrial tourism. Opportunities for city and enterprise*. Ashgate.

Palmer, M. (2005). Industrial archaeology: Constructing a framework of inference. In E. Conlin Casella & J. Symonds (Eds.), *Industrial archaeology. Future directions* (pp. 59–76). Springer.

Richards, G. (2018). Cultural tourism: A review of recent research and trends. *Journal of Hospitality and Tourism Management, 36*, 12–21.

Riviezzo, A., Garofano, A., & Napolitano, M. R. (2021a). *Corporate heritage marketing: Using the past as a strategic asset*. Routledge.

Riviezzo, A., Garofano, A., Mason, M. C., & Napolitano, M. R. (2021b). Italian corporate museums as industrial tourism destinations: A segmentation study based on strategic orientation. *European Journal of Tourism Research, 29*, 2906. https://doi.org/10.54055/ejtr.v29i.2425

Riviezzo, A., Garofano, A., & Napolitano, M. R. (2016). "Il tempo è lo specchio dell'eternità". Strategie e strumenti di heritage marketing nelle imprese longeve italiane. IL CAPITALE CULTURALE. *Studies on the Value of Cultural Heritage, 13*, 497–523.

Riviezzo, A., Mason, M. C., Garofano, A., & Napolitano, M. R. (2022). Uncovering the worth of the past: The relationship between corporate museums' strategic orientation and dual performance. *Management Decision, 60*(7), 1852–1874. https://doi.org/10.1108/MD-01-2021-0037

Romagosa, F. (2020). The COVID-19 crisis: Opportunities for sustainable and proximity tourism. *Tourism Geographies, 22*(3), 690–694.

Rosato, P. F., Caputo, A., Valente, D., & Pizzi, S. (2021). 2030 agenda and sustainable business models in tourism: A bibliometric analysis. *Ecological Indicators, 121*, 106978. https://doi.org/10.1016/j.ecolind.2020.106978

Saarinen, J. (2006). Traditions of sustainability in tourism studies. *Annals of Tourism Research, 33*(4), 1121–1140.

Savoja, L. (2011). La visita d'impresa. Da attrazione a prodotto turistico. *ROTUR: Revista de Ocio y Turismo, 4*(1), 33–56.

UNWTO. (2017). *Tourism and the sustainable development goals—Journey to 2030*. World Tourism Organization (UNWTO).

UNWTO. (2018). *Report on tourism and culture synergies*. UNWTO.

Waligo, V. M., Clarke, J., & Hawkins, R. (2013). Implementing sustainable tourism: A multi-stakeholder involvement management framework. *Tourism Management, 36*, 342–353.

Xie, P. F. (2015). *Industrial heritage tourism*. Channel View Publications.

Yang, X. S., Xu, H., & Wall, G. (2019). Creative destruction: The commodification of industrial heritage in Nanfeng Kiln district, China. *Tourism Geographies, 21*, 54–77.

Documentation Activities for Historic Railways: Detection of Potential and Weaknesses for a Sustainable Cultural Heritage Resource

Luca Rossato

Abstract This contribution intends to emphasise a possible methodology for the documentation and enhancement of historic railways, seen both as socio-economic drivers and touristic resources. The research activities were mainly focused on regions facing development issues in a low-income scenario, taking into consideration two important case studies in Asia and South America. In this research framework, it is crucial to highlight that the European Commission has decided to make 2021 the European Year of Rail. Despite the pandemic period, a series of actions and initiatives in 2021 promoted rail as a sustainable and cultural mode of transport, highlighting the important role played by railways in not only connecting the EU countries physically but also improving the political cohesion among members. On the one hand, railways are one of the most sustainable and safest modes of transport, and, on the other hand, these railways that cross boundaries, mountains, rivers, small villages, and big cities are capable of creating an innovative formula for rail tourism. For these reasons, detailed and effective documentation activities need to be carried out to bring these tracks from the past into the future. The research analyses digital tools such as GIS platform to create affordable and effective dataset useful for the site management and improvement, trying to define a possible methodology for the implementation of knowledge and information. It is only by preserving the memory of these lines that it is possible to promote still hidden areas of many regions and, at the same time, the idea of slow and sustainable tourism as an important economic resource.

1 Introduction

The year 2021 was chosen by the European Commission as the "European Year of Rail" to emphasise one of the most sustainable, innovative, and safe modes of transport. A series of activities have been planned for the whole year across the

L. Rossato (✉)
Department of Architecture, University of Ferrara, Ferrara, Italy
e-mail: luca.rossato@unife.it

E. Borin et al. (eds.), *Cultural Leadership in Transition Tourism*, Contributions to Management Science, https://doi.org/10.1007/978-3-031-14121-8_7

continent to encourage the use of rail by both citizens and businesses and thus contribute to achieve the European Green Deal's goal of making the European Union climate neutral by 2050.

Railways have always played a crucial role in connecting areas and in rebalancing the existing gaps between the different regions with intermodal solutions (Senes, 2004). Beside the structural function, the oldest railways are also examples of outstanding ingenuity. In some cases, their design applies audacious and ingenious engineering solutions to the problem of establishing an effective rail link across mountainous landscapes.

The mix of technical creativity and beauty of natural environment makes several historic railways a cultural resource that could represent huge possibilities in terms of tourism potential and economic development for vast territories (Chun et al., 2020). Currently, the international discussion among experts underlines the great impact of reusing neglected historic rail paths for the valorisation of the territories and abandoned areas (Ahern, 1995; Ryan et al., 2001).

This contribution aims at emphasising the potentials of railway heritage considering two main aspects: on the one hand, it is considered for its social and economic functions in history, since railways are of great importance for enhancing cultural identity, and, on the other hand, it could be a crucial asset to foster regional sustainable development at socio-economic level (Zhang, 2013). Furthermore, historic railways could also act as green circuits to link separate sites in a sustainable way, improving regional economy and communication (Howard, 2003; Wang, 2011).

The text analyses two case studies located in India and Brazil, the Darjeeling and Curitiba-Paranaguá railways (Rossato, 2007), respectively, in West Bengal and Paraná states. The research activities performed by the author in these two sites show how historic rail paths are affected by abandonment and mismanagement with great underestimation of their cultural, economic, and sustainability values (Mennella, 2004). On this basis, the documentation of this cultural heritage must be improved and encouraged at regional level for the implementation of effective management and preservation plans.

From this perspective, if disused railway tracks represent potential routes for a new use of the territory, disused stations, usually located in a strategic position in the territory, provide spaces available for new destinations related to the objectives of promotion and enhancement. The topic perfectly fits into the work developed by the International Committee for the Conservation of the Industrial Heritage (ICCIH), an international group dedicated to industrial archaeology and cultural heritage, bringing to the discussion many sites worthy of documentation and conservation (ICOMOS – TICCIH, 2011) in terms industrial heritage (Douet, 2013). As an industrial heritage, historic railways include both tangible (equipment, coaches, stations) and intangible dimensions (such as technical know-how, the organisation of work and workers).

The redeveloped routes can therefore become a vehicle for the fruition of the historical, artistic, architectural, cultural, environmental, and landscape heritage and enhance quality production chains linked to traditional agriculture (Oppido, 2010).

For instance, in Italy, more than 600 km of railways are protected and respond to the new aim of re-evaluating the territories they cross by creating a widespread museum of small villages and landscape jewels. This was possible thanks to the 2017 law on historic railways, which aimed at recovering lines that were abandoned because they were not very attractive for the regular transport of people and goods, but which still have a lot to offer: travelling along their route, you can admire unique views in areas of Italy to be rediscovered and opened up to tourism. Only with a careful documentation of all the historic features it will be possible to properly conserve these ingenious structures for future generation and at the same time to foster local economy of low-income areas.

2 Historic Railways

The phenomenon of tourist railways and historical trains is now widespread throughout the world, with very different contexts, but all highlighting the interest focused both on the places to be reached (and visited) and the journey (considering the vehicles used to complete it). Historically, the railway was created for the transport of goods, followed by the transport of people, mainly for reasons related to work and economic activities. The train was used for tourism only by a small part of the population, belonging to the wealthier social classes (Flinn, 1984). From the end of the nineteenth century onwards, the train became a key element in the development of tourism, when the great European capitals were linked to the beautiful Mediterranean coasts of France and Italy, as well as the famous Italian cities of art, the destinations of the Grand Tours of writers and poets, accessible in a few days' travel. Trains such as the Orient Express, the Calais-Méditerranée Express (later the Train Bleu), and the Rheingold Express are still part of the collective imagination (Von Buch, 2007). At the same time, short but very audacious lines were built for tourism purposes, the famous cogwheels climbing up mountains and glaciers, which can still be found in the French and Swiss Alps, such as the Tramway du Mont Blanc, the Chamonix-Montenvers Railway, the Jungfrau Railway, and the Achensee Railway. Many railway lines that originated with economic functions, for the transport of passengers and goods, have gradually lost their capacity to carry out this task effectively, due to various factors that have made other modes of transport preferable. Indeed, many of these railways were built in difficult orographic conditions, minimising costs, which resulted in tortuous routes and, consequently, low commercial speeds, with villages often far from stations and sometimes subject to depopulation.

In other cases, it was the construction of a modern road infrastructure that "condemned" the railways, together with the increase in the rate of individual motorisation and the lower cost of public road transport. Some of these lines were modernised, but still for many of them more efforts are needed. Recently, the United Kingdom, a country where the railway network is extremely widespread and capillary, played a lead role towards the preservation of the most attractive railway lines.

In the United Kingdom, many associations were created with the aim of safeguarding part of this immense historical and technical heritage that was being lost, because not only were the lines being cut, but also the entire heritage of steam locomotives and antiquated rolling stock was in danger of disappearing completely. This had already been known for years, so much so that the first of the "preserved railways" was the Talyllyn Railway (in central Wales), for which the first association for the recovery for tourist purposes was formed on 11 October 1950. Obviously, the most beautiful lines from a landscape point of view were chosen, as well as probably those where the operating costs could be reasonably contained, with an average length of just over 10 km up to a maximum of 35 km. Ultimately, a tourist railway is defined as a line on which passenger services have been discontinued (and in some cases a sporadic freight service may coexist) and on which passengers use the train for recreational purposes, with circulations during limited periods and days, mostly concentrated in the summer season.

The United Kingdom currently operates around 100 tourist railways, with 2000 historic steam, diesel, and electric locomotives. Since the ban on steam locomotives on the mainline network was lifted in 1971, many of Britain's historic and tourist trains also run on sections of line open to normal rail traffic. The British model has spread across the European continent since the 1970s, with many railways being taken out of normal operation and converted into tourist lines. As far as historical trains are concerned, it should be stressed that most of them are used for tourist purposes, precisely because the charm of vintage railway vehicles is itself an element of attraction for the public (Muriel, 2006). Of course, there are also tourist trains with modern equipment, such as those of the Rhaetian Railway, which also operates various historic rolling stocks, such as carriages, steam engines, electric locomotives, and electric trains, for special occasions. The numbers of travellers carried and the direct turnover (without considering the ancillary business, which is estimated to be about 2.7 times what is spent on railways) of European tourist railways are very respectable: data from the European Federation of Tourist Railways (FEDECRAIL), which brings together about 500 associations involved in tourism and railway history, indicate about 5000 km of tourist railway lines, 20 million travellers a year, 500 million euros income, and about 3700 paid staff, supported by at least 26,000 volunteers. To sum up, tourist railways and historical trains nowadays represent a tourism offer that allows people to immerse themselves in history and discover a particular sector of industrial archaeology. This touristic approach seems to be very sustainable because it doesn't assume new infrastructures enhancing something that already exists. Furthermore, we must consider the possibility of new forms of tourism with positive effects on the induced activities of the territories they pass through, with investments and costs that are often very low, thanks also to the work of volunteers. This could also bring an optimisation of the offer of local resources in terms of promotion of centres of cultural, naturalistic, and gastronomic interest with opportunities for the promotion and sale of typical products. Obviously, the management of such complicated systems also has to deal with local peculiarities such as the historic railways' different scenarios, considering their growth in distinct historical moments in each country, as well as the different purposes (in some countries, railways were mainly built to transport goods, whereas in other areas they were meant to carry passengers). These two factors are actually part of the overall approach and must be taken into consideration for the preservation and enhancement of these cultural heritage sites.

3 Research Design and Presentation of Case Studies

The research is based on the analysis of two case studies, the Darjeeling Himalayan Railway and the Curitiba-Paranaguá historic railway located in India and Brazil, respectively. The two examples were selected based on the criteria of variety and representativeness (Patton, 2002), since they are outstanding examples of historic railways located in Global South countries facing common issues but also presenting peculiarity in terms of equipment and technology. The cases have been analysed providing first an introduction to the main particularities and then reflecting on the potential of creating a system able to record resources and manage risks and potentials. Thus, the paper discusses the potential methodology that could be used to conceive and improve digital technology application balancing cultural preservation/enhancement and development of the local economy of remote areas (Bottero et al., 2018). The following sections present the results of this analysis, starting from the consideration of the peculiarity of each site aiming at a possible future vision of the management that integrates virtual tours of 360° panoramas, 3D models from photo modelling of pictures taken by drones, multimedia contents, and 2D/3D historic evolution schemes (De Fino et al., 2020).

3.1 From East to West: Extra European Railways

All over the world, historic railways were recently taken into consideration as an important factor for the economic development of territories and communities. Many examples show us how an effective enhancement master plan and management plan could improve tourist flows in depressed and isolated areas. For instance, the Vietnamese train Ho Chi Minh City-Hanoi travels 1726 km between Ho Chi Minh City and Vietnam's capital Hanoi. It takes 2 full days to cover the distance whilst admiring stunning coastlines and seas and historic cities.

The Lake Titicaca Railway, which connects Puno to Cuzco, allows passengers to admire the splendour of the Andes – not only geological and geographical, with magnificent views of snow-capped mountains and phenomenal valleys, but above all sociological, as you pass through remote villages and tiny towns. It takes 10 h for 388 km: a space/time relationship that invites slowness. Again, there is no historic railway more impressive in its change of scenery than the one that covers 3750 km in 40 h and links the imperial architecture of China's capital Beijing with the mountains and monasteries of Lhasa, Tibet. The train climbs more than 5000 m from the bustling city to the fragrant silence of Tibet. It is possible to find historic railways also in Africa, the Tazara Railway which links Dar es Salaam in Tanzania with Kapiri Mposhi in Zambia. It leaves twice a week on a route of almost 1900 km and takes about 46 h, at least on paper. The splendour is both the flora and fauna and everyday life that is experienced between stops. Globalisation and digital marketing strategies are making what were previously unknown attractions a reliable economic resource for many communities, having a good impact on the quality of the life of residents and not impacting too negatively on the local environment (Matthews & Thakkar, 2012). Unfortunately, some historic railways with huge potential are still abandoned or underestimated as economic drivers. Lack of funding and local management skills are affecting the preservation of some beautiful lines that are in bad state of conservation and in some cases already dismissed.

3.2 Darjeeling Himalayan Railway, India

At the end of the eighteenth century, it was first proposed to build a steam tramway from Siliguri to Darjeeling. The idea was then better conceived in 1879 and the construction began. Gillanders Arbuthnot Company was appointed to build the line, and by March 1880, the track was already operating. Given the steepness of the road in 1882, four loops and four zigzags were built between Sukna and Gayabari, and later, in 1886, the line was extended to Darjeeling Bazar using a small train lately called by the locals the "toy train" (Fig. 1). After a short time, a period of station renovation began: the Darjeeling station was renovated in 1891, and Kurseong got a new station building in 1896. Over the years, the railway had to face several impactful events, such as earthquakes in 1897 and 1934 and a cyclone in 1899. In 1910, the line was able to carry more than 174,000 passengers and 47,000 ton of goods, whilst during World War II, it transferred military troops to the nearby camps. In 1951, the railway was purchased by the Indian government and put under the

Fig. 1 Colonial buildings along the Darjeeling Himalayan Railway (left) and the old train still in use, the so-called toy train (right). (Images by the author)

government railway organisation. In 1962, the railway was straightened at Siliguri and expanded nearly 6 km to New Jalpaiguri (Baker, 2014). The extension began freight service that year and passenger service in 1964. It was declared a World Heritage Site by UNESCO in 1999. The line nowadays represents probably one of the most interesting examples of slow travel by train (Roy & Hannam, 2013).

3.3 *Curitiba-Paranaguá Railway, Brazil*

The Curitiba-Paranaguá is a railway that connects the cities of Paranaguá and Curitiba in Paraná state, Brazil. Designed in 1870 by brothers André, Antônio, and José Rebouças and built between 1880 and 1885 by Antônio Ferrucci and João Teixeira Soares, it has 110 km of track, with 14 tunnels, the largest 457 m long, as well as 10 intermediate stations and 30 bridges and viaducts (Fig. 2).

The road is currently managed by Rumo Logística, with daily freight transport, and a tourist train operated by Serra Verde Express. Before the railway, for more than 200 years, the only connection route between the coast and the plateau of the interior of Paraná State was the Graciosa Road, initially a trail opened by the Indians and only later paved with stones. Compagnie Génerale, under the direction of Antonio Ferrucci, an Italian who had participated in the construction of the Suez Canal, began construction in 1880 with a ceremony attended by Emperor Dom Pedro II. Nine thousand workers took part in the construction under challenging working conditions as the area was swampy and prone to flooding, and there was no road to facilitate transport. The road was built with Belgian steel, produced by Dyle et Bacalan and brought by ship to Brazil. In 1883, the Paranaguá-Morretes stretch was inaugurated in a ceremony with Princess Isabel (Damas & Bettega, 2020).

At the time, it was one of the most daring engineering works in the world, completed in 1884 under the direction of Soares the Serra do Mar, with 500-ton iron viaducts built without the use of machinery and under constant rainfall (Castro, 2014). The complete road was inaugurated on 2 February 1885, less than 5 years

Fig. 2 Old bridges built along the Curitiba-Paranaguá railway line. (Images by the author)

after work had begun. There are currently no more passenger transport lines on the Curitiba and Paranaguá track, other than a tourist line maintained by the Serra Verde Express company with a wide offer of leisure solutions.

4 Documenting Historic Railways: A Possible Methodology

Nowadays, the main challenges that these historic railways face are different. First, the presence of the railway in growing urban contexts or the high cost for the maintenance of the railway and related structures does not allow the municipalities and public bodies to effectively manage the systems. Very often local authorities that also have many difficulties in terms of coordination due to the complicated administrative divisions are unable to properly maintain the track and the buildings along it. Furthermore, some historical stations have already been destroyed or significantly modified, as they were not recognised as valuable cultural heritage. This highlights a lack of awareness regarding railway heritage in both stakeholders and residents.

Furthermore, the conflict between old (and expensive to maintain) and new (more comfortable and sustainable) coaches is still an aspect under investigation. On one side, the recovery of old coaches seems to be a sustainable activity, delivering to future generations traditional techniques and machines. On the other side, some could argue that using contemporary trains these tracks could benefit from great improvement in terms of costs and pollution reduction.

Presently, the discussion is still open, and many research studies have been carried out on historic railways, from the perspectives of history and innovation (Dann, 1994), tourism (Fan, 2011), and cultural heritage protection (Coulls, 1999). From these experiences, it seems that GIS systems have often been applied in the study of railways and their landscape towards the cultural heritage regeneration (Girard & Gravagnuolo, 2017).

Additionally, in order to help solve these problems, the creation of a heritage GIS geodatabase seems to be essential for further analysis, planning, and tourism management as already implemented in some international case studies (Sang & Piovan, 2019). The databases, usually created on reliable maps, include a wide set of data that should be inserted into the platform. These data could be topographic maps, digital elevation models, on-site survey data, digital historical documents, and topographic and thematic historical maps (Sang & Piovan, 2019). Some RSD (remote sensing data) can be found from the satellite covering and should be integrated with on-site digital documentation (photos, direct survey, 3D survey). In addition, historical documentation (historical pictures, paintings, and documents) plays a crucial role in understanding the main modification; for this, broad research in national and international archives, libraries, foundation, and museums is required, and its outputs should be part of the platform dataset and can be organised for dissemination purposes (Fig. 3). Data are the fundamental component of a spatial information system, and they are distinguished in digital geographic data (vector, raster, tables, databases) and analogue data, which can be used through digitisation

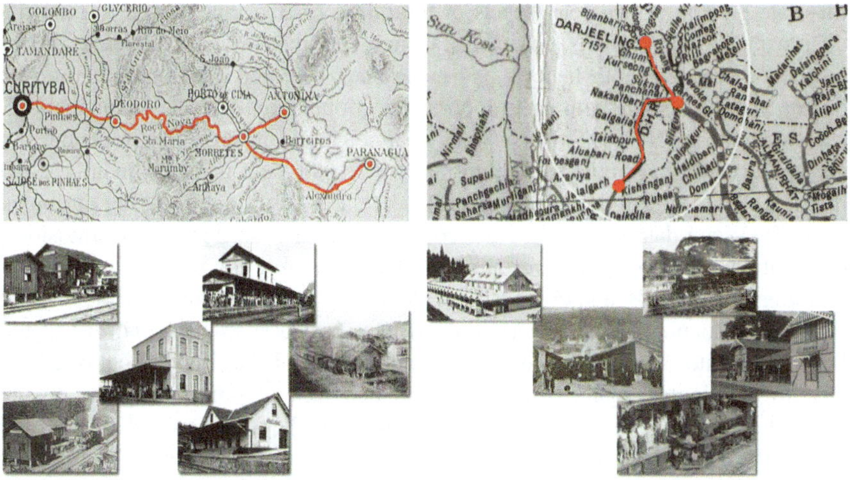

Fig. 3 Paths and main stations of the Curitiba-Paranaguá railways (left) and Darjeeling Himalayan Railway (right). (Graphic by the author)

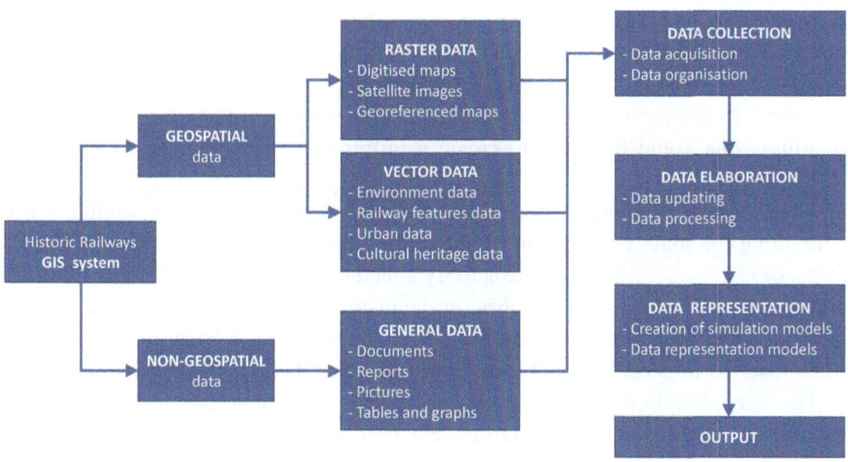

Fig. 4 Workflow for the creation, organisation, and elaboration of a GIS dataset for historic railways. (Graphic elaboration by the author)

such as historical maps in paper format, photographs, etc. For our methodology, two types of data were considered, geospatial data and non-geospatial data (Fig. 4). These data can be expressed by means of cartograms or tables and can refer to a more or less extensive portion of the territory, depending on the different applications.

The first group consists of vector and raster data representing features or objects on the earth's surface (man-made or natural) and describing them spatially through CAD (computer-aided design) drawings and maps (traditional, satellite, or georeferenced ones); the second group comprises all the document conceived to illustrate general or peculiar aspects of the railway management such as financial documents, pictures, charts, and tables. Concerning the two case studies described by these pages, for both historic paths, the data were collected starting from the documentation search with the support and cooperation of local municipalities, stakeholders, and NGOs. The first step was aimed at organising the vector and raster content divided by typology of information. Vector data included mainly CAD maps, sometimes very basic but useful to serve as a base for any further attributed data. The vector datasets in both cases comprised features such as the land divisions, distribution of environmental resources (fields, lakes, forests), residential areas and buildings, main roads, and obviously the rail track (along with tunnels, bridges, and stations). Also, point cloud data acquired by terrestrial laser scanners or drones could be inserted into the platform for specific purposes (Ilvitskaya & Prihodko, 2018). Due to the limited sources of information and lack of time during the on-site research, very few historic maps were found, but they were georeferenced and analysed in close relation to the dataset. In this case, the database works as a time machine: it is possible to switch from the current layout of the track to the old one described in the historic maps and see track alteration and structural improvement (despite the low level of information and accuracy of old maps).

This "time machine concept" could be very useful if we want to analyse the urban pressure along the lines, which, in the case of Darjeeling, is one of the most crucial points of the last part of the track (where buildings were built very close to the railways and some more old houses strongly hybridised with a great loss of cultural heritage). In Darjeeling, the dataset was enriched by the cataloguing of historical buildings that could be listed to create a buffer zone all along the railway track (Fig. 5).

Here, the on-site quick documentation was carried out using digital forms containing the main information of each built structure worthy of consideration in terms of historic significance. The activity aimed to start the survey campaign for the conservation of urban colonial fabric in the historic centre of Darjeeling in connection with the railway system. The research was based on a basic physical survey (data inserted in excel file) and local interviews (for the general report).

For each building, a dataset was collected, describing date of construction, number of storeys, structure materials, external wall materials, roof typology, roof material, state of conservation, and ground floor use. The result of this action was a first interoperable GIS map which identified almost 80 colonial period buildings. Beside technical data, the evaluation of historical documentation can provide information about the works and funds related to railway construction and maintenance, and this dataset could also be useful to assess the past financial sustainability of each line. As stated by Purcar, "Historic studies should be conceived as planning tools, destined to inform decision-making in conservation and spatial policies in both town and countryside, in both exceptional and ordinary places" (Purcar, 2012). Similarly, Sang asserts, "After creating and populating the geodatabase, different operations can be performed, such as visualisation and display, management and analysis of the data, data overlay and hierarchical browsing, 3D roaming, data retrieval and thematic map customisation" (Sang & Piovan, 2019). The maps of the case studies were

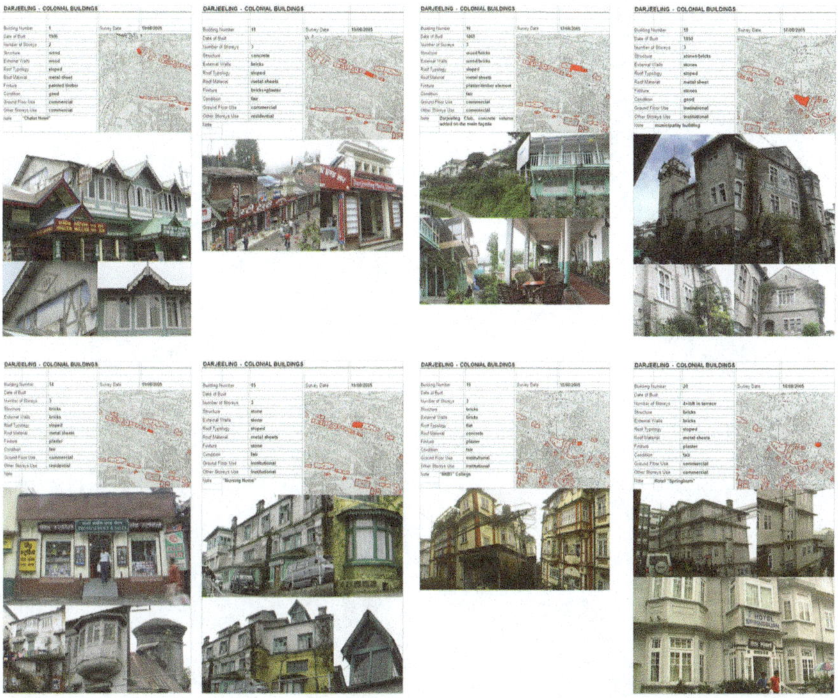

Fig. 5 Cataloguing forms of some of the 80 buildings documented in Darjeeling close to the railway track

conceived using ArcView, a GIS software by Esri company able to provide a complex set of cartographic tools and instruments for using, visualising, analysing, editing, and geoprocessing data.

Thanks to the acquired database and using an effective workflow, it is possible to view, query, analyse, and integrate geographic data, as well as create and edit geographic information. Furthermore, integrating the GIS platform with GPS data, it is possible to geolocate both contemporary and historical pictures and compare the transformation of the lines during the last decades. To increase the effectiveness, in terms of management, of the GIS platform for historic railways, it seems to be crucial to establish a proper identification of both natural and cultural resources available along the lines to foster possible reuse and increase the impact on local economy (Bowitz & Ibenholt, 2009).

The case study of the historic railway Curitiba-Paranaguá shows us how the main focus of the managers of the tourism companies is very often the line itself or the trains, not taking into consideration the net of attraction that could be linked to the railway. It is possible, using very accurate identification of natural and cultural assets, to conceive an itinerary where it could be possible to visit local point of interest, as these are slow trains able to stop frequently (Fig. 6). Some of the

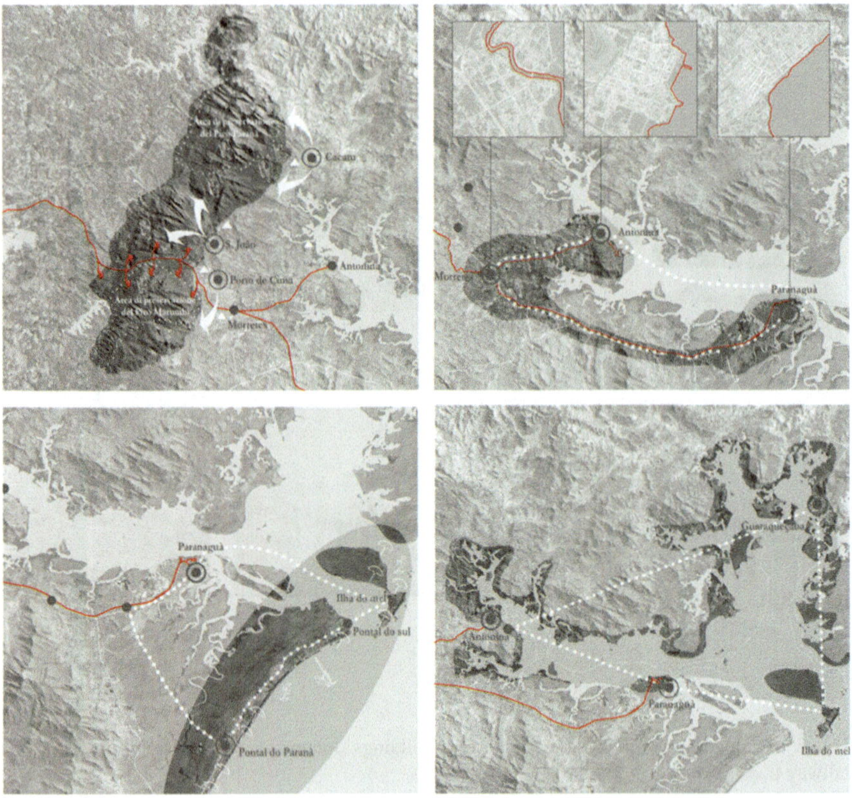

Fig. 6 During the research activities for the Curitiba-Paranaguá railway case study, satellite maps and vector maps were used as a base for physical survey of natural and cultural resources documented on-site by the research teams and then located into the GIS system

abandoned stations could be starting points for trekking activities or house a museum space which could involve the local community and revitalise arts and crafts skills. These places along the railway line have a strategic value, and they can play a decisive role both in the production and in the organisation of the local cultural and tourist opportunities (Cerquetti, 2007).

The documentation activity of these resources is thus a key factor for the improvement of these services and also for the redistribution of income due to the station's recovery (Bottero et al., 2019). All along the track, the development of tourism could be seen in the form of small hostels, local crafts shops, or simple leisure places (Lisienkova & Lisienkova, 2020). In this framework, the cultural heritage along the railway line can work both as a buffer zone and tourism resource, protecting the historic track and at the same time attracting visitors, if properly valorised and managed.

An effective management digital archive can be very useful if specifically designed for maintenance and activity control within the boundaries of historic area (Fig. 7).

Fig. 7 Using the GIS system, it is possible to create detailed maps based on highlighted features such as the building age or priority of intervention. (Database elaboration by the author)

5 Conclusions

GIS and other spatial tools are playing a pivotal role for the analyses and master plan of large-scale heritage sites like those shown in these last two railway case studies. This essay aimed to illustrate the process of identification of weaknesses and strengths by means of digital tools. By the integration of information, knowledge, and skills (that nowadays geodatabase requires) which can be realised by the stratification of raster, documentation, and vector data such as *dwg maps, it is possible to provide administrators and stakeholders with effective tools for the management of railway sites.

For instance, the case study of Antonina-Paranaguá railway showed how important it could be to think in a more regional way when working for the preservation of historic railways. This means taking into consideration the value of historic areas by also documenting private properties that are actually part of the local cultural heritage and, in many ways probably, the most difficult to be maintained and included in the tourism system (Fig. 8).

By their structures, these GIS projects represent something much more than a CAD applied to cartography; they make it possible to integrate multi-source data, perform advanced spatial analyses, examine phenomena, perform network analyses, create thematic maps, etc. for the management of these tracks. Furthermore, the development of these databases will provide technical support for effective management of all the spatial and non-spatial information for the tourism activities, the cultural heritage preservation, and the enhancement and regional planning. The importance of an adequate methodology of digital documentation activity is thus crucial to achieve.

Fig. 8 The digital documentation of the historic centres linked by the Curitiba-Paranaguá railway (in the map, façades survey of Antonina village) is a crucial factor towards the protection and management of the whole area. (Graphic elaboration by the author)

References

Ahern, J. (1995). Greenways as a planning strategy. *Landscape Urban Planning, 33*(1), 131–155.

Baker, J. C. (2014). Mobility, tropicality and landscape: The Darjeeling Himalayan Railway, 1881–1939. *Journal of Historical Geography, 44*, 133–144.

Bottero, M., Dell'Anna, F., & Nappo, M. (2018). Evaluating tangible and intangible aspects of cultural heritage: An application of the PROMETHEE Method for the Reuse Project of the Ceva–Ormea railway. In G. Mondini, E. Fattinnanzi, A. Oppio, M. Bottero, & S. Stanghellini (Eds.), *Integrated evaluation for the management of contemporary cities* (pp. 285–295). Springer.

Bottero, M., D'Alpaos, C., & Oppio, A. (2019). Ranking of adaptive reuse strategies for abandoned industrial heritage in vulnerable contexts: A multiple criteria decision aiding approach. *Sustainability, 11*(3), 785.

Bowitz, E., & Ibenholt, K. (2009). Economic impacts of cultural heritage—Research and perspectives. *Journal of Cultural Heritage, 10*(1), 1–8.

Castro, W. R. D. (2014). *As ações das instituições públicas e privadas sobre o patrimônio cultural ferroviário da linha Paranaguá-Curitiba (Estrada de Ferro do Paraná)* [Master degree thesis dissertation]. UEPG.

Cerquetti, M. (2007). The cultural component of integrated tourist product: The value creation for context through local museums. *Sinergie, 73–74*, 421–438.

Chun, Z., Dai, S., & Xia, H. (2020). Reuse of abandoned railways leads to urban regeneration: A tale from a rust track to a green corridor in Zhangjiakou. *Urban Rail Transit, 6*(2), 104–115.

Coulls, A. (1999). *Railways as world heritage sites. Occasional papers for the World Heritage Convention* (pp. 1–23). ICOMOS.

Damas, M. T., & Bettega, M. C. (2020). Transformação X Abandono: um olhar sobre a Estação Ferroviária de Paranaguá, Paraná. *Cenário: Revista Interdisciplinar em Turismo e Território, 8*(15), 100–112.

Dann, G. M. (1994). Travel by train: Keeping nostalgia on track. In A. Seaton (Ed.), *Tourism: The state of the art* (pp. 775–782). Wiley.

De Fino, M., Ceppi, C., & Fatiguso, F. (2020). Virtual tours and informational models for improving territorial attractiveness and the smart management of architectural heritage: The 3d-impact PROJECT. In *International Archives of the Photogrammetry, Remote Sensing & Spatial Information Sciences* (Vol. XLIV-M-1, pp. 473–480).

Douet, J. (2013). *Industrial heritage re-tooled: The TICCIH guide to industrial heritage conservation*. Routledge.

Fan, Y. (2011). An introduction of the development of tourism resources of the Yunnan-Vietnam railway. *Human and Inhabitant, 5*, 64–65.

Flinn, M. (1984). *History of the British Coal Industry 1700–1830*. Oxford University Press.

Girard, L. F., & Gravagnuolo, A. (2017). Circular economy and cultural heritage/landscape regeneration. Circular business, financing and governance models for a competitive Europe. *BDC. Bollettino Del Centro Calza Bini, 17*(1), 35–52.

Howard, P. (2003). *Heritage: Management, interpretation, identity*. Continuum.

ICOMOS – TICCIH. (2011). Principles for the conservation of industrial heritage sites, structures, areas and landscapes. In *Proceedings of 17th ICOMOS General Assembly on 28 November 2011* (p. 7).

Ilvitskaya, S., & Prihodko, V. (2018). Innovative technologies in the field of topography, land management, territorial planning, construction and architecture. *IOP Conference Series: Materials Science and Engineering, 365*(2), 022030.

Lisienkova, L., & Lisienkova, T. (2020). Evaluation of railway stations as cultural heritage objects. In *IOP Conference Series: Materials Science and Engineering* (Vol. 918).

Matthews, L. C., & Thakkar, B. S. (2012). The impact of globalization on cross-cultural communication. In H. Cuadra-Montiel (Ed.), *Globalization—Education and management agendas*. IntechOpen.

Mennella, V. G. G. (2004). *Greenway per lo sviluppo sostenibile del territorio*. Il Verde Editoriale.

Muriel, M. (2006). Industrial heritage and cultural economics: The re-use of historic railways in Western Andalusia. *Cuadernos de Economía de la Cultura, 4*(5), 141.

Oppido, S. (2010). Renewal and enhancement of natural and built heritage in rural areas: A greenways network for a sustainable management strategy. In R. Amoêda, S. Lira, & C. Pinheiro (Eds.), *Proceedings of Heritage 2010. 2nd International Conference on Heritage and Sustainable Development*, Évora, Portugal, June 22-26, 2010 (Vol. I, pp. 601–609).

Patton, M. Q. (2002). Two decades of developments in qualitative inquiry: A personal, experiential perspective. *Qualitative Social Work, 1*(3), 261–283.

Purcar, C. (2012). Along the tracks. Interdisciplinary challenges in the research of historic railways. *Acta Technica Napocensis: Civil Engineering & Architecture, 55*(4), 384–404.

Rossato, L. (2007). Un viaggio in treno tra le nuvole, Darjeeling Himalayan Railway. *Siti, 3*(2), 44–47.

Roy, S., & Hannam, K. (2013). Embodying the mobilities of the Darjeeling Himalayan Railway. *Mobilities, 8*(4), 580–594.

Ryan, R., Fábos, J. C., Csemez, A., Machado, J., & Toccolini, A. (2001). Balancing culture and nature through greenway planning. In *American Society of Landscape Architects Annual Meeting Proceedings* (pp. 118–121).

Sang, K., & Piovan, S. E. (2019). The application of GIS in railway heritage management: The case of Yunnan-Vietnam Railway. In *Proceedings of the International Cartographic Association, 29th International Cartographic Conference (ICC 2019)*, 15–20 July 2019, Tokyo, Japan.

Senes, G. (2004). *Il censimento delle linee dismesse in Italia, Ferrovie, territorio e sistema di greenways* (pp. 100–107). ISFORD.

Von Buch, A. (2007). In the image of the grand tour: Railway Station Embellishment and the Origins of Mass Tourism. *The Journal of Transport History, 28*(2), 252–271.

Wang, Y. (2011). The traditional regional cultural landscape protection model based on landscape fragmentation analysis—Taking Zhifu in Zhuji, Zhejiang Province as an example. *Geographical Research, 1*, 10.

Zhang, M. (2013). The protection and value of YunnanVietnam Railway. *Thinking, 2*, 100–101.

Digital Documentation for Enhancement and Conservation of Minor or Inaccessible Heritage Sites

Federica Maietti

Abstract This contribution aims to explore some aspects of current research avenues in the field of digital documentation of cultural heritage, starting with two topical challenges: on the one hand, the heritage at risk, due to the lack of conservation or development strategies affecting small historic centres or barely accessible heritage sites, and, on the other hand, the impact that digital technologies can have in this direction in terms of knowledge and accessibility improvement.

Compared to the overcrowding that has often affected tourist attractions in major art cities, many small historic centres offer huge potential for alternative sustainable tourism, thanks to their cultural heritage, history, arts and traditions. The impact of COVID-19 increased this challenge, also enhancing digital applications as a tool towards new inclusive and sustainable way for knowledge-sharing, interdisciplinary collaborations and possible diversification of tourist flows. But there is still plenty of room in the use of digital tools as a support for conservation actions and as a means for knowledge, dissemination, enhancement and accessibility.

The research analyses possible applications of digital documentation starting with two emblematic case studies, opening up the analysis to international approaches and scenarios.

1 Introduction

We are living in a time when events related to climate change, new emergencies resulting from the pandemic crisis and new social trends increasingly require actions and responses in which the built environment and society are at the forefront. Moreover, the digital is becoming pervasive at every level of management and across all disciplines. In this overall framework, cultural heritage, its knowledge and accessibility, in all possible ways becomes essential.

F. Maietti (✉)
Department of Architecture, University of Ferrara, Ferrara, Italy
e-mail: federica.maietti@unife.it

E. Borin et al. (eds.), *Cultural Leadership in Transition Tourism*, Contributions to Management Science, https://doi.org/10.1007/978-3-031-14121-8_8

Among the purposes of the Strategic Plan of Horizon Europe, the most recent funding programme for research and innovation, the commitment to maximise the full potential of cultural heritage, arts and cultural and creative sectors and industries is stated.[1] According to this plan (Horizon Europe Strategic Plan 2021–2024, n.d.), research and innovation will support the access to our common heritage through new technologies, high-quality digitisation and curation of digital heritage assets and by developing solutions for sustainable and inclusive cultural tourism in Europe. New, participatory management models, including those for museums and cultural institutions, will be developed, and new technologies will be identified to increase the international competitiveness of Europe's cultural production.

Investments will also contribute to safeguarding endangered cultural heritage from natural hazards and anthropogenic disasters by preventive interventions, as well as ensuring equal and wide access to cultural assets and heritage sites.

In this policy document, the priorities related to the enhancement of cultural heritage are clear. Three key issues can be pinpointed: endangered cultural heritage due to natural hazards and anthropogenic disasters, access to our common heritage and new technologies, high-quality digitisation for sustainable and inclusive cultural tourism.

This contribution aims to explore the most recent research avenues and international guidelines in the field of digital documentation for knowledge, enhancement and conservation of cultural heritage. A focus on minor or inaccessible heritage sites through the analysis of two case studies located in the northeast of Italy will also be presented. In fact, more and more, small historic centres, or cultural assets outside the routes characterised by large tourist flows, are the focus of redevelopment and networking actions, all over the world. And this trend further developed as a result of the crisis in the tourism sector during the COVID-19 pandemic. At national level, initiatives aimed at enhancing and rediscovering local identities are growing, covering heritage conservation aspects and actions to prevent abandonment and encourage repopulation. Just to mention one example, the "Progetto Borghi",[2] launched in September 2021, is an initiative supported by the Italian Ministry of Culture and other institutions, as a digital forum aimed at developing a socio-economic mapping of the territory to enhance and regenerate small historic centres, developing an ecosystem capable of combining tradition and innovation. Also, the recent DPCM (Prime Ministerial Decree) of 23 July 2021, which identifies the 5518 municipalities that will benefit from the measures provided for by the Realacci Law (Law 158/2017), aims to enhance the value of small municipalities and the redevelopment of their historic centres. This law introduces simplifications for the recovery of historic centres either abandoned or at risk of depopulation to be reconverted into widespread hotels. It provides for the instigation of maintenance work with priority

[1]Horizon Europe Strategic Plan (2021–2024), CLUSTER 2 (Culture, Creativity and Inclusive Society). https://ec.europa.eu/info/sites/default/files/research_and_innovation/funding/documents/ec_rtd_horizon-europe-strategic-plan-2021-24.pdf, accessed 10 May 2022.

[2]https://progettoborghi.host-b2b.com/, accessed 10 May 2022.

given to the protection of the environment, the securing of roads and schools, the energy efficiency of the public building stock and the interventions supporting residents and productive activities established in smaller centres.

Nevertheless, there is still plenty of room to slow down the process of degradation and alteration of historic urban fabrics, giving back to the communities a heritage very often underused and not accessible to the public.

Italy is rich in small historic centres and a widespread heritage that is often not included in the major tourist circuits. Often, this condition of marginality or difficult accessibility creates conditions of abandonment, degradation or lack of maintenance (Francini et al., 2012; Maietti, 2008).

The 2030 Agenda[3] (United Nations, n.d.) already foresees a focus on cities in the sustainability goals, including challenges related to inclusive and sustainable urbanisation and capacity for participatory, integrated and sustainable human settlement planning and management, and efforts to protect and safeguard the world's cultural and natural heritage (Goal 11: Make cities and human settlements inclusive, safe, resilient and sustainable).

Moreover, the recent National Recovery and Resilience Plan (Piano Nazionale di Ripresa e Resilienza, 2021)[4] highlights, under the section related to tourism and culture, the need to improve culture and tourist accessibility through digital investments and investments to remove physical and knowledge barriers to heritage, and regenerate small historic centres through promoting participation in culture, and the enhancement of sustainable tourism. Investments to achieve these goals include the "digital strategy" and platforms for cultural heritage and regeneration of small cultural sites, to foster the development of new tourism/cultural experiences and balance tourist flows in a sustainable way.

These targets address two important issues: the small, "minor" or barely accessible heritage and the impact that digital technologies can have in this direction. A careful assessment of the impact that COVID-19 had on human activities related to accessing, using and enjoyment of heritage (Kono, 2020) is part of this scenario. This strategy foresees collaboration with the Regions and ANCI (Associazione Nazionale Comuni Italiani – National Association of Italian Municipalities) so that sites with specific characteristics can be identified—very small villages, almost completely uninhabited—to which a prevailing aptitude (or function) can be attributed on which the State can deploy resources for interventions on the public and private heritage. The planned functions also include tourism, such as widespread hotels.

The paper aims to deepen the state of the art in the use of digital technologies for the documentation and dissemination of historic sites (Lo Turco et al., 2019; Trizio

[3] Agenda 2030 per lo Sviluppo Sostenibile (https://unric.org/it/wp-content/uploads/sites/3/2019/11/Agenda-2030-Onu-italia.pdf). https://unric.org/it/agenda-2030/, accessed 10 May 2022.

[4] https://www.mef.gov.it/focus/Il-Piano-Nazionale-di-Ripresa-e-Resilienza-PNRR/, accessed 10 May, 2022.

https://www.governo.it/sites/governo.it/files/PNRR.pdf, accessed 10 May 2022.

et al., 2019). The purpose is to explore the current research avenues focused on digital documentation and digital applications and platforms (Myers et al., 2016) as a tool towards new inclusive and sustainable way for knowledge-sharing and promoting interdisciplinary collaboration and the diversification of tourist flows.

The two selected case studies are aimed at showing an example of a scenario that is widespread in Italy and beyond and at analysing the potential of digital documentation to support conservation actions, using simple or more complex tools.

2 Heritage in Emergency Scenarios and Small Historic Centres

> Disasters and catastrophes in consequence of natural hazards, anthropogenic effects and extreme climate change events pose risks to the physical conservation requirements of cultural heritage assets, their social, cultural, historic and artistic values, the safety of occupants and users, and to local employees whose economies are linked to tourism. In addition, urban expansion into coastal areas and flood plains, coupled with inappropriate building practices and overburdened infrastructure, increase the risks of putting enormous additional pressures on heritage sites, especially those in vulnerable urban locations. (Maxwell et al., 2018)

We are all increasingly aware of the high number of threats affecting our territories, built heritage and cultural assets. In particular, the impact of climate change on monuments includes air pollution combined with frequency of events of intense wind-driven rain, floods and landslides, wind risks and hurricanes, earthquakes, fire risks as well as armed conflicts and terrorism and anthropic actions.

In addition to these risks (Machat & Ziesemer, 2020) and in some regions perhaps to a greater extent, small historic centres are affected by different types of damage, even in non-emergency conditions, due to several forms of abandonment, degradation or lack of maintenance.

This means that many small historic centres, representing cultural and historic values or rich in traditions, lose their testimonial significances every day due to a lack of redevelopment or enhancement or due to barriers that compromise their accessibility. Often, these already unfavourable conditions are combined with damage due to territorial vulnerability or events such as earthquakes (Markham et al., 2016).

In this overall scenario, the Sendai Framework for Disaster Risk Reduction 2015–2030[5] provides UN Member States with concrete actions to protect development gains from the risk of disaster. The Sendai Framework (UNDRR, n.d.) works hand in hand with the other 2030 Agenda agreements, including the Paris Agreement on Climate Change, the Addis Ababa Action Agenda on Financing for Development, the New Urban Agenda and ultimately the Sustainable Development Goals.

[5] United Nations Office for Disaster Risk Reduction – https://www.undrr.org/implementing-sendai-framework/what-sendai-framework, accessed 10 May 2022.

The Sendai Framework sets out four priorities for action:

- Priority 1. *Understanding disaster risk*, by developing and disseminating awareness, allocating EU funding for specific projects
- Priority 2. *Strengthening disaster risk governance to manage disaster risk*, through collaboration among competent authorities, documentation also by using IT tools, such as the digitalisation of archival records and on-site laser scanning, ensuring easy access to required information, stakeholder engagement
- Priority 3. *Investing in disaster risk reduction for resilience*, through the establishment of priorities for protecting cultural heritage assets and the enhancement of education
- Priority 4. *Enhancing disaster preparedness for effective response and to "Build Back Better" in recovery, rehabilitation and reconstruction*, drafting European Standards, adopting recovery measures by cultural institutions, etc.

It is particularly relevant that, under Priority 2, strategies to manage disaster risks include digitisation of cultural heritage, collecting data to ensure information and data access. The digital documentation can be a way to reach the so-called resilience that can be defined as the ability of a system, community or society exposed to hazards to resist, absorb, accommodate, adapt to, transform and recover from the effects of a hazard in a timely and efficient manner.[6] This definition includes the preservation and restoration of structures and functions through risk management.

The application of digital surveying technologies is becoming increasingly widespread in documenting historic centres, in particular following disaster events such as earthquakes (Vernizzi, 2016; Oreni et al., 2014), making it possible to create databases and three-dimensional models that can be used for different types of analysis of the historic fabric (Balzani et al., 2013) and as support for restoration work. Resilience and identity of historic cities are additional and very topical issues where documentation and the use of digital tools are playing an increasingly crucial role (Puma, 2020). Digital documentation is therefore increasingly seen as a strategic action to ensure better access to, understanding of and engagement with cultural heritage—as well as developing sustainable and inclusive cultural tourism and sustainable development (Council of Europe, 2018)—improving the protection, enhancement, conservation and more efficient restoration through high-quality digitisation and curation of digital heritage assets.

[6] United Nations Office for Disaster Risk Reduction - https://www.undrr.org/terminology, accessed 10 May 2022.

3 Documentation, Analysis, Enhancement and Conservation of Small Historic Centres: Two Case Studies

The state of abandonment in which small historic centres often lie has a twofold negative effect: the progression of the state of decay, which can lead to the risk of losing historic but underused buildings, and the lack of networking and exploitation of this heritage, which could instead create new forms of tourism and thus drive resources and new economic opportunities.

The two case studies described below are examples of possible methodological approaches in which digital documentation plays an essential role as a knowledge base for implementing not only conservation actions but also enhancement strategies for their reuse.

These two small historic centres have been the focus of analysis and documentation as they are both:

- Characterised by a historic-architectural landmark and a set of "minute" architectures, mostly unknown or barely enhanced and accessible
- Destinations for cycle tourism routes and hit-and-run tourism, aimed at the single architectural landmark without offering alternative visitor routes or strategies to enhance other historic buildings
- Mainly characterised by a condition of abandonment, degradation and lack of maintenance, in addition to some environmental risk conditions due to their proximity to rivers

3.1 Landscape and Historic Fabric: Stellata

The small historic centre of Stellata di Bondeno, in the province of Ferrara, in the northeast of Italy, has been the focus of studies and research aimed at documenting in terms of knowledge and safeguarding and defining conservation strategies that are not limited to historic buildings but extend to the urban fabric (Fig. 1).

The historic centre extends along the right bank of the Po River (Fig. 2), following a linear development. In this position, the borders of the three most important domains in the history of the lower Padana area intersected: Gonzaga, Este and the Republic of Venice. Today, the administrative borders of the regions of Lombardy, Emilia-Romagna and Veneto meet at this point.

Its border position made Stellata a military outpost of essential importance, and it was fortified during the fourteenth century with an impressive fortress.

This outstanding example of military architecture seems to be the only heritage trace in a place of great cultural and urbanistic-architectural value. Quite the opposite, the fabric of the small historic centre still shows the imprint of the frontier nucleus of the late sixteenth century and includes several significant buildings and an urban layout, imposed by the presence of the Po riverbank, which has remained almost unaltered.

There is a relationship of "necessary connection" between the riverside landscape, the settlement pattern and the architecture (Pini, 2004). It is a historic centre whose value is the relative integrity that has remained over the centuries, which is more the result of a condition of territorial marginality, of a lack of economic and functional motivation, which is expressed either by trivial interventions or by abandonment. These small provincial historic centres are indeed often forgotten and seem frozen in time.

The historic urban fabric is dotted with buildings of historic and architectural interest, less studied and preserved than Stellata's monumental landmark, the Rocca Possente (fortress), recently restored following the 2012 earthquake.

The research has deliberately tackled small-scale historic architectures. The documentation methodology started from the territorial scale, retracing the historic cartography in order to recognise the identifying features, up to the architectural scale, examining the Oratory of San Francesco as a symbol of the state of abandonment in which are almost all the historic-architectural buildings of great value, but which are almost unknown, unused or difficult to access.

Fig. 1 Overview of the historic centre of Stellata developed along the river, showing some of the historic buildings that dot the historic fabric and the houses along the embankment. In the centre, the connecting fabric that extends from the oratory of San Francesco down to the river is highlighted. (Graphic elaboration by the author)

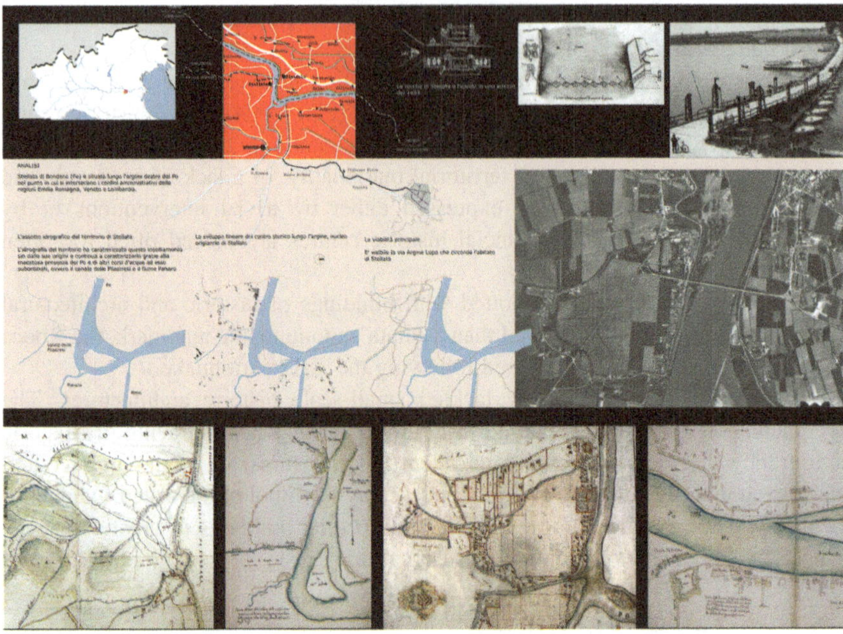

Fig. 2 Analysis and documentation at the territorial scale, highlighting the relationship between the river and the historic centre. Below, a selection of historic cartography found in various archives, which formed the basis for the analysis of urban development. (Graphic elaboration by the author)

In order to make the study on the historic centre a starting point for widespread recovery strategies—and not limited to the single building chosen as the focus of the methodological proposal—a survey of the heart of the small historic centre was performed, in this case through a two-dimensional documentation (Fig. 3). This choice was adopted in order to provide the municipal administration with more easily accessible and usable drawings.

In addition to the proposal for the restoration and re-functionalisation of the oratory, the research focused on several possible recovery strategies and new functions in unused historic buildings potentially acting as attractors for new forms of cultural tourism, proposing a new urban pavement—with a connective function—extending from the oratory to the embankment to reconnect buildings.

The proposed intervention enhanced the relationships between historic buildings and context and the testimonial and symbolic features, proposing actions that make the historic buildings an active presence, available for public use.

The redesigning of the public spaces and of the path connecting the oratory's churchyard, the square and the river and the restoration of the small oratory, a minor monument strongly connected to the context, are part of an overall strategy.

In this case, a 2D digitisation has been realised in order to document the state of conservation and overall conditions of historic buildings and to have digital representation supporting analysis, conservation and restoration projects. This work has

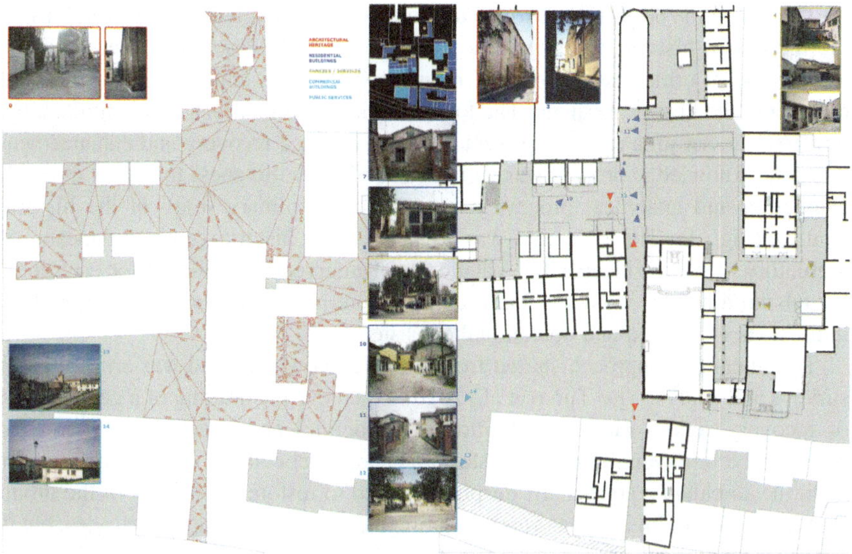

Fig. 3 Survey, representation and documentation of the main features of the heart of the historic centre of Stellata as a starting point for the proposal of conservation and redevelopment strategies. (Graphic elaboration by the author)

also been used as the starting point for restoration work on artefacts damaged or completely lost due to the earthquake in 2012. Other historic buildings are indeed now under restoration, such as the main church and an annexed oratory by implementing seismic improvement, and now the San Francesco Oratory is in the initial state of conservation analysis with the aim of setting up a restoration project. The analysis and digital documentation developed are being used as a starting point for this new restoration work.

3.2 Regeneration and Enhancement of the Historic Centre of Mesola

The small historic centre of Mesola, in the province of Ferrara, is also characterised by the presence of the river, being a municipality in the Po Delta, and by its border position. Mesola is indeed located south of the branch of the river called Po di Goro, which represents both the municipal and the regional border between the Veneto and Emilia-Romagna regions.

The history of the area is linked to its position as an extreme eastern sea border and to the presence of the Este family, who enhanced the territory by undertaking large-scale land reclamation work, designing the Alcina port on the Po di Ariano and building Mesola Castle.

The aim of the project was to draw up a methodological guide for interventions on the facades of historic buildings. The historic centre of Mesola is indeed an interesting case of critical assessment of regulatory practices aimed at recovering and enhancing the historic heritage. The local administration's decision to approach the university to set up a research project aimed at heritage recovery and enhancement through a knowledge process was an indicator of the willingness to trigger in-depth evaluation and analytical processes for a multidisciplinary study of the historic context (Fig. 4). The objective was to protect the historic centre by means of regulatory measures identifying "categories of intervention" tailored for each individual building, far from the logic of rigid, pre-packaged plans, but suggesting indications of method and operational consistency.

The integrated approach started from historic knowledge and was articulated in surveys and analysis (Lo Turco et al., 2020) to achieve the drafting of a general plan of recovery actions and the definition of guidelines and operational criteria for conservative intervention on the historic facades. The concept of the project is far from the so-called colour charts despite the need to restore the colours of the urban facades and is not limited to norms regulating the choice of colours for facades, but is based on a methodology that considers the urban fabric in its entirety: from the study of public spaces to the elements of street furniture, from the typological features to the relationships between monumental architecture and small historic buildings to the study of construction techniques, materials and the state of conservation of architectural surfaces.

The research was conceived as a project for the recovery of historic and aesthetic values, structuring a plan to propose operational-technical tools for interventions on historic facades and urban quality on a historic-critical basis.

The survey campaign was based on an integration of methodologies (Parrinello, 2019), including the direct method based on a critical and targeted assessment of the

Fig. 4 Some views of the historic context around Mesola Castle. Here, too, the relationship with the river has been and still is crucial

Fig. 5 3D point clouds of the historic centre of Mesola. At an urban scale, the 3D database allowed to investigate features of the small-scale historic architectures around the Mesola Castle

contexts to be surveyed, integrating the topographical method with three-dimensional laser scanner survey (Centofanti et al., 2014; Maietti et al., 2018).

Laser scanning consists of generating three-dimensional digital models, often very accurately, of an object, a building and a site, through non-contact and non-invasive procedures. In this case study, a pulsed instrument based on time-of-flight technology was used, producing the so-called point cloud 3D models (Fig. 5), as they are made up of millions of points (three-dimensional coordinates). The model can be navigated and explored in its three dimensionality and can be, more often, the basis for producing two-dimensional drawings (plans, elevations and sections) to be used for many purposes (documentation, conservation, monitoring, etc.). Today, a large number of technologies and procedures are available for documentation and mapping purposes and digital recording of cultural heritage. These tools are often used in an integrated way depending on the aims of the survey and documentation (Remondino, 2011).

The digital recording of the historic centre of Mesola allowed the creation of metric-morphological models to be used at different levels of data interpretation and extraction (Bianchini et al., 2015), also for the development of thematic representations for the study of surfaces and for in-depth analysis of colours, materials and the state of conservation (Fig. 6). Indeed, by using time-of-flight laser scanning, the measure of the position of each surveyed point is based on the time elapsed between the emission of the laser beam and its reflection by the object to be surveyed. Reflectance is the intensity of the laser beam at the moment of the "impact" with the object to be surveyed, and it is related to surface features, such as the material and possible conditions of degradation.

The guidelines aim to indicate to operators in the field of restoration and recovery of historic buildings how to proceed in the various stages of defining the intervention

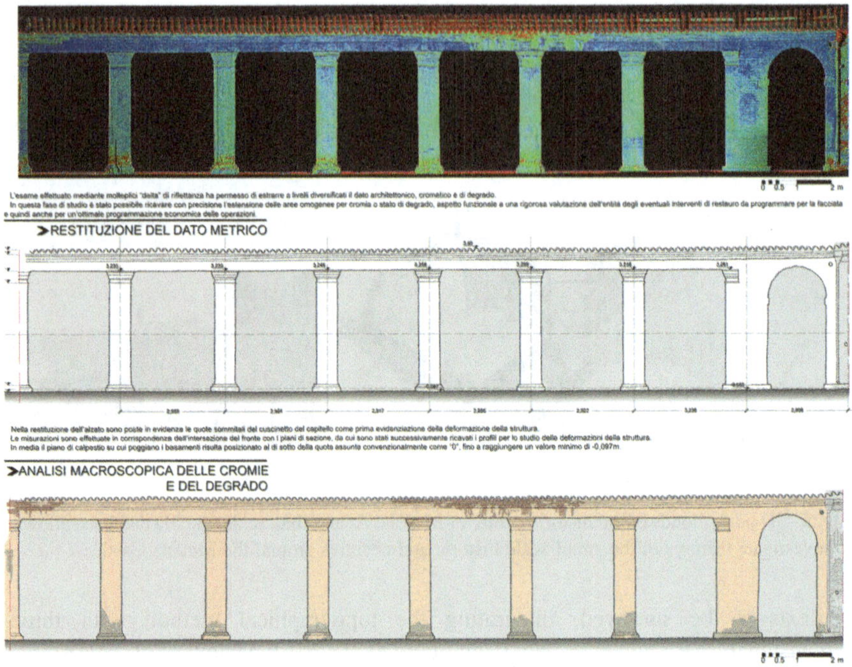

Fig. 6 Mesola, 3D survey. Example of a thematic representation. The data obtained by the 3D survey is used as a support for the analysis of morphological and surface features. (Graphic elaboration by the author)

project, from the survey to the more complex conservative interventions concerning external finishes and the colouring layers.

The study carried out in the first phases indicated the parameters for assessing the quality of the survey, design and implementation of the colouring work on the façades of the historic centre, in order to allow the preparation of management and operational procedures concerning all the phases of the colouring work on the façades of historic buildings.

With the conclusion of the research, the procedures were linked to the Municipal Building Regulations in order to address the restoration and recovery interventions on the external facades of historic buildings.

Following the conclusion of the research project, other initiatives were launched by the municipal administration, such as competitions for ideas to redevelop public spaces and the square in front of the church.

Today, the guidelines developed as a basis for the regeneration and enhancement of small historic buildings are used by the public administration to define interventions and make choices on the finishes to be adopted in the areas of the historic centre that have been studied in depth.

4 European Strategies on Heritage Digital Documentation

As already mentioned, more and more actions are being developed at international level to define guidelines and procedures for the digitisation of cultural heritage. This process started in particular in the framework of the Horizon 2020 programme, where a number of calls were focused on advancing digitisation. Horizon 2020 recognised the changing role of digital models in the representation of heritage and its analysis for digital reconstructions, preservation and conservation, supporting information integration/linking, shape-related analysis and providing semantic information for in-depth studies by researchers and users.

Many research experiences in this direction had already been developed, but the programme gave rise to several related initiatives, which increasingly pushed towards the willingness to draw up guidelines and develop the huge potential of digitisation for heritage.

In addition to a series of funded projects aimed at developing research and innovation towards digital heritage documentation,[7] in 2019, 27 European countries signed a Declaration of *Cooperation on advancing digitisation of cultural heritage*. Among the premises to this declaration (European Commission, 2019), the following statement summarises the purpose of the initiative: "Europe has a very rich cultural heritage that embodies our past and inspires us with a sense of a shared history and identity. Digital transformation can play an essential role in enabling cultural experiences, knowledge creation, preservation, and use and re-use of cultural heritage across borders. Digitised cultural objects moreover unlock the potential for broader societal, cohesive and economic benefits of sectors such as tourism, education and creative sectors".

General recommendations set by the European Commission are gathered under three main pillars:

- A pan-European initiative for 3D digitisation of cultural heritage artefacts, monuments and sites, elaborating common standards, methodologies and guidelines and defining the framework conditions for an open European repository for storing, managing and reusing interoperable 3D models
- Reuse of digitised cultural resources to foster citizen engagement, innovative use and spillovers in other sectors
- Enhancing cross-sector, cross-border cooperation and capacity-building in the sector of digitised cultural heritage, creating and fostering partnerships between culture and tourism ministries and organisations

After this initiative, an additional step was achieved in 2020, when the expert group on Digital Cultural Heritage and Europeana was tasked by the member states

[7] Inception, i-MareCulture, Gravitate, Time Machine and ViMM, to mention just some of the funded projects on this topic. See also https://cordis.europa.eu/article/id/413473-how-digital-technologies-can-play-a-vital-role-for-the-preservation-of-cultural-heritage, accessed 10 May 2022.

to contribute to the development of guidelines on 3D cultural heritage assets. The outcome is a document (European Commission, 2020) including a list of basic principles and tips for 3D digitisation of tangible cultural heritage.[8]

This document highlights how 3D digitisation can be valuable for many purposes, including conservation and preservation, reproduction, research, education, exploration and creative or tourism-related reuses, or for increasing the exposure of cultural heritage and reusing it in various applications for different sectors. 3D digitisation is also important where access is limited or impossible or to ensure better protection of physical cultural heritage sites at risk.

Moreover, the very recent recommendation published by the European Commission concerning a "European data space for cultural heritage"[9] is aimed to accelerate the digitisation of all cultural heritage monuments and sites, objects and artefacts for future generations, to protect and preserve those at risk and to boost their reuse in domains such as education, sustainable tourism and cultural creative sectors. As stated in the recommendation document, "Cultural tourism represents up to 40% of all tourism in Europe, and cultural heritage is an essential part of cultural tourism. Advanced digitisation of cultural heritage assets and the reuse of such content can generate new jobs not only in the cultural heritage sector but also in other cultural and creative sectors, including for instance the video game and film industries". Digitisation is strongly linked to preservation, and it is particularly important for cultural heritage at risk. Even where the risk of deterioration or destruction is low, it should be a priority given the high reuse potential of digitised assets, such as for innovative experiences and sustainable tourism.

This recommendation is also based on the results of the *Tourism and Culture Synergies* report (World Tourism Organisation, 2018), where several experiences in using new technologies to develop innovative relationships between tourism and culture are analysed (Poland, Spain, Israel, Romania and Serbia). In addition to specific applications that can aid on-site interpretation or navigation around cultural attractions in the destination, there is also the potential to develop systems that make the large amounts of data stored by public authorities available to visitors. The development of new systems should also stimulate the local economy through the participation in the creation of new technologies by small- and medium-sized enterprises. In some countries, there are specific funding schemes aimed at stimulating such developments.

[8] Basic principles and tips for 3D digitisation of cultural heritage, https://digital-strategy.ec.europa.eu/en/library/basic-principles-and-tips-3d-digitisation-cultural-heritage, accessed 10 May 2022.

[9] Commission proposes a common European data space for cultural heritage, https://digital-strategy.ec.europa.eu/en/news/commission-proposes-common-european-data-space-cultural-heritage, accessed 10 May 2022.

4.1 Digitisation and Post-pandemic Scenarios

The global crisis due to the COVID-19 pandemic had a strong impact on the cultural heritage sub-sector, which is in most cases venue- and visitor-based, due to the closing of museums and heritage sites and strict bans on internal movement and tourism, which is the first source of income for many heritage sites (IDEA Consult et al., 2021).

In general, during the pandemic, several initiatives were launched, and numerous studies were drafted on specific impacts of the crisis on museums, cultural heritage sites and stakeholders (Europa Nostra, 2020; ICOM, 2020; NEMO, 2021). Responses to the post-pandemic crisis by cities suffering from overtourism were analysed as well (Pasquinelli et al., 2021).

Also, the Horizon Europe Work Programme 2021–2022, *Culture, creativity and inclusive society*, highlights that COVID-19 pandemic crisis is continuously demonstrating the extent to which science and innovation policy needs to be at the core, fostering the use of digital means to engage citizens.

Although, even before the pandemic, the debate about the impact of mass tourism on historic cities was a hot topic, the crisis has prompted additional needs to diversify the tourist flows by moving them from the large attractors to different urban contexts.

Although in this scenario digital tools and devices were especially used by museums to engage citizens in new forms of off-site access and fruition,[10] other initiatives began to consider digital applications to foster real and virtual tourist routes to enhance the value of villages, artistic assets and natural heritage.[11]

5 Conclusions and Future Visions

As stated by the Commissioner for Internal Market, Thierry Breton:

> *We owe the preservation of our European cultural heritage to future generations. This requires building and deploying our own technological capabilities, empowering people and businesses to enjoy and make the most of this heritage. We must take advantage of the opportunities brought by artificial intelligence, data, and extended reality. The European data space for cultural heritage will promote creation and innovation within the cultural heritage sector, and beyond, in education, tourism, and cultural and creative sectors.*[12]

---

[10] Digital Ideas for Museums in a Post-Pandemic World – https://www.museumnext.com/article/digital-ideas-for-museums-in-a-post-pandemic-world/, accessed 10 May 2022.

[11] https://www.cisam.org/connessioni-museali-tra-valli-e-monti-borghi-e-citta/, accessed 10 May 2022.

[12] *Commission proposes a common European data space for cultural heritage*, https://digital-strategy.ec.europa.eu/en/news/commission-proposes-common-european-data-space-cultural-heritage, accessed 10 May 2022.

The requirement for an efficient use of digital technologies is coupled with challenges such as deterioration due to climate change, pollution, natural or man-made disasters, lack of finance or insufficient valorisation of cultural heritage.

Future challenges include the use of digital and cutting-edge technologies, an active engagement with stakeholders and citizens, a robust plan for how projects will use or build on outputs, results from research already undertaken and technology already available and lessons learnt from the COVID-19 crisis in view of a sustainable management of the post-crisis society.

A recent national survey on the situation of historic centres[13] shows that the priorities to be followed in order to implement a new vision that rebalances management and transformation processes are:

- Strengthening heritage protection within the framework of regional urban planning laws
- Regulating tourist uses of the residential heritage
- Managing and planning tourist flows in cases of excessive pressure
- Promoting the development of compatible creative and productive activities
- Promoting the use of advanced technologies for management
- Providing risk prevention and crisis response plans
- Developing systematic tools for understanding the dynamics of change affecting historic centres

The possibilities of enhancing the interpretation of cultural assets through augmented reality and other techniques are also explored by several initiatives.

The direction in which the application of digital tools is heading is clear, although obstacles such as financial resources to achieve massive digitisation or digital gaps in share and use of collected data still remain. In order to reach new frontiers for the application of digital technologies, there is a need for awareness of the significance of documentation and knowledge and the development of a critical approach, which also adopts digital tools to support the onsite fruition of cultural heritage.

References

Balzani, M., Boscaro, C., & Medici, M. (2013). Rilievo integrato, rilievo in emergenza. *Paesaggio urbano, 2*, 48–59.

Bianchini, C., Ippolito, A., & Bartolomei, C. (2015). The surveying and representation process applied to architecture: Non-contact methods for the documentation of cultural heritage. In S. Brusaporci (Ed.), *Handbook of research on emerging digital tools for architectural surveying, modeling, and representation* (pp. 44–93). IGI Global.

[13] Centri storici e futuro del paese. Associazione nazionale centri storico artistici - Indagine nazionale sulla situazione dei Centri Storici. http://www.cresme.it/doc/rapporti/Centri-storici-e-futuro-del-Paese.pdf, accessed 10 May 2022.

Centofanti, M., Brusaporci, S., & Lucchese, V. (2014). Architectural heritage and 3D models. In P. Di Giamberardino, D. Iacoviello, R. N. Jorge, & J. M. R. S. Tavares (Eds.), *Computational modeling of objects presented in images* (pp. 31–49). Springer.

Council of Europe. (2018). *European cultural heritage strategy for the 21st century*. Retrieved January 15, 2021, from https://www.coe.int/en/web/culture-and-heritage/strategy-21

Europa Nostra. (2020). *Covid-19 & beyond. Challenges and opportunities for cultural heritage*. Retrieved January 10, 2021, from https://www.europanostra.org/wp-content/uploads/2020/10/20201014_COVID19_Consultation-Paper_EN.pdf

European Commission. (2019). *Cooperation on advancing digitisation of Cultural Heritage*. Retrieved December 17, 2021, from https://ec.europa.eu/digital-single-market/en/news/eu-member-states-sign-cooperate-digitising-cultural-heritage

European Commission. (2020). *Basic principles and tips for 3D digitisation of cultural heritage*. Retrieved January 18, 2022, from https://digital-strategy.ec.europa.eu/en/library/basic-principles-and-tips-3d-digitisation-cultural-heritage

Francini, M., Colucci, M., Palermo, A., & Viapiana, M. F. (2012). *I centri storici minori. Strategie di rigenerazione funzionale*. FrancoAngeli.

Horizon Europe Strategic Plan 2021–2024. (n.d.). Retrieved December 17, 2021, from https://ec.europa.eu/commission/presscorner/detail/en/ip_21_1122

ICOM - International Council of Museums. (2020). *Museums, museum professionals and COVID-19*. Retrieved December 28, 2021, from https://icom.museum/en/news/museums-museum-professionals-and-covid-19-survey-results/

IDEA Consult, Goethe-Institut, Amann, S., & Heinsius, J. (2021). *Research for CULT committee—Cultural and creative sectors in post-Covid-19 Europe: Crisis effects and policy recommendations*. European Parliament, Policy Department for Structural and Cohesion Policies.

Kono, T. (Ed.), Adetunji, O., Jurčys, P., Niar, S., Okahashi, J., & Rush, V. (2020). *The impact of COVID-19 on heritage: An overview of responses by ICOMOS National Committees and Paths Forward*. Retrieved January 20, 2021, from http://openarchive.icomos.org/id/eprint/2415/

Lo Turco, M., Piumatti, P., Calvano, M., Giovannini, E. C., Mafrici, N., Tomalini, A., & Fanini, B. (2019). Interactive digital environments for cultural heritage and museums. Building a digital ecosystem to display hidden collections. *Disegnarecon, 12*(23), 7–1.

Lo Turco, M., Giovannini, E. C., & Mafrici, N. (Eds.). (2020). *Digital & documentation. Digital strategies for cultural heritage* (Vol. 2). Pavia University Press.

Machat, C., & Ziesemer, J. (Eds.). (2020). *Heritage at risk. World report 2016–2019 on monuments and sites in danger*. ICOMOS, hendrik Bäßler verlag.

Maietti, F. (2008). *Centri Storici Minori. Progetti di recupero e restauro del tessuto urbano fra identità culturale e salvaguardia*. Maggioli Editore.

Maietti, F., Di Giulio, R., Balzani, M., Piaia, E., Medici, M., & Ferrari, F. (2018). 3D data acquisition and modelling of complex heritage buildings. In M. Ioannides (Ed.), *Digital cultural heritage* (pp. 1–13). Springer.

Markham, A., Osipova, E., Lafrenz Samuels, K., & Caldas, A. (2016). *World heritage and tourism in a changing climate*. United Nations Environment Programme/United Nations Educational, Scientific and Cultural Organization. Retrieved July 29, 2021, from https://whc.unesco.org/document/139944

Maxwell, I., Drdácký, M., Vintzileou, E., Bonazza, A., & Hanus, C. (2018). *Safeguarding cultural heritage from natural and man-made disasters. A comparative analysis of risk management in the EU*. European Commission Publication Office. Retrieved October 21, 2021, from https://data.europa.eu/doi/10.2766/224310

Myers, D., Quintero, M. S., Dalgity, A., & Avramides, I. (2016). The Arches heritage inventory and management system: A platform for the heritage field. *Journal of Cultural Heritage Management and Sustainable Development, 6*(2), 213–224.

NEMO - Network of European Museum Organizations. (2021). *Follow-up survey on the impact of the COVID-19 pandemic on museums in Europe*. Retrieved December 15, 2021, from https://

www.nemo.org/fileadmin/Dateien/public/NEMO_documents/NEMO_COVID19_FollowUpReport_11.1.2021.pdf

Oreni, D., Brumana, R., Della Torre, S., Banfi, F., & Previtali, M. (2014). Survey turned into HBIM: The restoration and the work involved concerning the Basilica di Collemaggio after the earthquake (L'Aquila). *ISPRS Annals of the Photogrammetry, Remote Sensing and Spatial Information Sciences, 2*(5), 267–273.

Parrinello, S. (Ed.). (2019). *Digital & documentation databases and models for the enhancement of heritage* (Vol. 1). Pavia University Press.

Pasquinelli, C., Trunfio, M., Bellini, N., & Rossi, S. (2021). Sustainability in overtouristified cities? A social media insight into Italian branding responses to Covid-19 crisis. *Sustainability, 13*, 1848. https://doi.org/10.3390/su13041848

Piano Nazionale di Ripresa e Resilienza. (2021). Retrieved December 11, 2021, from https://www.governo.it/it/articolo/piano-nazionale-di-ripresa-e-resilienza/16782 and https://www.governo.it/sites/governo.it/files/PNRR.pdf

Pini, D. (2004). Il sito, il tessuto urbano e le emergenze: un rapporto di necessità. In F. Maietti (Ed.), *Dalla grammatica del paesaggio alla grammatica del costruito. Territorio e tessuto storico dell'insediamento urbano di Stellata* (pp. 15–17). Nardini Editore.

Puma, P. (2020). The historical cities in transition in the global trend: Some issues of architecture's identity survey and representation of the genius loci. In F. Nepravishta & A. Maliqari (Eds.), *IFAU19 - 3rd International Forum for Architecture and Urbanism. Modernisation and globalization. Challenges and opportunities in architecture, urbanism, cultural heritage* (pp. 82–89). Flash Publishing.

Remondino, F. (2011). Heritage recording and 3D modeling with photogrammetry and 3D scanning. *Remote Sensing, 3*(6), 1104–1113.

Trizio, I., Brusaporci, S., Luigini, A., Ruggieri, A., Basso, A., Maiezza, P., Tata, A., & Giannangeli, A. (2019). Experiencing the inaccessible. A framework for virtual interpretation and visualization of remote, risky or restricted access heritage places. In *International Archives of the Photogrammetry, Remote Sensing and Spatial Information Sciences* (Vol. XLII-2/W15, pp. 1171–1178).

UNDRR. (n.d.). *Sendai framework for disaster risk reduction 2015-2030*. United Nations. Retrieved November 17, 2021, from https://www.undrr.org/publication/sendaiframework-disaster-risk-reduction-2015-2030

United Nations. (n.d.). *Agenda 2030 for sustainable development*. Retrieved November 17, 2021, from https://sdgs.un.org/2030agenda, accessed 10 May 2022.

Vernizzi, C. (2016). Integrated survey for the representation of urban landscape features: The case of Navelli and Civitaretenga (Aquila). In F. Capano, M. I. Pascariello, & M. Vison (Eds.), *Old and new media for the image of the landscape* (pp. 99–107). Circe

World Tourism Organization. (2018). *Tourism and culture synergies*. UNWTO. Retrieved December 4, 2021, from https://doi.org/10.18111/9789284418978

Digitalisation and Cultural Heritage Tourism in Asia: Stepping in Without Stepping Out

David Ocón

Abstract During the early years of the 2020s, large parts of the world came to a standstill instigated by COVID-19. The pandemic dramatically affected people's mobility, including intensive lockdowns, border closures, and prohibition of local and international travel. These restrictions had dramatic implications for tourism, particularly for destinations that live from it and need it to survive. Locations, particularly with cultural heritage attractions such as monuments, historical buildings, archaeological sites, landscapes, urban ensembles, and sites known for their customs, folklore, oral and performing traditions, and religious manifestations, suffered the consequences. While digital leisure experiences cannot replace in-person visits, recent developments in digitalisation, such as 360-degree technology, virtual and augmented reality, and gaming, both 'serious'—video games designed for educational objectives—and for pure entertainment, are increasingly impacting how we visit, interact with and consume cultural heritage. This chapter investigates the latest trends in digitalisation in its application to cultural heritage tourism in Asia, particularly in contexts of crisis like the COVID-19 pandemic, where in-person opportunities to engage with cultural heritage assets are significatively reduced. The chapter argues that, albeit initially largely unintentional, the push for these new technologies during the pandemic has generated more sustainable and equal access paths to cultural heritage assets, particularly for visitors with limited means or mobility.

For the purpose of this chapter, the use of Asia corresponds to the distinct UNESCO definition of the Asia and the Pacific region (APAC), which includes Northeast Asia, Southeast Asia, South Asia, Central Asia, Australia, and Oceania.

D. Ocón (✉)
Singapore Management University, Singapore, Singapore
e-mail: davidocon@smu.edu.sg

E. Borin et al. (eds.), *Cultural Leadership in Transition Tourism*, Contributions to Management Science, https://doi.org/10.1007/978-3-031-14121-8_9

1 Introduction

In 2019, tourism contributed to nearly 20% of Thailand's GDP, and the country earned around 3 trillion baht (approximately $89 billion) in tourism revenue, two-thirds from foreign visitors (Phoonphongphiphat, 2021). Close to 40 million international tourists flocked to the country's beaches of Phuket, Krabi, or Koh Samui. Many travelled to experience Thailand's cultural heritage sites that include the thousands of *wats* (temples) that dot the country, and popular sites such as the historic city of Ayutthaya and Chiang Mai 'Creative City', and iconic spots such as *The Grand Palace* and the *Temple of the Reclining Buddha* in Bangkok. Tourists also visited Thailand to explore popular festivals such as *Loi Krathong* festival of light and *Songkran* water festival, and indulge in world-famous dishes such as pad thai, green curry, and tom yam kung, the latter proposed as a candidate to enter UNESCO's list of intangible cultural heritage, where it would join the Thai dance form *Khon*, based on the *Ramakien* (Glory of Rama) epic, and traditional Thai massage. Access to most of these tangible and intangible cultural heritage attractions stopped at the beginning of the 2020s when the COVID-19 pandemic struck. As a result, in 2020 and 2021, tourism visits to Thailand declined 83% and 99%, respectively, from the 2019 numbers. The 40 million foreign visitors turned to barely 6.7 million in 2020 and 427,000 in 2021 (Trading Economics, 2022).

Forty million foreign visitors were also expected in Japan in 2020 when the country was scheduled to host the Olympic and Paralympic Games. One year earlier, close to 32 million tourists had travelled to Japan, many of them to visit the historic monuments of ancient Kyoto and Nara, the *Hiroshima Peace Memorial*, the *Hijemi Castle* in Hyōgo Prefecture, and the country's many shrines and temples, part of the country's rich cultural heritage. As in the case of Thailand, culinary experiences are central to Japan's identity, and *Washoku*, the name used to describe all types of traditional Japanese dishes, was added to the UNESCO intangible heritage list in 2013. Many visitors also engaged with other intangible cultural heritage manifestations, such as *Nogaku* and *Kabuki* theatre, the traditional Ainu dances in Hokkaido, or *Washi*, the craftsmanship art of traditional Japanese handmade paper. However, with the arrival of COVID-19, foreign visits to Japan plunged 87.1% in 2020 from a year earlier and barely 10% (4.12 million) of the projected 40 million tourists visited the country (Kyodo, 2021; Arba, 2021). That drop was 99.2% in 2021, when only 245,900 people visited Japan, the lowest since 1964, when comparable data became available (Kyodo, 2022).

Thailand and Japan were, in 2019, the second and third largest tourist markets in Asia, only behind China. These two examples illustrate how the restrictions derived from the handling of the COVID-19 pandemic impacted visits to Asia's rich cultural heritage in Asia. In 2019, Asia had recorded 360 million international tourist arrivals and earned $443 billion in tourism receipts, 25% and 30%, respectively, of the world's total (UNWTO/GTERC, 2020).[1] The closure of museums and heritage sites and the cancellation of performances was just the tip of the iceberg of consequences that affected every aspect of life. The COVID-19 pandemic caused unprecedented disruption to global tourism, triggering the most severe tourism crisis on record in the region. Asia was the first region to suffer the impact of the pandemic and is still the hardest hit. By the second half of 2021, far from improving, the situation had worsened. With high-impact measures to contain the spread of the virus, such as widespread border closures, as of August 2021, Asia had lost 95% of its international arrivals compared with the same period in 2019, resulting in sharp declines in demand for intra-regional and international travel. Southeast Asia led the world's tourism decline, with 97.9% fewer visitors than in 2019 (UNWTO, 2021).

In 2022, we start envisioning how post-COVID-19 contexts can look, largely thanks to widespread vaccination campaigns. Progressively, pre-pandemic travel resumes in some parts of the world, with tools such as extensive testing, travel corridors or 'bubbles' between countries with low transmission rates, and travel lanes for vaccinated travellers, enabling this. However, many domestic and international travel restrictions continue in Asia. Stay home notices, expensive hotel quarantines, and pervasive testing are still frequent in many Asian countries, reducing the appeal of travelling and, resultantly, significatively diminishing opportunities to experience cultural heritage first hand.

While digital leisure experiences cannot replace in-person visits and experiences, recent developments in digitalisation are increasingly diversifying how we visit, interact with and consume cultural heritage products. 360-Degree technology, virtual and augmented reality (VR and AR), 3D modelling and printing, 'edugames', interactive maps, and podcasts are examples of technological tools that have progressively joined our cultural heritage consumption habits. Traditionally, these technological offerings have supported our efforts to know more about cultural heritage sites, investigate their relevance, and create an alternative encounter with them. However, during the COVID-19 pandemic more than ever, the synergy between technology, cultural heritage, and tourism had been as intensive, stemming from the evident contextual necessity.

This chapter investigates the latest trends in digitalisation in its application to cultural heritage tourism, particularly when in contexts of crisis like the COVID-19 pandemic, in-person opportunities to engage with cultural heritage assets are significatively reduced. It uses a multiple case study research approach, in which several essential cases across Asia are analysed using multiple data collection methods, which allows for a holistic understanding of the current state of the matter. The chapter argues that, albeit initially largely unintentional, the push for these new technologies during the pandemic, most of them free of charge, has generated more

[1] This share had expanded continuously from 16% in 2000 to 25% in 2019. Asia is the second most visited region in the world after Europe and has seen the fastest growth in recent years. As of 2019, China was the top destination in Asia, with 66 million international arrivals, followed by Thailand with 40 million and Japan with 32 million. Other top Asian destinations are Malaysia, Hong Kong (China), Macao (China), Vietnam, and India.

sustainable and equal access paths to cultural heritage assets, particularly for visitors with limited means or mobility.

2 Cultural Heritage Tourism

Cultural Heritage refers to the contemporary society's use of resources inherited from the past conferred for the benefit of future generations (UNESCO, 2014; Graham et al., 2016). Today, cultural heritage comprises a multitude of items. While until the last part of the twentieth century, the term primarily referred to monuments, historical buildings, and archaeological sites (Vecco, 2010), in the last decades, it has substantially expanded its scope. Cultural heritage includes now tangible assets such as parks and gardens, landscapes, industrial and rural heritage, urban ensembles, works of art, cuisine and written texts. It also comprises intangible features, like folklore, language, memories, values, traditions and customs, in associations that evolve with time and often change from place to place (Timothy & Boyd, 2003; Nilson & Thorell, 2018).[2] A cultural heritage item is defined today by its historical relevance and artistic attributes, but also by its cultural values, its identity associations, and its capacity to trigger memories (Vecco, 2010), in a comprehensive approach that incorporates the heritage's context (Greffe, 2004).

Culture and heritage have long contributed to increasing the appeal of tourist destinations. We refer to cultural heritage tourism when people travel to a location to visit buildings, artefacts, or landscapes and experience activities that authentically represent the stories and people of the past, including aspects such as traditions, celebrations, music, clothing, and food. Cultural heritage tourism provides opportunities for people to experience other cultures, whether by visiting attractions, historical or culturally relevant sites, or taking part in cultural activities. Cultural heritage tourism has become increasingly popular, with 40% of current travellers identified as 'cultural tourists' (Global Heritage Fund, 2019) participating in a cultural visit or activity as part of their stay. Tourists interested in cultural heritage generally want to learn something about the beliefs, practices, struggles, and successes that shaped the shared identity of a group of people or culture (McNulty & Koff, 2014).

Cultural heritage tourism is important for multiple economically beneficial reasons, for instance, by injecting new money into the economy, boosting businesses and tax revenue, creating new sources of development and income, jobs, events, and attractions, thus helping diversify the local economies. Some of the social benefits of heritage tourism can include helping to build social capital by enabling culture as an instrument that increases exposure to other cultures and communities and facilitates

[2] In the 2000s, partly due to its integration into UNESCO frameworks and Conventions (2003 Convention for the Safeguarding of the Intangible Cultural Heritage, 2005 Convention on the Protection and Promotion of the Diversity of Cultural Expressions), the notion of cultural heritage expanded to include notions of immateriality and orality, such as customs, folklore, oral and performing traditions, religious or profane manifestations (Barrio et al., 2012).

understanding among people. Cultural heritage tourism can also promote the preservation of local traditions and customs, develop local crafts and other cultural expressions, enhance community pride and reinforce identity, and provide research and education opportunities. Furthermore, if conducted sustainably, cultural heritage tourism can help encourage a culture of preservation in communities, boosting awareness of a particular site and encouraging residents and visitors to be mindful of their impact on it.

Increasingly embedded in the so-called 'experience economy' (Nilson & Thorell, 2018), cultural heritage tourism can also be a double-edged sword or even a 'burden' for some sites (Caust & Vecco, 2017). When poorly managed, tourism can pose threats to fragile cultural sites and unwanted social and economic consequences. Overexploitation of cultural heritage resources can physically damage sites and erode unique atmospheres and charm of destinations, diluting identity traits. D'Eramo (2014) labels the, by and large otherwise celebrated, UNESCO World Heritage recognition, as 'the kiss of death', which, he posits, often marks the replication of 'non-places' deployed of authenticity. International attention can also lead to loss, damage, or looting of cultural treasures. Uncontrolled surges in visitation can result in increased waste, water and energy consumption, and the development of precarious facilities and services in and around popular sites. Increased crime and prostitution, and cultural tensions and clashes between tourists and residents can also happen (Porter & Salazar, 2005).

The travel and tourism industry is directly responsible for more than 300 million jobs worldwide, and it creates one in every five new jobs. The direct global value of culture and heritage tourism is over $1billion, with that of the Asia-Pacific region approximately $327 million. Cultural heritage tourism is directly responsible for more than 50 million jobs in APEC countries[3] (Childs, 2015). Cultural heritage tourists tend to stay longer (+22%), spend more (+38%), and typically travel more often than other types of tourists (OTTI 2013 Industry Sector Profile: Cultural Heritage). In 2020, however, the travel and tourism industry's total GDP accounted for just 5.5% of the global GDP, a massive decline from 10.4% in the previous year (Statista Research Department, 2021a, b). In Asia-Pacific, the total contribution of travel and tourism to GDP declined by 53.7% and to employment by 18.4%. India is one of the region's biggest markets. Until the arrival of the COVID-19 pandemic, tourism contributed 6.8% to the country's GDP and 8% to employment. In 2020, over 8 million jobs were lost in the industry, and the GDP contribution decreased to 4.7% (WTTC, 2021). The crown of the Indian cultural heritage tourism industry is the Taj Mahal, which attracts over 8 million visitors annually, 10% from overseas (Taj Mahal, 2021). In 2020, however, the monument saw a decrease of 76% in ticketed visits (Mint, 2021).

[3] APEC stands for *Asia-Pacific Economic Cooperation*, a regional economic forum established in 1989 to leverage the growing interdependence of the Asia-Pacific. Although the majority of APEC's 21 members are Asian countries, it also includes members in the Americas and Russia.

With these figures, the cultural heritage tourism sector had to innovate and turn to the digital world to provide alternative virtual access to their iconic sites. Digital leisure experiences cannot replace in-person visits and experiences and, in most cases, cannot repair the economic and social damage provoked by the pandemic. However, recent developments in digitalisation have provided unconventional, creative, and innovative ways to engage with cultural heritage. In some cases, technological advancements open up opportunities to complement and enhance access to cultural heritage assets, often in more sustainable and equal ways.

3 The Potential of Digitalisation as Support for Cultural Tourism

The range and accessibility of digital tools supporting cultural heritage tourism have expanded exponentially in the last decade. Digital archiving of existing documents such as technical data, blueprints and old pictures, together with digital capturing techniques that include full coverage photography and videography (for instance, with the help of 360-degree technology and drones), photogrammetry, 3D laser scanning, and reconstruction techniques such as digital drawing and 3D modelling, can be used today for virtual reality purposes. Digitalisation, in its intersections with tourism and cultural heritage, can perform a wide range of functions beyond its traditional role in documentation and archiving. One of its prominent roles is safeguarding visual reminiscences in anticipation of potential damage, irreversible destruction or degradation due to natural disasters or human-made actions or decisions. Digitalisation can help as a 'memory insurance policy', lengthening and even expanding the cultural heritage item's life once its original physical manifestation is gone or damaged (Ocón, 2021). Several recent examples illustrate this: fires damaged in 2015 Moscow's *Novodevichy Monastery*'s bell tower, a UNESCO World Heritage site, devastated over 90% of Brazil's *National Museum*'s 20 million artefacts in 2018, and 1 year later destroyed *Notre Dame of Paris*' iconic spire and most of its thirteenth-century wooden roof; an earthquake in 2015 reduced to piles of rubble many of Kathmandu valley's most historically significant buildings, some dating more than 1700 years. Hostilities and armed disputes were behind the destruction in 2001 of the emblematic Buddhas of Bamiyan by the Taliban in Afghanistan, and of parts of the 4000-year-old oasis city of Palmyra (Syria), often known as the *Venice of the Sands*, and Mosul's (Iran) 850-year-old *Great Mosque of al-Nuri* by Daesh.

Beyond their preservative and research roles, digital tools can also improve and diversify the way cultural heritage is accessed by the public, by complementing and enhancing the experiences offered. They can help to provide details challenging to see in a gallery or a heritage site, thus bringing the visitor closer to the essence of the works (Kersten et al., 2017; Styliani et al., 2009; Deggim et al., 2017; Tschirschwitz et al., 2019). This is particularly handy in cases where there is restricted access due to the site's vulnerability (Tanasi et al., 2019), removing traditional barriers such as money, location and opening hours, providing alternative environments of display and engagement beyond the confines of conventional museums (King et al., 2016). This aspect is particularly relevant in extraordinary circumstances such as the COVID-19 pandemic, which dramatically limited physical access to many heritage sites. For instance, Istanbul's iconic *Suleymaniye Külliye* (Mosque and Complex) was fully digitalised in 2019 and can now be experienced virtually via 3D modelling and Virtual Reality (Kan et al., 2019).

Digital tools are also essential when the cultural heritage is of problematic, restricted, or elusive access (Aiello et al., 2019; Pizziolo et al., 2012; Garaba, 2015). Through the *Te Ataakura* project, New Zealand's Māori organisation *Toi Hauiti* contributed to documenting and making accessible the heritage of their indigenous communities. With their ancestral *taonga* (artefacts) dispersed among collections throughout New Zealand and many other countries, physically visiting the *taonga* is challenging. With *Te Ataakura*'s 'digital taonga' and 'virtual repatriation' proposals, Māori cultural heritage is revisited, rekindled, restored innovatively and dynamically, and made available to the Māori community and visitors worldwide (Ngata et al., 2012).

Virtual visits can also help expand the reach of the cultural heritage, allowing for new visitor flows. For instance, through the project *Singosari: the origins of Majapahit*, Leiden's *Museum Volkenkunde* and Jakarta's *Museum Nasional Indonesia* developed an app that showcased augmented reality statues from the thirteenth-century syncretic Hindu-Buddhist Singosari temple in East Java Singhasari-Majapahit kingdom. This digital collaboration allowed tourists and visitors to these two museums separated by over 11,000 km to interact with the remains of this unique civilisation and access its distinctive historical and religious contexts (Brink, 2013).

Digitalisation can open innovative ways to engaging virtually with a heritage site. One of them is through educational tools that can help enhance teaching and learning experiences, for instance, through *edutainment* materials designed for educational objectives, which allow for a new kind of knowledge dissemination, often targeting younger generations. These so-called *serious games*' aims include anchoring the intended information, developing decision-making skills, and increasing strategic thinking abilities. Edugames comprise a wealth of typologies, including historical, patrimonial, or even archaeological. Since *The Oregon Trail*, an educational game that familiarised US elementary students with the nineteenth-century American pioneer life, was introduced in the 1970s, *serious games* have become increasingly popular. One of the latest arrivals is *The Sericum* Via, which stimulates interest in Iran's cultural heritage by proposing a virtual encounter with eighteenth century's Safavid Caravanserais, the network of roadside inns where travellers could rest and recover from the journey across the *Silk Roads* routes that traversed Persia (Eshaghi et al., 2021). Beyond educational games, the world of video games also provides numerous examples that offer detailed access to digital versions of tangible and intangible cultural heritage. For instance, through real-time and strategy games such as the *Age of Empires* and the *Civilisation* series, or action-adventure video games like Ubisoft's *Assassin's Creed* saga, celebrated for providing realistic versions of major cities and landscapes across the world.

Today, the drop in prices (Tschirschwitz et al., 2019; Aiello et al., 2019) and the gains in accuracy in 3D printing technologies (Balletti et al., 2017) offer new possibilities of display and analysis of heritage assets. This new 'digital cultural heritage', also known as 'cultural heritage information resource' (Lor & Britz, 2012), can lengthen heritage's lifespan and expand its reach. When the *Pekeliling Flats* in Malaysia's capital were demolished in 2015, a team of university researchers had virtually reconstructed them to preserve them for the future. The *Flats*, an iconic sight in Kuala Lumpur, were the first project in the 1960s that introduced *Industrialised Building System* (using precast concrete) to Malaysia, providing affordable housing to low-income groups. Although the original *Pekeliling Flats* have now disappeared, their digital restoration lives and the 3D models produced using 3D printing technology can be used for exhibition and educational purposes (Esmaeili et al., 2014).

The digitalisation of cultural heritage is not always an easy task. For instance, the lack of available data and sources, or misalignments between contextual information and the digital content can create confusion and misunderstandings (Chowdhury, 2010). Moreover, despite the recent drop in prices, digital tools are still expensive and time consuming (European Commission, 2015), which often goes hand in hand with adequate professional, technical skills or knowledge (Cignoni & Scopigno, 2008; Bruno et al., 2010). This is something that this chapter's author has experienced first-hand: when in 2021, I decided to use UAV filming to digitalise the aerial view of a Muslim cemetery in Singapore, months of waiting for permits and costly licences significantly delayed and challenged the project. With respect to this, some authors have also stressed the risks associated with universal virtual accessibility to sensitive spaces of cultural or religious nature, as they can hold personal memories and connections with the internees.

Whether tangible or intangible, cultural heritage celebrations are often seen as vehicles for expanding economic interests, as actual income and tourism generators (Waterman, 1998; Getz, 2009; Hughes, 1999; MacCannell, 1992). When this happens, the boundaries between culture, tourism, and profit-making become increasingly blurred, with these festivities acting as producers of tourism experiences (Szmigin et al., 2017). Once encompassed as part of a city's tourism supply, powerful economic forces can turn the cultural heritage in ways that undermine local traditions and decision-making.

A longer-term concern for using digital tools to showcase cultural heritage is the utilitarian usage of the results by decision makers, for instance, to justify demolishing a site to use the space for other developments, as has happened in some large Asian cities.

4 Digitalising Cultural Tourism in Asia: Case Studies from the Region

4.1 360-Degree Technology, 3D Modelling and Photogrammetry: Reimagining Visits to Shurijo Castle, Okinawa, Japan

In October 2019, a fire caused by an electrical fault devastated the fourteenth-century *Shurijo Castle* in Japan's Okinawa Islands. *Shurijo Castle* sat on top of a hill overlooking the city of Naha, Okinawa's capital. Between 1429 and 1879, the wooden castle was the palace of the Ryukyu dynasty. The Ryukyu kingdom thrived on maritime trade, connecting countries in the region. The castle had burned down three times during the Ryukyu dynasty and was entirely destroyed by the American forces in 1945 during World War 2's *Battle of Okinawa*. After the war, the site became a campus for the *University of the Ryukyus* until the 1970s. In 1992, commemorating the 20th year since Okinawa's reversion to Japan, *Shurijo Castle* was restored, modelled after the castle from the eighteenth century, and opened as a National Park. In 2000, it was designated a UNESCO World Heritage Site as part of the Gusuku Sites and Related Properties of the Kingdom of Ryukyu, the 11th in Japan, and appeared in commemorative 2000-yen notes. The castle, the largest wooden building in Okinawa, was one of the most popular attractions in the region, visited by over two million tourists every year. The castle was central to understanding Ryukyu's history, culture, and identity, 'a beautiful symbol of resilience and endurance' and of 'the splendor of Okinawa' (Denny Tamaki, Governor of Okinawa, as quoted in Callot, 2020).

Shortly after the blaze, a team of volunteers that included researchers, students, and engineers with expertise in Computer Vision (CV), Human–Computer Interaction (HCI) and Virtual Reality (VR), joined forces to virtually reconstruct this essential element of Okinawa's cultural heritage. They created *Our Shurijo* project, a 3D digital reconstruction of *Shurijo Castle*'s shapes and textures using 3D modelling technology and photogrammetry.[4] The team collected over 80,000 photos and videos shared by 3000 people across the globe who wanted to see the digital reconstruction of this site and preserve the legacy of Ryukyu culture for generations to come. Over the months, the team created several 3D reconstructed models with annotations. Since then, the virtual Shurijo Castle Reconstruction has received over 83,000 visitors, with the 3D *Seiden* (the castle's main temple) visited by over 30,000 visitors and *Shurijo*'s throne by close to 7000 visitors. In addition, more than 1700 visitors have appreciated the detail provided by the 3D digital reconstruction of the *Shurijo*'s King's Crown, also destroyed in the blaze. While these numbers are far from the two million visitors that visited Shurijo before its destruction, they symbolically speak of the power of virtual tools to preserve memories, as educational

[4]For more information about the project: www.our-shurijo.org/en/.

Fig. 1 Shujiko Castle 3D Digital model (Shuri Castle Digital Reconstruction, 2019)

tools, and to attract tourists until the physical castle is rebuilt, something that could take decades (Kaneko, 2019; Machemer, 2020). Beyond the *Our Shurijo* project, the Okinawa Prefectural government collaborated with Google's Arts & Culture platform to highlight the castle's heritage through the digital project *Reconstructing Shurijo Castle* (Callot, 2020). The project compiles materials that include 360-degree visits, recordings made by drones flying through and above the castle park, a virtual historical compilation of paintings and photographs that contribute to keep the memory of the site alive (Figs. 1 and 2).

In other parts of Asia, the use of digital tools such as 360-degree technology, 3D modelling, and photogrammetry is also multiplying the possibilities of accessing and visiting cultural heritage sites. As a result, national heritage bodies, regional governments, municipalities, and private entities are increasingly exploring the possibilities of technologies to enhance and increase access to valuable cultural heritage sites. Religious buildings such as temples are among the most popular choices. For instance, eighteenth century's *Lukang Longshan* is Taiwan's best-preserved Buddhist temple dating to the *Qing* Dynasty (1644–1912). The temple, designated a national monument in 1983, exhibits impressive architectural elements that include intricately painted murals, woodcarvings, a caisson or spider web ceiling, and granite columns with carved dragons. *Lukang Longshan* was affected by a large earthquake in 1999 and suffered extensive structural damage to the roof, walls and decorated pillars. Although the site's reconstruction was completed in 2008, it remains vulnerable to tectonic activity and is threatened by smog air pollution. A recent public–private partnership was formed to provide *Lukang Longshan* with

Fig. 2 Use of photogrammetry to create Shujiko Castle's 3D Digital model (Capturing Reality, 2021)

digital life (CyArk, 2021).[5] Using 3D scanning technology and drones to capture data aerially, the team created a 360-degree virtual tour and 3D models of several temple sections, including its intricate spider web roof and a carved stone pillar, that can be viewed in VR using the free *Sketchfab* platform. Now virtual visitors can 'tour' *Lukang Longshan* as if they were there.

Hindu heritage is also being digitalised and made progressively accessible to virtual visitors. For example, in the Indonesian island of Bali, *Melanting* and *Pulaki* temples were digitalised using VR technologies. Bali, Indonesia's tourist crown and its most visited destination with 38.47% of the country's international tourists in 2019 (Fadli, 2020), suffered tremendously due to the COVID-19 pandemic. In 2020, the decrease of foreign tourist arrivals to Bali was 83.26%.[6] VR experiences enable the discovery of objects inside the temples in 3D, complemented with narrations and background music (Darmawiguna et al., 2020). Virtual visitors can now discover the island's unique religious architecture and celebrate its historical importance from any location in the world.

[5]Between Taiwan's *Bureau of Cultural Heritage*, non-profit organisation *Cyark*, dedicated to digitally recording, archiving and sharing cultural heritage worldwide, and *Iron Mountain*, a global company that provides storage and information management services.

[6]As of the first half of 2021, the foreign tourist arrivals to Bali decreased by 100% compared to the same period the year before. In mid-October 2021, Indonesia finally reopened Bali for fully vaccinated international tourists (Statista Research Department, 2021b).

4.2 Virtual Exhibitions: Experiencing Strategically Curated Asian Cultural Heritage

At the end of 2020, amid the COVID-19 pandemic and given the impossibility of opening a physical exposition, India's *National Museum* in New Delhi launched the *SCO Virtual Exhibition on Shared Buddhist Heritage*.[7] The exhibition displays 400 digital copies of rare Buddhist artworks from India, Kazakhstan, China, Kyrgyzstan, Pakistan, Russia, Tajikistan, and Uzbekistan.[8] The virtual display showcases several fundamental events of Buddha's life, Buddha's teachings, and the philosophy and rituals of Buddhism. The participating museums and institutions provided 3D photographed objects and artefacts,[9] and India's *National Museum* team of curators integrated them into a collective digital exhibition that replicates a physical presentation. This virtual exhibition complements, enhances, and augments *National Museum*'s offerings and provides a platform that transcends national boundaries and encourages trans-national dialogues through regional and cultural aesthetics. In addition, the exhibition circumvents traditional limitations when physically visiting a museum: money, the virtual display is free of charge, and can be visited from anywhere in the world; time, the show opened in 2020 and has no end date; location, as it occupies a virtual space, it does not hinder the arrangement of other exhibitions; and scope, as it displays in virtual format valuable objects from several countries for a prolonged time, something not possible for a physical exhibition. Admittedly, the SCO virtual exhibition also has flaws: for instance, to enter most countries' halls, the loading times are long even with a high speed internet connection, which diminishes visitors' interest; some of the virtual spaces created are rudimentary, which decreases the exhibition's attractiveness; lastly, there are notable discrepancies when visiting the different countries' resources: while India seems overrepresented, Uzbekistan and Kyrgyzstan barely display a couple of dozens of objects each, and China's resources don't appear integrated into the exhibitions' main website but located in an external webpage belonging to the *Dunhuang Academy* in Gansu (Figs. 3 and 4).

Another significant example to note, more for its potential than its current achievements, is the *ASEAN Cultural Heritage Digital Archive* (ACHDA).[10]

[7] *SCO Virtual Exhibition on Shared Buddhist Heritage*: https://nmvirtual.in/.

[8] These eight countries are full members of the *Shanghai Cooperation Organisation*, a Eurasian political, economic and security alliance established in 2001.

[9] The exhibition includes 3D Buddhist artworks from India (*National Museum New Delhi*, *Indian Museum* in Kolkata), including the Gandhara and Mathura Schools, Nalanda, Amaravati and Sarnath. Objects depicting the life of Gautam Buddha from Karachi, Lahore, Taxila, Islamabad, Swat, and Peshawar museums in Pakistan, *National Museum* of Kazakhstan, *National Historical Museum* of the Kyrgyz Republic, *National Museum* and *National Museum of Antiquities* in Tajikistan, and archaeological sites of Uzbekistan, are also displayed. Objects from Moscow's *State Oriental Art Museum* and China's *Dunhuang Academy* are also included (Journals of India, 2020; The Hindu, 2020).

[10] *ASEAN Cultural Heritage Digital Archive* (ACHDA): https://heritage.asean.org/.

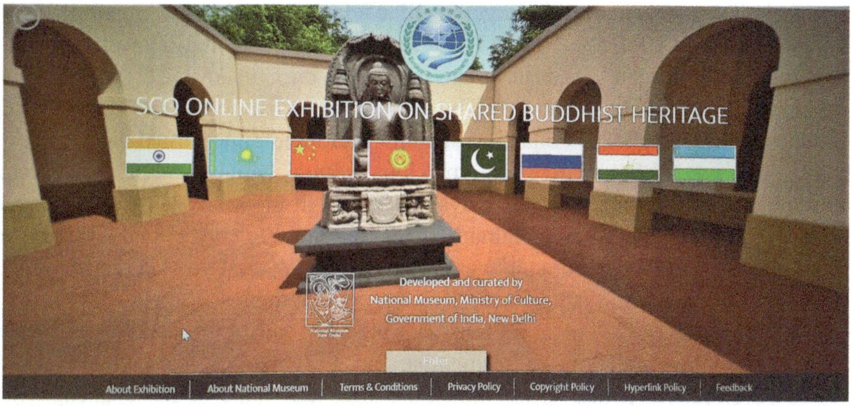

Fig. 3 Web captures of *SCO Virtual Exhibition on Shared Buddhist Heritage* (National Museum, New Delhi, India, 2022)

Fig. 4 Web captures of *SCO Virtual Exhibition on Shared Buddhist Heritage* (National Museum, New Delhi, India, 2022)

Announced at the beginning of 2018, the ACHDA project aims to present, celebrate and raise awareness of Southeast Asia's rich and diverse cultural heritage on one website, including artefacts, old documents and paintings (ASEAN, 2018; ACHDA, 2022). The *ASEAN Secretariat*[11] implements the ACHDA project with the financial

[11] ASEAN stands for *Association of Southeast Asian Nations*. Created in 1967 ASEAN aims to promote cooperation and facilitate integration among its members and other countries in Asia. ASEAN comprises ten members: Thailand, Singapore, Malaysia, Philippines, Indonesia, Brunei, Vietnam, Laos, Myanmar, and Cambodia.

and technical support of the Government of Japan through the *Japan-ASEAN Integration Fund* (JAIF). NTT DATA, a Japanese multinational information technology service company, leads the archiving system through AMLAD® technology, which manages digital content owned by institutions such as museums, libraries, and archives worldwide.[12] The collaboration allows for three-dimensional digital accessibility to some of Southeast Asia's precious cultural heritage assets (NTT Data Corporation, 2020), until now only available through physical visits to a selected number of collections. These cultural heritage treasures are conveniently accessible free of charge to visitors from all over the world. However, 4 years down the road, the ACHDA project still has a long way to go, with a modest number of resources currently on display (274 as of mid-2022), from only half of the ASEAN countries (Thailand, Malaysia, Indonesia, Cambodia, and Myanmar), and uncertainty in its future development plans. These limitations illustrate some of the challenges to making Southeast Asia's cultural heritage widely available.

4.3 Podcasts: 'Travelling by Ear' to Dream Southeast Asia

An alternative method of digitally engaging with cultural heritage sites is listening to travel podcasts that offer 'virtual' tourists immersive auditory travelling experiences. Travelling by ear has significant demand—during the COVID-19 pandemic, the *Podcast Consumer Tracking Report* shared a 30-min increase in listening times, while 'home' remains the top location for podcast listening (Sachs, 2020). For mobile users, there has been longer time engagement with deep, digital long-form news content (Mitchell et al., 2016), an area podcasts can satisfy.

Contrasting the spatial and time restrictions of physical travel during COVID-19, podcasts are permanently within the 'grasp' of tourists no matter the place. At the same time, listening can be complemented by other activities (like commuting or exercising), maximising time use (Schlütz & Hedder, 2021). Furthermore, in dealing with financial barriers, podcast travel requires low monetary investment due to the ready availability of smartphones and podcast apps (Schlütz & Hedder, 2021).

Most importantly, podcasts offer intense, immersive travelling experiences that somewhat replicate actual travelling. Podcasts offer 'the driveway moment', where we listen to a story so 'inherently dramatic' and 'compelling that it keeps you in your car listening to the radio' (Schlütz & Hedder, 2021). The screen-free format of consumption reduces distractions and allows for focused concentration on the auditory experience. By extension, the imaginative aspect allows for varying and unique interpretations of a specific place for every person that listens. The first

[12] In 2018, using the AMLAD technology, NTT DATA completed the digitisation of 3000 historical handwritten documents held by the *Vatican Library*. The technology has also been used by other cultural institutions, such as Japan's *National Diet Library* and *Koyasan University* (NTT Data Corporation, 2020).

method of auditory travel offers this immersive storytelling format, packed with ambient noises to imitate on-the-ground atmospheres and expressions of drama in the narrations. Notable ones include 'Armchair Explorer', in which travellers recount their 'hardcore' and adrenaline-pumping adventures in navigating challenging terrains, while episodes are edited in a cut documentary and cinematic style (Armchair Explorer, 2021). Scott Gurian's '*Far from Home*' is recorded while he is at the tourist site itself—meaning the episodes are not post-edited in a studio—therefore making the listener feel like he is there alongside Gurian. In an episode where he visits the *Angkor Wat* in Cambodia, we hear the motorcycles from the busy Cambodian traffic buzzing in the background, which slowly fades into ancient temple music as he gets closer to the entrance of the temple, describing the long line of locals waiting to enter (Far from Home, n.d.). If a listener prefers to focus on a site's white noise without narration, '*Field Recordings*' offers soundscapes from global contributors who document their country's immediate surroundings. In an episode by Colin Hunter, he recorded the songs of ticket sellers at the Alor Setar bus station in Kedah, Malaysia, trying to attract customers in a competitive private bus service industry. Additional context is provided as text on the website:

> There are many different private bus companies operating in Malaysia, each offering slightly different routes, destinations, comfort and price. Competition between these companies is high and the right sales technique can be what makes the difference in ensuring the bus parts with no empty seats. In Alor Setar bus station – in order to attract customers – ticket sellers from each of the bus companies announce the destinations at the top of their voices. But the result is far from a cacophony. Instead, they try to sing and rhyme the names of cities. And each seller has their own particular style which they use to try to cut through the other noise. (Hunter, 2021)

The second method of auditory travel is investigating a specific socio-cultural issue characteristic of the country, offering a more sustained form of tourism, where tourists embrace the complexities of a country beyond pretty photographs. This could be exploring the local music, covered by BBC Radio 3's 'Music Planet: Road Trip', or specific design structures of a place, covered by Roman Mars' '99% Invisible'. 'Far Flung with Saleem Reshamwala' explores tales of how countries adapt to problems in their societies. In an episode on Bangkok, the host explores how the country responds to its complex traffic congestion problem, which stunts emergency services from reaching those in traffic who are in need. To solve this problem, some local motorcycle police are trained as mechanics for when vehicles break down, while others are trained as midwives to medically help pregnant women who go into labour in their cars (Reshamwala, 2020). In leveraging on the role of digital nomads in activating digital tourism, podcasts that retell their reflections regarding their interactions with local society also make up auditory travel.

4.4 Augmented Reality (AR): Reviving Ancient Glories in South Korea

Hwangnyongsa Temple was a Buddhist Temple in the city of Gyeongju, today's South Korea. Gyeongju was the capital of the ancient kingdom of Silla (57 BC–935 AD) that ruled over Southern and Central parts of the Korean Peninsula for close to one thousand years. At its peak during the seventh and ninth centuries, Gyeongju had an estimated population of one million, then the fourth city in the world, next to Constantinople, Baghdad and Chang'an (Adams, 1989). *Hwangnyongsa Temple*, whose name means 'Imperial Dragon Temple' (Chung, 2001), was the cultural centre of Silla's state-sponsored Buddhism and represented a beacon of ancient Korean culture (Lee, 2017). Hwangnyongsa was designed to be a place where monks prayed for the nation's welfare and impressed foreign dignitaries. The temple featured a 47-metre-long main hall and an enormous 68-high nine-story wooden pagoda with an interlocking design that needed no iron nails, making it one of the tallest structures in East Asia at the time of its construction (Adams, 1983). The temple was entirely destroyed by Genghis Khan's Mongolian invasion in 1238, and only the massive foundation stones of the temple remain today.

In 2019, South Korea attracted 17.5 million international visitors, most of them from other parts of East Asia, and had over 40 million domestic tourists. Until the arrival of COVID-19, travel and tourism contributed to 4.2% of the country's GDP and provided 4.8% of the total employment. Cultural heritage tourism is becoming increasingly important in South Korea. The historic city of Gyeongju, with its vast number of ancient ruins and archaeological sites, is one of its most popular tourist destinations, visited by 6 million tourists annually, including 750,000 foreigners. Among its historical treasures are the *Seokguram Grotto* and *Bulguksa Temple*, the *Gyeongju Historic Areas* and the *Yangdong Folk Village*, all UNESCO World Heritage Sites.

One of the latest additions to the list of sites to visit is the *Hwangnyongsa Temple*, not in its physical form since it was gone 800 years ago, but in augmented reality (AR). AR is an interactive experience of a real-world environment where the real-world objects are enhanced by computer-generated perceptual information, sometimes across multiple sensory modalities. Unlike VR, which creates its own cyber environment, AR adds to the real world and allows the user to interact with it. AR uses a tablet or a mobile device's camera and enhances reality by adding contextual information to the surroundings. Thus, AR brings the digital world into the real world. The most mainstream examples of AR are Nintendo's game *Pokémon GO*, the instant messaging app *Snapchat*, and the furniture retailer IKEA's *Place* app. Since 2020, part of *Hwangnyongsa Temple*'s cultural heritage, its inner gate and South corridor, can also be experienced in AR. The *National Research Institute of Cultural Heritage* (Nrich) has digitally restored the temple to portray every life-size detail of the structures and their interiors in their original locations. Using a tablet loaned on-site, tourists can experience the ancient temple and engage with various programmes like a treasure hunt game utilising excavated artefacts, experience the buildings with different seasons' backgrounds, and animations about the construction process. Nrich plans to digitally restore *Hwangnyongsa*'s main building, hall, and wooden pagoda in the years to come.

The *Donuimun Gate* of the Joseon dynasty (1392–1897) was one of the *Four Great Gates* in the City Wall of Seoul, today's South Korea's capital. It was demolished in 1915 during the Japanese occupation, and in its place, stands now a busy road junction. In 2019, it was digitally restored thanks to a public–private partnership.[13] Visitors can now visit the digitally restored *Donuimun Gate* using a mobile app that places it in its original location in modern Seoul and can be experienced with several lights based on different day moments. Also in Seoul, a recent partnership[14] recreates the *Changdeok Royal Palace* with 5G AR technology, offering unconventional ways for visitors to appreciate the cultural heritage. The fifteenth-century monument was one of the *Five Grand Palaces* of the Joseon dynasty in Seoul. Despite being heavily damaged during the Japanese occupation (only 30% of its structure survives), it was inscribed as a *UNESCO World Heritage Site* in 1997 for being an 'outstanding example of Far Eastern palace architecture and garden design' (UNESCO World Heritage Centre, 2021b). 'Changdeok ARirang' is the latest development to allow visitors and tourists to interact with the *Changdeok Royal Palace*. When visitors launch the AR app, a *haechi* sculpture (an imaginary horned creature that fights off catastrophe and is Seoul city's mascot) comes to life on the screen. *Haechi* serves them as a guide through 12 main spots, including some off-limits to the physical visitors, such as the *Huijeongdang Hall* and the *Juhamnu Pavilion* in the garden where they find the royal library. Visitors can also take pictures with virtual Joseon kings and queens and experience virtual archery and kite-flying. In front of the *Nakseonjae Hall,* users can watch a performance of *Chunaengmu*, a type of Joseon court dance (Kwon, 2020). As a complement to the on-site AR experience, at the end of 2021, 'Changdeok Arirang at Home' was launched, which offers AR and VR experiences to virtual tourists from any location worldwide. This emphasis underscores the potential of technological developments to provide access to cultural heritage, particularly in pandemics and other emergencies, for those living far from the sites and vulnerable social groups without the means to travel in person (Figs. 5 and 6).

While South Korea is leading the way, other parts of Asia are also increasingly using AR to revive, enhance accessibility, and increase interactive encounters with cultural heritage. For example, in Hong Kong, a new tourism project featuring AR enables smartphone users to enjoy panoramic views of how Hong Kong used to be in the past. This initiative is designed by the city's Tourism Commission to bolster its appeal ahead of the return of foreign visitors once the COVID-19 pandemic eases (Sun, 2021). The *City in Time* project features 360-degree historical panoramas

[13]Between South Korea's *Cultural Heritage Administration* (CHA), the *Seoul Metropolitan government*, real estate and construction company *Woomi Construction*, and marketing company *Cheil Worldwide*.

[14]Between South Korea's *Cultural Heritage Administration* (CHA) and Google Korea.

Fig. 5 AR *Donuimun Gate* in Seoul (MuseumNext, 2019)

Fig. 6 AR *Hwangnyongsa Temple* in Gyeongju (National Research Institute of Cultural Heritage, 2020)

created from heritage photos and illustrations by young Hong Kong artists, along with lively animations and special sound effects. The public can download the project's mobile phone app and scan markers installed at designated locations around town to view the historical sights. The app is also equipped with a selfie function that allows users to share the experience on social media.

In Singapore, The *National Museum* showcases an immersive installation called *Story of the Forest*.[15] The exhibition focuses on 69 images from the *William Farquhar Collection of Natural History Drawings*. The Japanese digital art collective *teamLab* created an AR project that brings the drawings to life by turning them into three-dimensional animations that visitors can interact with. Visitors download an app and then use the camera on their device to explore the paintings. Much like *Pokémon Go*, virtual visitors can hunt for and 'catch' items, in this case the plants and animals within the paintings. They can then add them to their personal virtual collection as they walk around the museum. The app shows more information about them as they are collected, and visitors can learn facts such as habitat, diet and how rare the species are.

5 Conclusion

No digital tool or experience can yet completely replace an in-person visit to a cultural heritage site. However, recent developments in digitalisation in their application to cultural heritage tourism, paired with the critical necessity prompted by the COVID-19 pandemic, are impacting the way we access, interact with and consume cultural heritage. In the last few years, tools such as 360-degree technology, interactive visits, VR and AR, and even podcasts, have naturally integrated our cultural tourism habits. This chapter has explored cases from Asian locations in Japan, South Korea, India, Singapore, Indonesia, Taiwan, and Hong Kong, that helped illustrate and better understand this trend.

Across Asia, regional bodies, national governments, local entities, and private companies are exploring ways to integrate digital technology with cultural heritage tourism, often through original public-private partnerships. Many of these synergies have safeguarded the (virtual) accessibility to sites that have remained locked to in-person tourism for prolonged periods of time. Often, they have provided unconventional, sometimes enriched, ways to explore and interact with the sites. Admittedly, these efforts may have removed some of the traditional barriers to visitors, such as money, time, and location: wherever we are in the world, we now can visit a cultural heritage site listening to podcasts (that make us dream we are at the site), exploring it via virtual visits (that help us experience the site from the comfort of our homes), or using VR and AR technologies (that help us explore it in depth, often in an interactive manner).

Of course, it could be argued that digital cultural tourism can add other barriers. Improved digital access to sites, and its free of charge nature, if not integrated well with the in-person experiences, can cause losses to the secondary economic activities connected to cultural tourism. Some of the technological applications discussed in

[15] For more information: https://www.nhb.gov.sg/nationalmuseum/our-exhibitions/exhibition-list/story-of-the-forest.

this paper, such as AR and VR, require a steady and fast Internet connection and powerful devices to guarantee decent virtual tourism experiences. As discussed in this chapter, these public–private collaborations do not always ensure success; in others, the potential is yet to be confirmed. Some are asymmetric ad hoc collaborations that lack continuity and long-term commitments. With the exception of ASEAN, which operates with numerous limitations such as financial and human resources, no regional organisation is taking the lead to offer virtual tourists and visitors consolidated access to the region's fascinating sites of cultural heritage. Another challenge is the lack of available data. Across the board, public and private organisations, from local governments and museums to profit-making companies, tend to not public disclose numbers and statistics, including the number of online visitors, which raises questions about their virtual initiatives' effectiveness and reach.

However, the wide variety of countries and cases featured in this chapter and their proposals' diversity speak of a trend difficult to ignore: in Asia, digitalisation and its application to cultural heritage tourism are here to stay, with or without crises like the COVID-19 pandemic. It is free and often easier now to virtually visit a well-curated online exhibition than physically travel to a museum, and explore recently or long-gone heritage sites recreated by VR and AR technologies than imagine their relevance by solely visiting their physical sites. Albeit initially largely unintentional, the push for these new technologies during the COVID-19 pandemic has arguably generated more sustainable and equal access paths to cultural heritage assets, particularly for visitors with limited means or mobility. Practically speaking, in today's Asian context, stepping into a cultural heritage site without stepping out of the comfort of our homes has never been easier.

Bibliography

ACHDA. (2022). *ASEAN cultural heritage digital archive*. ASEAN Cultural Heritage Archive. Retrieved from https://heritage.asean.org/

Adams, E. B. (1983). *Korea's Kyongju: Cultural Spirit of Silla in Korea*. Seoul International Tourist Publishing Company.

Adams, E. B. (1989). The legacy of Kyongju. *The Rotarian*, 28–31.

Aiello, D., Fai, S., & Santagati, C. (2019). Virtual museums as a means for promotion and enhancement of cultural heritage. *The International Archives of the Photogrammetry, Remote Sensing and Spatial Information Sciences, XLII-2/W15*, 2019. https://doi.org/10.5194/isprs-archives-XLII-2-W15-33-2019

Arba, A. (2021, December 3). Japan: Foreign visitor numbers 2020. *Statista*. Retrieved August 17, 2021, from https://www.statista.com/statistics/654263/japan-number-foreign-visitors/

Armchair Explorer. (2021). *About*. Armchair Explorer. Retrieved October 11, 2021, from https://www.armchair-explorer.com

ASEAN. (2018, February 22). *ASEAN launches cultural heritage digital archiving project*. Association of Southeast Asian Nations. Retrieved from https://asean.org/asean-launches-cultural-heritage-digital-archiving-project/

Balletti, C., Ballarin, M., & Guerra, F. (2017). 3D printing: State of the art and future perspectives. *Journal of Cultural Heritage, 26*, 172–182. https://doi.org/10.1016/j.culher.2017.02.010

Barrio, M. J., Devesa, M., & Herrero, L. C. (2012). Evaluating intangible cultural heritage: The case of cultural festivals. *City, Culture and Society, 3*(4), 235–244. https://doi.org/10.1016/j.ccs.2012.09.002

Brink, J. A. (2013). Singosari: The origins of Majapahit. In *Digital Heritage: International Congress 2013*. Retrieved from http://www.digitalheritage2013.org/singosari/

Bruno, F., Bruno, S., De Sensi, G., Luchi, M.-L., Mancuso, S., & Muzzupappa, M. (2010). From 3D reconstruction to virtual reality: A complete methodology for digital archaeological exhibition. *Journal of Cultural Heritage, 11*(1), 42–49. https://doi.org/10.1016/j.culher.2009.02.006

Callot, E. (2020, October 30). Rebuilding Japan's Shurijo Castle in the Cloud. *The Keyword*. Retrieved from https://blog.google/outreach-initiatives/arts-culture/rebuilding-shurijo-castle/

Capturing Reality. (2021). Our Shurijo Project. *Capturing reality*. Retrieved from https://www.capturingreality.com/Article-our-shurijo

Caust, J., & Vecco, M. (2017). Is UNESCO world heritage recognition a blessing or burden? Evidence from developing Asian countries. *Journal of Cultural Heritage, 27*, 1–9. https://doi.org/10.1016/j.culher.2017.02.004

Childs, C. (2015, April 18). *How culture and heritage tourism boosts more than a visitor economy*. MyTravelResearch.Com. Retrieved from https://www.mytravelresearch.com/culture-and-heritage-tourism-boosts-visitor-economy/

Chowdhury, G. (2010). From digital libraries to digital preservation research: The importance of users and context. *Journal of Documentation, 66*(2), 207–223. https://doi.org/10.1108/00220411011023625

Chung, D. (2001). *Syncretism: The religious context of Christian beginnings in Korea*. SUNY Press.

Cignoni, P., & Scopigno, R. (2008). Sampled 3D models for CH applications: A viable and enabling new medium or just a technological exercise? *ACM Journal on Computing and Cultural Heritage, 1*(1), 1–23. https://doi.org/10.1145/1367080.1367082

CyArk. (2016a, June 10). Carved Stone Pillar at Lukang Longshan Temple. *Sketchfab*. Retrieved from https://sketchfab.com/models/a3db4955d951485d9053843de6518336/embed?autostart=1

CyArk. (2016b, June 11). Intricate roof of Lukang Longshan Temple. *Sketchfab*. Retrieved from https://sketchfab.com/models/46a41b805175412994ea82da3c3ce7c3/embed?autostart=1

CyArk. (2021). *Lukang Longshan Temple: In depth*. CyArk. Retrieved from https://cyark.org/projects/lukang-longshan-temple/in-depth

D'Eramo, M. (2014). UNESCOcide. *New Left Review, 88*, 47–53.

Darmawiguna, I. G. M., Pradnyana, G. E., Sindu, I. G. P., Karimawan, I. P. P. S., & Dwiasri, N. K. R. A. (2020, April 26). Bali Temple VR: The virtual reality based application for the digitalization of Balinese temples. *ArXiv:2004.13853 [Cs]*. Retrieved from http://arxiv.org/abs/2004.13853

Deggim, S., Kersten, T. P., Tschirschwitz, F., & Hinrichsen, N. (2017). Segeberg 1600—Reconstructing a historic town for virtual reality visualisation as an immersive experience. *The International Archives of the Photogrammetry, Remote Sensing and Spatial Information Sciences, XLII-2*(W8), 87–94. https://doi.org/10.5194/isprs-archives-XLII-2-W8-87-2017

Eshaghi, S., Afshar, S. V., & Varinlioglu, G. (2021, March). The Sericum Via: A serious game for preserving tangible and intangible heritage of Iran. In *9th International Conference of the Arab Society for Computer-Aided Architectural Design: Architecture in the Age of Disruptive Technologies, ASCAAD 2021*, Cairo, Egypt. https://doi.org/10.6084/m9.figshare.14384318

Esmaeili, H., Woods, P. C., & Thwaites, H. (2014, December). Realisation of virtualised architectural heritage. In *2014 International Conference on Virtual Systems Multimedia (VSMM)* (pp. 94–101). https://doi.org/10.1109/VSMM.2014.7136676

European Commission. (2015, April 17). *Horizon 2020—Work Programme 2014–2015*. European Commission decision C (2015) 2453. European Commission.

Fadli. (2020, January 16). Riau Islands province ranks second after Bali as most-visited destination for foreign tourists. *The Jakarta Post*. Retrieved from https://www.thejakartapost.com/

travel/2020/01/16/riau-islands-province-ranks-second-after-bali-as-most-visited-destination-for-foreign-tourists.html
Far from Home. (n.d.). *Season two*. Retrieved November 10, 2021, from https://farfromhomepodcast.org/season-two
Garaba, F. (2015, August). The Timbuktu manuscripts: A model for preservation in Africa. In *81st. IFLA World Library and Information Congress*.
Getz, D. (2009). Policy for sustainable and responsible festivals and events: Institutionalization of a new paradigm. *Journal of Policy Research in Tourism, Leisure and Events, 1*(1), 61–78. https://doi.org/10.1080/19407960802703524
Global Heritage Fund. (2019, September 27). *Why we all need to travel with purpose*. Global Heritage Fund. Retrieved from https://globalheritagefund.org/2019/09/27/why-we-all-need-to-travel-with-purpose/
Google Arts & Culture. (2021). *Reconstructing Shurijo Castle*. Google Arts & Culture. Retrieved from https://artsandculture.google.com/project/castles-of-japan
Graham, B. J., Ashworth, G. J., & Tunbridge, J. E. (2016). *A geography of heritage: Power, culture and economy* (1st ed.). Routledge. https://doi.org/10.4324/9781315824895
Greffe, X. (2004). Is heritage an asset or a liability? *Journal of Cultural Heritage, 5*(3), 301–309. https://doi.org/10.1016/j.culher.2004.05.001
Hughes, G. (1999). Urban revitalization: The use of festive time strategies. *Leisure Studies, 18*(2), 119–135. https://doi.org/10.1080/026143699374998
Hunter, C. (2021, October 7). Alor Setar Bus Station, Kedah, Malaysia on 26th December 2009. *Field recordings*. Retrieved from https://fieldrecordings.xyz/2021/10/07/alor-setar-bus-station-kedah-malaysia-on-26th-december-2009-by-colin-hunter/
Iron Mountain. (2021). *CyArk and Iron Mountain provide virtual reality access to 18th century Buddhist Temple in China-Taiwan*. Iron Mountain. Retrieved from https://www.ironmountain.com/about-us/corporate-responsibility/stories/s/step-inside-a-buddhist-temple-with-virtual-reality
Jeong, Y. & Park, S. (2020, September 8). The recovery of Hwangnyongsa. *The Dankook Herald*. Retrieved from http://dkherald.dankook.ac.kr/news/articleView.html?idxno=12132
Journals of India. (2020, December 3). Virtual exhibition on shared Buddhist heritage of SCO. *Journals of India*. Retrieved from https://journalsofindia.com/virtual-exhibition-on-shared-buddhist-heritage-of-sco/
Kan, T., Buyuksalih, G., Ozkan, G. E., & Baskaraca, P. (2019). Rapid 3D digitalization of the cultural heritage: A case study on Istanbul Suleymaniye Social Complex (Kulliye). *The International Archives of the Photogrammetry, Remote Sensing and Spatial Information Sciences, XLII-2*(W11), 645–652. https://doi.org/10.5194/isprs-archives-XLII-2-W11-645-2019
Kaneko, R. (2019, December 25). Shuri Castle to rise from ashes in virtual resurrection. *Kyodo News*. Retrieved from https://english.kyodonews.net/news/2019/12/c896c4ae16c8-feature-shuri-castle-to-rise-from-ashes-in-virtual-resurrection.html
Kersten, T. P., Tschirschwitz, F., & Deggim, S. (2017). Development of a virtual museum including a 4D presentation of building history in virtual reality. *The International Archives of the Photogrammetry, Remote Sensing and Spatial Information Sciences, XLII-2/W3*, 361–367. https://doi.org/10.5194/isprs-archives-XLII-2-W3-361-2017
King, L., Stark, J. F., & Cooke, P. (2016). Experiencing the digital world: The cultural value of digital engagement with heritage. *Heritage & Society, 9*(1), 76–101. https://doi.org/10.1080/2159032X.2016.1246156
Korea JoongAng Daily. (2020, September 21). [Living tradition] Hwangnyongsa Temple restored through AR. *Korea JoongAng Daily*. Retrieved from https://koreajoongangdaily.joins.com/2020/09/21/culture/koreanHeritage/heritage-hwangnyongsa-temple/20200921164900398.html
Kwon, M.-Y. (2020, July 27). Haechi guides way in AR App for Changdeok Palace. *The Korea Times*. Retrieved from https://www.koreatimes.co.kr/www/art/2021/11/398_293484.html

Kyodo. (2021, January 20). Visitors to Japan in 2020 plunge 87%; biggest fall since 1964. *Nikkei Asia*. Retrieved from https://asia.nikkei.com/Business/Travel-Leisure/Visitors-to-Japan-in-2020-plunge-87-biggest-fall-since-1964

Kyodo. (2022, January 19). Foreign visitors to Japan in 2021 fell to record low 245,900. *The Japan Times*. Retrieved from https://www.japantimes.co.jp/news/2022/01/19/national/japan-2021-visitor-numbers/

Lee, H.-J. (2017). Phenomenological classification of cultural heritage: Role of virtual reality. *Virtual Archaeology Review, 8*(16), 69. https://doi.org/10.4995/var.2017.5962

Lim, S. (2019, September 17). South Korea restores historic gate with augmented and virtual reality to promote lost cultural assets. *The Drum*. Retrieved from https://www.thedrum.com/news/2019/09/17/south-korea-restores-historic-gate-with-augmented-and-virtual-reality-promote-lost

Lor, P. J., & Britz, J. J. (2012). An ethical perspective on political-economic issues in the long-term preservation of digital heritage. *Journal of the American Society for Information Science and Technology, 63*(11), 2153–2164. https://doi.org/10.1002/asi.22725

MacCannell, D. (1992). *Empty meeting grounds: The tourist papers*. Routledge.

Machemer, T. (2020, January 2). Volunteers digitally revive Japan's Shuri Castle following October fire. *Smithsonian Magazine*. Retrieved from https://www.smithsonianmag.com/smart-news/after-october-fire-volunteers-are-reviving-japanese-castle-online-180973890/

McNulty, R., & Koff, R. (2014). *Cultural heritage tourism*. Partners for Livable Communities.

Mint. (2021, January 17). Footfall to Taj Mahal dropped by 76% in 2020 due to COVID-19 situation. *Mint*. Retrieved from https://www.livemint.com/news/india/footfall-to-taj-mahal-dropped-by-76-in-2020-due-to-covid-19-situation-11610881057401.html

Mitchell, A., Stocking, G., & Matsa, K. E. (2016, May 5). Long-form reading shows signs of life in our mobile news world. *Pew Research Center's Journalism Project*. Retrieved from https://www.pewresearch.org/journalism/2016/05/05/long-form-reading-shows-signs-of-life-in-our-mobile-news-world/

MuseumNext. (2019, September 21). Augmented reality brings Korean heritage site back to life. *MuseumNext*. Retrieved from https://www.museumnext.com/article/augmented-reality-brings-korean-heritage-sit-back-to-life/

National Museum, New Delhi, India. (2022). *SCO virtual exhibition on shared Buddhist heritage*. Retrieved from https://nmvirtual.in/Sco.php

National Research Institute of Cultural Heritage. (2020). *Capture of the middle gate of Hwangnyongsa Temple in augmented reality*. National Research Institute of Cultural Heritage (NRICH). Retrieved from https://www.nrich.go.kr/kor/index.do

National Trust for Historic Preservation. (2021). Issues in preservation. *Preservation Leadership Forum*. Retrieved from https://forum.savingplaces.org/learn/issues#.Uz7tU1flvqE

Ngata, W., Ngata-Gibson, H., & Salmond, A. (2012). Te Ataakura: Digital Taonga and cultural innovation. *Journal of Material Culture, 17*(3), 229–244. https://doi.org/10.1177/1359183512453807

Nilson, T., & Thorell, K. (2018). *Cultural heritage preservation: The past, the present and the future*. Halmstad University Press.

NTT Data Corporation. (2020, February 27). NTT DATA creates and makes public a digital archive of the historical cultural heritage of the ASEAN region. *NTT DATA*. Retrieved from https://www.nttdata.com/global/en/media/press-release/2020/february/ntt-data-creates-and-makes-public-a-digital-archive-of-the-historical-cultural-heritage

Ocón, D. (2021). Digitalising endangered cultural heritage in southeast Asian cities: Preserving or replacing? *International Journal of Heritage Studies, 27*(10), 975–990. https://doi.org/10.1080/13527258.2021.1883711

Pacific Asia Travel Association. (2015, March 31). The connected visitor economy: The role of culture and heritage tourism in building the visitor economy—And beyond. *Flipsnack*. Retrieved from https://www.flipsnack.com/pata2014/the-connected-visitor-economy-bulletin-3rd-edition-2015.html

Phoonphongphiphat, A. (2021, January 29). COVID wave crushes Thailand's already Hobbled Tourism Sector. *Nikkei Asia*. Retrieved from https://asia.nikkei.com/Business/Business-trends/COVID-wave-crushes-Thailand-s-already-hobbled-tourism-sector2

Pizziolo, G., Pirisino, D., Tessaro, C., & Volante, N. (2012). Intra-site analysis and photogrammetry: The case study of the 'Buca Di Spaccasasso' (Grosseto, Italy) an Eneolithic Funerary Site. In *CAA2012 Proceedings of the 40th Conference in Computer Applications and Quantitative Methods in Archaeology,* Southampton, United Kingdom.

Porter, B. W., & Salazar, N. B. (2005). Heritage tourism, conflict, and the public interest: An introduction. *International Journal of Heritage Studies, 11*(5), 361–370. https://doi.org/10.1080/13527250500337397

Reshamwala, S. (2020, May). Bangkok. *Far Flung with Saleem Reshamwala*. Retrieved from https://open.spotify.com/episode/63KLTwhU99F6gILg0vsO4G

Sachs, A. (2020, December 11). With travel podcasts, explore the world through your ear buds. *Washington Post*. Retrieved from https://www.washingtonpost.com/lifestyle/travel/best-travel-podcasts-pandemic-listening/2020/12/10/44912020-364e-11eb-8d38-6aea1adb3839_story.html

Schlütz, D., & Hedder, I. (2021). Aural parasocial relations: Host–listener relationships in podcasts. *Journal of Radio & Audio Media*, 1–18. https://doi.org/10.1080/19376529.2020.1870467

Shiraishi, S. (2019, October 31). Shuri Castle: Fire destroys 500-year-old world heritage site in Japan. *BBC News*. Retrieved from https://www.bbc.com/news/world-asia-50244169

Shuri Castle Digital Reconstruction. (2019). みんなの首里城デジタル復元プロジェクト | Shuri Castle Digital Reconstruction. *Shuri Castle Digital Reconstruction*. Retrieved from https://www.our-shurijo.org/en/

Statista Research Department. (2021a, October 7). Share of the total gross domestic product (GDP) generated by the travel and tourism industry worldwide from 2000 to 2020. *Statista*. Retrieved from https://www.statista.com/statistics/1099933/travel-and-tourism-share-of-gdp/

Statista Research Department. (2021b, October 27). Growth of foreign tourist arrivals to Bali, Indonesia from 2011 to first half 2021. *Statista*. Retrieved from https://www.statista.com/statistics/976863/growth-foreign-tourist-arrivals-bali-indonesia/

Styliani, S., Fotis, L., Kostas, K., & Petros, P. (2009). Virtual museums, a survey and some issues for consideration. *Journal of Cultural Heritage, 10*(4), 520–528. https://doi.org/10.1016/j.culher.2009.03.003

Sun, F. (2021, March 27). New augmented reality tourism app gives users a panoramic view of Hong Kong's past. *South China Morning Post*. Retrieved from https://www.scmp.com/news/hong-kong/society/article/3127265/new-augmented-reality-tourism-app-gives-users-panoramic-view

Szmigin, I., Bengry-Howell, A., Morey, Y., Griffin, C., & Riley, S. (2017). Socio-spatial authenticity at co-created music festivals. *Annals of Tourism Research, 63*, 1–11. https://doi.org/10.1016/j.annals.2016.12.007

Taj Mahal. (2021). Views of Taj Mahal. *Taj Mahal*. Retrieved from http://www.tajmahal.gov.in/views-of-maj-mahal.aspx

Tanasi, D., Gradante, I., & Hassam, S. (2019). Best practices for 3D digital recording and global: Sharing of catacombs from late Roman Sicily. *Studies in Digital Heritage, 3*(1), 60–82. https://doi.org/10.14434/sdh.v3i1.25290

The Hindu. (2020, November 30). Virtual exhibition on shared Buddhist heritage of SCO countries launched. *The Hindu*. Retrieved from https://www.thehindu.com/news/national/virtual-exhibition-on-shared-buddhist-heritage-of-sco-countries-launched/article33214191.ece

The National Thailand and Asia News Network. (2021, March 23). Thailand serves Tom Yam for UNESCO Cultural Heritage 2021 list. *The Star*. Retrieved from https://www.thestar.com.my/aseanplus/aseanplus-news/2021/03/23/thailand-serves-tom-yam-for-unesco-cultural-heritage-2021-list

Timothy, D. J., & Boyd, S. W. (2003). *Heritage tourism*. Prentice Hall.

Trading Economics. (2022). *Thailand tourist arrivals: January 2022 data*. Retrieved from https://tradingeconomics.com/thailand/tourist-arrivals

Tschirschwitz, F., Buyuksalih, G., Kersten, T. P., Enc, G., Kan, T., & Baskaraca, P. (2019, January). Virtualising an Ottoman Fortress—Laser scanning and 3D modelling for the development of an interactive, immersive virtual reality application. *The International Archives of the Photogrammetry, Remote Sensing and Spatial Information Sciences, XLII-2/W9*, 723–729. 8th International Workshop 3D-ARCH "3D Virtual Reconstruction and Visualization of Complex Architectures", 6–8 February 2019, Bergamo, Italy. https://doi.org/10.5194/isprs-archives-XLII-2-W9-723-2019

Twining Ward, L., & McComb, J. F. (2020). *COVID-19 and tourism in South Asia: Opportunities for sustainable regional outcomes*. World Bank. https://doi.org/10.1596/34050

UNESCO. (2003). *Convención Para La Salvaguardia Del Patrimonio Cultural Inmaterial*. UNESCO. Retrieved from http://portal.unesco.org/es/ev.php-URL_ID=17716&URL_DO=DO_TOPIC&URL_SECTION=201.html

UNESCO. (2005, October 20). *Convention on the protection and promotion of the diversity of cultural expressions*. UNESCO. Retrieved from http://portal.unesco.org/en/ev.php-URL_ID=31038&URL_DO=DO_TOPIC&URL_SECTION=201.html

UNESCO. (2014). UNESCO *culture for development indicators: Methodology manual*. United Nations Educational, Scientific and Cultural Organization. Retrieved from https://en.unesco.org/creativity/sites/creativity/files/cdis_methodology_manual_0_0.pdf

UNESCO World Heritage Centre. (2019, October 31). *UNESCO expresses solidarity with Japan as fire rips through world heritage site in Okinawa*. UNESCO World Heritage Centre. Retrieved from https://whc.unesco.org/en/news/2053/

UNESCO World Heritage Centre. (2021a). *Seokguram Grotto and Bulguksa Temple*. UNESCO World Heritage Centre. Retrieved from https://whc.unesco.org/en/list/736/

UNESCO World Heritage Centre. (2021b). *Changdeokgung Palace Complex*. UNESCO World Heritage Centre. Retrieved from https://whc.unesco.org/en/list/816/

UNWTO. (2021). *UNWTO world tourism barometer and statistical annex, May 2021*. UNWTO World Tourism Barometer (English Version), *19*(3). Retrieved from https://www.e-unwto.org/doi/abs/10.18111/wtobarometereng.2021.19.1.3

UNWTO/GTERC. (2019). *UNWTO/GTERC Asia tourism trends—2019 edition. 6. UNWTO/GTERC Asia tourism trends*. UNWTO. Retrieved from https://www.e-unwto.org/doi/abs/10.18111/9789284420360

UNWTO/GTERC. (2020). *UNWTO/GTERC Asia tourism trends—2020 edition, executive summary. 7. UNWTO/GTERC Asia tourism trends*. UNWTO. Retrieved from https://www.e-unwto.org/doi/epdf/10.18111/9789284422272

Vecco, M. (2010). A definition of cultural heritage: From the tangible to the intangible. *Journal of Cultural Heritage, 11*(3), 321–324. https://doi.org/10.1016/j.culher.2010.01.006

Waterman, S. (1998). Carnivals for elites? The cultural politics of arts festivals. *Progress in Human Geography, 22*(1), 54–74.

World Travel & Tourism Council. (2021). *Economic impact reports*. World Travel & Tourism Council. Retrieved from https://wttc.org/Research/Economic-Impact

Part III
Reshaping the Relationship Between Cultural Assets, Cities and Territories

Girona Episcopal: An Analysis of a New Cultural Brand

Marta Crispí and Judith Urbano

Abstract In recent years, Girona has become a cultural centre of reference in Catalonia. The Girona City council's creation of cultural and tourist brands presenting Girona as a city of cultural experiences actively contributed to the development of this centrality. These cultural and tourist promotion strategies are based on the network of events and festivals that consolidate the city as a local, national, and international cultural reference. However, before the adoption of these strategies, Girona was already a tourist destination due to the appeal of the rich and diverse historical and artistic heritage, concentrated, above all, in the Barri Vell (Old Quarter).

The diocese of Girona and the Generalitat de Catalunya signed an agreement in 2016 to create a cultural brand called Girona Episcopal, which included Girona's cathedral, the church of Sant Feliu, and the Museu d'Art de Girona. This new brand enriches and further strengthens the city's attractive heritage by creating a coherent circuit that links the buildings and collections of the powerful bishopric of Girona. The network of cultural institutions generated, as has happened in other countries, contributes to offering content, resources, and services that, in addition to the single entrance ticket for the three heritage sites, allow feedback from visitors between them and give visibility to the collections of the Museu d'Art de Girona. This chapter aims to analyse the new cultural and tourist brand, according to the brand management and the brand orientation models. In addition, the brand's strengths and weaknesses will be identified.

M. Crispí (✉) · J. Urbano
Faculty of Humanities, Universitat Internacional de Catalunya, Barcelona, Spain
e-mail: mcrispi@uic.cat; jurbano@uic.cat

E. Borin et al. (eds.), *Cultural Leadership in Transition Tourism*, Contributions to Management Science, https://doi.org/10.1007/978-3-031-14121-8_10

1 Introduction

Girona is the capital of the province of the same name and is located about 100 km north of Barcelona, therefore about 60 km from France. It is also important to note that it is 40 km from the Costa Brava, one of the most important tourist destinations in Europe. Its population slightly exceeds 100,000 inhabitants.

Girona is a city with a Roman foundation and 2000 years of history. Its monumental and historical centre within the medieval walls, known as Barri Vell, is of great artistic and cultural value. It includes the Gothic Cathedral of Santa Maria (with some Romanesque elements and the Baroque façade), the Banys Àrabs (Arab Baths), the basilica of Sant Feliu, and the Jewish Quarter. Today, the city is also home to seven museums: el Museu d'Art de Girona and the pharmacy of the old Hospital de Santa Caterina, the Museu d'Història de la Ciutat, the Museu del Cinema, the Tresor de la Catedral, the Museu d'Arqueologia de Catalunya (inside the romanesque monastery of Sant Pere de Galligants), the Museu dels Jueus, and the Casa Masó.

Tourist flows in the Barri Vell of Girona have been growing in recent years. Romanticism fixed the tourist image of the city in the nineteenth century, based on the monumental ensemble around the cathedral, which had already appeared in the guides of the time. After some years of degradation of the Barri Vell between the 1950s and 1980s of the twentieth century, when the beach tourism of the Costa Brava was more important, this model began to be questioned, and tourism was moved toward the interior. The *Pla Especial de Rehabilitació Interior per al Barri Vell* (Special Interior Rehabilitation Plan for the Old Quarter) of 1982 and some actions of the Girona City Council recovered the historic centre as a point of reference. During the 1990s, tourism increased, helped by a bigger offer of accommodation and different cultural events that attract more visitors every year (Temps de Flors, Fires de Sant Narcís, Temporada Alta etc.).[1] Thus, that romantic image of the nineteenth century around the cathedral, the Arab Baths and the basilica of Sant Feliu has been consolidated in the collective imagination (Galí & Donaire, 2006a).

In 2016, the cathedral began a collaboration with the Basilica of Sant Feliu and the Girona Art Museum (md'A) to have a joint entrance and establish synergies for the promotion and dissemination of the ensemble with the name of Girona Episcopal. This chapter aims to analyse this new cultural and tourist brand and detect its strengths and weaknesses following two models: Cerquetti's (2010) branding management and Evans et al. (2012) brand orientation.

[1] *Temps de Flors* (Flower Time) is an event that has been run since 1954. For 10 days Girona's historical buildings, courtyards, and gardens are adorned with floral arrangements and art installations related to nature and flowers with a programme of cultural events. *Fires de Sant Narcís* (Saint Narcís fairs) is the annual festival of the city which includes a trade fair, a carnival, musical performances, street fairs and markets, a parade of "giants", a street fireworks festival (*correfoc*), etc. *Temporada Alta* (High Season) is an international performing arts festival mainly in the cities of Girona and Salt between the months of October and December.

2 Theoretical Background

The academic literature on cultural marks is recent and sparse, as Baumgarth et al. pointed out in their state-of-the-art study (2016). The birth of the concept of "cultural branding" and the application of commercial marketing techniques to cultural institutions dates to the 1990s, when a real concern arose about the relationship between audiences and cultural enterprises. Then, the need to reach new segments of the public became peremptory. As a result, the means were put in place to understand and analyse these audiences with methodologies developed for marketing.

Now several scholars study, from a theoretical profile, the concept and singularities of cultural brands as well as other related concepts such as brand identity, brand personality, brand equity, brand orientation, brand management, or brand audit (Caldwell, 2000; Scott, 2000; Wallace, 2006; Baumgarth & O'Reilly, 2014; Kladou & Kehagias, 2014; Baumgarth et al., 2016). As Baumgarth and O'Reilly have pointed out, cultural brands have their own identity and parameters. They are endowed with a complexity that must be considered when establishing brand strategies and subsequently evaluating them. These authors emphasise, among other aspects, that creators need to know the history and traditions of the area where the new product is generated to create consistent symbolic associations with the past and present and that cultural brands are products of diverse typologies where sensory and experiential factors play a prominent role in the product and its consumption (Baumgarth & O'Reilly, 2014).

Among the cultural sectors in which brand studies have had the greatest development are, in particular, museums and, tangentially, temporary exhibitions (Izquierdo et al., 2008; Camarero et al., 2010, 2012). Thus, more theoretical and generalist works have been published on museums, analysing the concept of the brand and pointing to some first evaluation methodologies (Caldwell, 2000; Wallace, 2006). However, above all, there are many case studies that apply analytical concepts, tools, and methodologies to different museums, such as brand creation and brand positioning (Scott, 2000; Phillips & O'Reilly, 2007), brand identity and brand values (Caldwell, 2003; Pusa & Uusitalo, 2014), brand orientation (Baumgarth, 2009; Evans et al., 2012), brand audit (Scott, 2000; Caldwell & Coshall, 2002), and the impact of the brand on visitors (Recuero et al., 2017). Between these different tools and methodologies, there are two that can fit the analysis of the recent brand of Girona Episcopal according to two parameters: brand management and brand orientation.

Brand management is related to the strategies and tools used to develop and spread the brand. Cerquetti (2010) studied the brand management of a network of Italian museums localised in the Marche Region (Italy), in the framework of the enhancement of the local heritage through thematization of cultural contents. According to the Cerquetti, this thematization can be considered as a brand for the cultural product and its management is one of the ways to put in value the local heritage. The author designed a model based on brand management structured on three levels: the *stock of cultural capital* in the place (that are the cultural resources of the territory), the *cultural product* created (that can be the result of the thematization of a group of resources related to historical or artistic subjects) and, above all, the *communication* (particularly the naming) of the brand. This typology of cultural products has the potential to generate experiences that constitute a particular attraction for local and foreign audiences and contribute to the value of local heritage (Cerquetti & Montella, 2015).

Brand orientation is the second concept that we will use to analyse Girona Episcopal. According to Evans et al. (2012), this concept results from a fusion of brand concept and business orientation. The definitions that have been given are multiple and oscillate between those that start from a more philosophical view (Hankinson, 2001a, b), on which the proposal of Bridson and Evans (2004) is based: the degree to which the organisation value brands and its practices are oriented towards building brand capabilities, and the definitions that prioritise the marketing approach (Wong & Merrilees, 2008). The authors propose a brand orientation model based on two philosophical aspects, *culture* and *compass*, and four behaviours (or capabilities) that they define as *distinctiveness*, *functionality*, *augmentation*, and *symbolism*. Both philosophical concepts, *culture* and *compass*, are the basis and the background of the whole brand strategy. As a compass, the brand and its values guide the business organization and all the processes for decision-making and these actions are only possible because the stakeholders—the staff, the marketing, communication team etc.—have turned the brand into an element that identifies the organizational culture of the company (Evans et al., 2012, p. 438). By the *distinctiveness* of the brand, the authors understand the aspects that make the brand unique, whether it is the logo, The slogan or, by extension, elements of the product itself, such as the historical monuments. The *functionality* has to do with the ability to target the needs and profiles of the public and to manage their expectations. Brand *augmentation* refers to the attitude of the promoters to develop new products or improve the existing ones to increase the quality, scope, and impact of the product. The last behaviour, *symbolism*, is related to the symbolic values that brands generate in consumers' minds, causing an emotional bond or a more personal connection with the brand. These symbolic values correspond to what other authors call brand values or brand associations (Camarero et al., 2010; Kladou & Kehagias, 2014), which may be linked to various aspects of the brand.

3 Methodology

The analysis of the case study is based on a dual source of quantitative and qualitative information: first of all, the actions taken, the management, and the visitor data for the three heritage spaces have been analysed thanks to the published reports of the project (*Girona Episcopal. Report* 2016, 2017, 2018, 2019 and 2020) that systematically and exhaustively collect the information on dissemination actions, the number of visitors with details of the public of each site, the economic management, and the future strategies to be implemented in the next course.

Secondary data consist of semi-structured interviews conducted with five people, three of them directly responsible for the project: Joan Piña, director of the Tresor de la catedral de Girona, Gustau Torres, technician at the cathedral and Carme Clusellas, director of the md'A and Àngels Miralles, responsible for the press and communication for the museum and communications and distribution for Girona Episcopal, and Carme Sais, head of services of Museus, Patrimoni Cultural i Arts Visuals (Museums, Cultural Heritage, and Visual Arts) of the Girona City Council, an external person to the project, but with an in-depth knowledge of the city's museum and heritage network, providing an outside view not directly linked to the product (*Interview with Àngels Miralles*, 2021; *Interview with Carme Clusellas*, 2021; *Interview with Carme Sais*, 2021; *Interview with Joan Piña and Gustau Torres*, 2021).

These data have been analysed according to the brand management model proposed by Cerquetti (2010) and the brand orientation model suggested by Evans et al. (2012) in the previous section of the chapter.

4 The Case Study

4.1 Girona Historical Background

At the beginning of the nineteenth century, Alexandre Laborde's *Voyage Pittoresque et Historique de l'Espagne* is already referring to the history of Girona, its cathedral, and the Arab Baths (Laborde between 1806 and 1820). It includes a view of Girona from the north and two of the Arab Baths. Other foreign romantic authors also gave their views of the city at that time. However, Gironian journalist and historian Narcís Blanch's *Gerona histórico monumental*, published in 1853, was the most prominent publication which reviews the history of Girona and describes its monuments (Galí & Donaire, 2006a).

Also noteworthy is the *Revista de Gerona*, which began in 1876 with a series of varied contributions to the city and its history, archaeology, literature, and science up to 1895 (Pla, 1965). The Diputació de Girona continued it in 1955 with the title *Revista de Girona*, which today reaches number 329.

Santiago Sobrequés i Vidal created the Institut d'Estudis Girondins in 1946, from which congresses, informative itineraries, and conferences are organised, preferably focused on the city of Girona. They published *Anales del Instituto de Estudios Gerundenses* from 1946 to 1978, which continues to this day with the title in Catalan. The annals aim to publicise the research carried out by members of the organisation and scholars from other scientific and cultural institutions, who are invited to participate in tributes and miscellany. *Quaderns de la Revista de Girona* is a collection that began in 1985 and presents a series of local monographs and guides.

For the specific historiography of the cathedral of Girona, we refer to Marc Sureda's article "La Catedral de Girona, matèria històrica. Historiografia a l'entorn de la seu (ss.XVII–XXI)" in *Annals de l'Institut d'Estudis Gironins*, that reviews the bibliography of the building from the modern era to 2002 (Sureda, 2004). Thus, starting with some seventeenth-century publications such as those of Joan Gaspar Roig or Onofre Relles, the article culminates with the choral publication of Joaquim Nadal in 2002, a monograph on the cathedral (Nadal et al. 2002).

The basilica of Sant Feliu has also been the subject of several publications. We highlight the monographic book of Clara and Marquès from 1992 and the doctoral dissertation of Chamorro from 2004. More recent works are from Nolla and Palahí (2011) and Freixas (2016), as well as one dedicated to the ecclesiastical monuments of Girona by Ferrer et al. (2020).

As for the city's museums, the first publications were *Gerona. La Catedral y el Museo Diocesano* by Font (1952) and *Los museos de Gerona* by Pla in 1956. Although later, we have books and monographic articles: Reixach (1982), Soler (1984, 1988), Llorens (2011), and Servitja and Burch (2014), among others.

The topic of Girona as a tourist city has also been extensively covered in literature, especially in the last few years. However, we must go back to the creation of the Societat d'Atracció de Forasters de Girona (Society of Attraction of Foreigners) in 1914, which meant having an association that favoured tourism in different aspects, as was already the case in other European cities. The architect Rafael Masó was one of its founders (Crous-Costa, 2010). For this subject, we highlight the review made by Galí and Donaire from the beginning of the phenomenon in the nineteenth century to 2006 (Galí & Donaire, 2006a).

Recently, critical discourses on the issue of tourism have also been analysed. This fact has arrived at Girona from some groups in Barcelona, who are critical of overtourism in the capital of Catalonia. This research conducted by Blasco, Palou, and Poch has been done through qualitative interviews (2021).

We must remember that the Girona Episcopal is a cultural product closely linked to the city. Over the last two decades, it has been the subject of the construction of a tourist brand designed by the public administration (Pla Estratègic de Turisme de Girona, 2018). Thus, the different plans for urban development and beautification of the city have been decisive strategies to turn it into a tourist city, also an object of attention by researchers (Crous-Costa & Vidal-Casellas, 2013) as well as the analysis of the tourist image in the city and the behaviour of its visitors (Galí & Donaire, 2003, 2006b; Blasco et al., 2021). The city has several cultural events, some of which are from the private sector, which have turned it into an important tourist and cultural destination (Jiménez & San Eugenio, 2009; Kammerer, 2019). Today, Girona, temps de flors (Jiménez & San Eugenio, 2009; Comas et al., 2013), Girona, ciutat de festivals (Camprubí & Planas, 2020), and events such as the Temporada Alta theatre festival (Iglesias, 2013; Vidal-Casellas et al., 2013) are recognised benchmarks for a segment of the local public and tourists interested in cultural consumption and practices.

4.2 Girona, a Cultural City

Transformation of Girona into a cultural city has been a slow process. The Centre for Initiatives and Tourism was set up for its promotion in 1964. The first incentive was when the Barri Vell was proclaimed a Conjunt Històric-artístic (Historic-Artistic Site) in 1967. The Patronat de Turisme de Girona (Girona Tourist Board) dates to 1976. It would be necessary to wait until 1983 to carry out the Pla Especial de Conservació i Reforma Interior del Barri Vell (Special Plan for the Conservation and Interior Reform of the Barri Vell), set up by the city council.[2] Commercial and residential activity in the area was gradually promoted. The installation of the University of Girona in 1981 in the old Convent of Sant Domènec and in the building called Les Àligues was also an important push. Another important event was restoring the houses on the banks of the river Onyar in 1983. In 1992, the Girona Convention Bureau was founded within the Cambra de Comerç de Girona (Girona Chamber of Commerce) to promote congresses and other commercial and business activities. However, it was not possible to have an Auditorium-Congress Hall until 2006. In 1994, the Comissió de Comerç i Turisme de Girona (Girona Trade and Tourism Commission) was created, as a municipal initiative, and in 2004 the Àrea de Desenvolupament local i promoció (Area of local development and promotion) within the city council. We must add transport to all these initiatives: the low-cost company Ryanair arrived at Girona airport in 2000, and in 2013 the high-speed train also arrived (Crous-Costa & Vidal-Casellas, 2013).

The tourist image of the city has been shaped by several monuments, as we have seen, and has attracted visitors to the Barri Vell. The Arab Baths, a monument that has aroused the interest of romantics since the nineteenth century, currently belongs to the Diputació de Girona. We have known of their existence since the eleventh century. However, Jaume II restored them in the late thirteenth century after being partially destroyed by a siege of the city by "Philippe le Hardi" of France. They have been open to the public since 1932.

On the other hand, the Jewish Call began to be promoted in the 1990s. The Patronat Municipal del Call (Municipal Call Board) was founded, and the Museu d'Història dels Jueus was opened in 2000. It had 75,805 visitors in 2019 (*Memòria* 2019, 2020.

However, the cathedral has always been the epicentre of the tourist experience in Girona. It is the most visited site (free until 2005). Ninety percent of itineraries include the religious building (Galí & Donaire, 2005). In 2019, there were 243,837 visitors, which represent 9% more than the previous year, consolidating the upward trend (*Catedral de Girona. Memòria,* 2019). We can see the visitors of the Museu d'Història de Girona, also in the Barri Vell, to make a comparison: 35,307; or those of the md'A, when there were 21,369.[3]

[2] http://ptop.gencat.net/rpucportal/AppJava/cercaExpedient.do?reqCode=veureDocument&codintExp=211863&fromPage= (2nd November 2022).

[3] https://cultura.gencat.cat/ca/temes/museus/dades/visitants/museus-registrats/ (2nd November 2022).

It should be added that the city's public museums (Museu d'Arqueologia de Catalunya-Girona, Museu d'Art, Museu d'Història de la Ciutat, Museu d'Història dels Jueus i Museu del Cinema) established a collaboration network in 2003 with the launch of a combined ticket (Girona Museus) with a 50% discount for the user from the second museum visited (the first museum is paid 100%) with a duration of 6 months.

In regard to the actions of the city council, it is necessary to emphasise the previously mentioned *Pla Estratègic de Turisme* from 2018 and the recent *Mapa Cultural* and *Girona Cultura* of 2021 (*Mapa Cultural*, 2021; *Girona Cultural*, 2021). The first meant the analysis of the cultural programming of the city to carry out strategic and operational decisions for improving them and increasing the diversity of the audiences involved. The second is a website with information for the public and the "targeta (card) Girona Cultura" initiative, which allows you to enjoy discounts and benefits on tickets and special activities (theatre, concerts, workshops, guided tours etc.).

4.3 The Origins of the Project "Girona Episcopal"

The project "Girona Episcopal" has its origin dating back to January 14, 2016, when the Bishopric of Girona and the Agència Catalana de Patrimoni Cultural[4] (Catalan Cultural Heritage Agency) signed a collaboration agreement for the promotion and dissemination of the episcopal monumental site of Girona, focused on three important heritage centres: the cathedral of Santa Maria, the basilica of Sant Feliu,[5] and the md'A (*Girona Episcopal. Memòria 2016*. (2017)). The first stage of this agreement was a single joint entrance to the three spaces for citizens and tourists. The name of the new cultural product is Girona Episcopal, evoking the relationship between the three buildings: the cathedral as the bishop's own church and the diocese's main temple, the basilica of Sant Feliu, which developed the functions of a cathedral, at least from the Carolingian period and, later, it was closely linked to the Girona main Christian site until it became a parish in the nineteenth century (Freixas, 2016; Nolla, 2020) and, finally, the md'A, located in the old episcopal palace,[6] which gathers the

[4] *Girona Episcopal. Memòria 2016*. (2017). The agreement is signed by the Bishopric of Girona, on which the cathedral and the basilica of Sant Feliu depend, and the Agència Catalana de Patrimoni Cultural (Catalan Agency for Cultural Heritage), an entity created in 2014, under the Department of Culture of the Generalitat de Catalonia, which manages Catalan monuments and museums, including the Girona Art Museum.

[5] A joint itinerary and entrance ticket was created between the cathedral and the basilica of Sant Feliu in 2012, thanks to an agreement between the rector and the cathedral chapter. The management of the visit is delegated to the technicians of the Treasury of the Cathedral.

[6] The origins of the Episcopal Palace of Girona date back to the tenth and twelfth centuries; the building was later expanded with new additions in the last centuries of the Middle Ages and Modern Era. It was an episcopal residence until 1973.

collections of the Museu Diocesà, the Diputació de Girona and the Generalitat de Catalunya.[7]

The creators and promoters of the new project had previously forged a series of collaborations—a loan of works for the organisation of the exhibition *Vitralls. La llum de la Catedral de Girona* (2015–2016) and the organisation of joint school activities—which paved the way for the formalisation of the agreement. The key elements of the project are:

1. The creation of a brand: Girona Episcopal, which expresses the essence of the project and is easily converted into an image that can be visualised in the collective imagination of visitors.
2. The brand rationale: It is a solid and rigorous story, which is the basis of the joint activities developed.
3. The communication and dissemination of the new brand through promotional campaigns.

The functional operation is specified in the definition of a pricing policy and the subsequent distribution of profits according to the agreed criteria and also with the leadership of the md'A in the implementation of consensually agreed communication strategies.

5 Research Results

5.1 Brand Management Analysis

5.1.1 The Stock of Cultural Capital

The stock of cultural capital refers to the numerous heritage resources of the Barri Vell, a historical, artistic complex endowed with a rich and diverse cultural capital of a heritage nature that includes many buildings from different eras and, above all, from different cultures and religions. This cultural diversity is one of the main hallmarks and attractions of the city.

The monumental site of Barri Vell, with its intricate urban plot and the excellent state of preservation of the main historic buildings, also preserves the atmosphere that allows visitors to relive its history with authenticity, generating a particularly attractive experience for them. This character of the visit-itinerary is a common phenomenon in monumental cities. In fact, Galí describes Girona as an "itinerary city" (Galí, 2004).

[7]El Museu d'Art de Girona was created in 1976 thanks to the agreement between the Diputació de Girona and the Bishopric of Girona. It has been managed by the Generalitat de Catalunya since 1992. The md'A was integrated into the Agència Catalana de Patrimoni Cultural in 2014. It has a significant collection of more than 13,753 works from the fourth to the twentieth centuries.

The cultural stock of the Barri Vell, apart from the mentioned historical buildings and monuments, also includes a variety of public and private institutions with a lot of different activities.

There are two important centres for the arts: *Bòlit, Centre d'Art Contemporani* and *CaixaForum*. The first develops research programmes, and the production and exhibition of contemporary art projects carried out by creators and professionals from other areas of local, national and international knowledge, with a strong involvement with their context. Bòlit offers a programme of integrated and linked activities, such as workshops, cycles of conferences, urban interventions, cycles of screenings, specific proposals, and work in progress, seeking the participation and involvement of civil society since its inception. The second, *CaixaForum*,[8] offers temporary exhibitions on diverse topics: art, photography, cinema, heritage, science, and cooperation. The exhibitions are itinerant and can be seen in more than one CaixaForum all over Spain. They design other activities for families and schools: workshops, music shows, dance theatre, visual arts, and science.

La *Mercè Cultural Center* is a multipurpose and multidisciplinary municipal facility that manages a varied programme of cultural activities and hosts various initiatives. The aim of the centre is to provide the city with a space open to cultural activity in which knowledge is facilitated and the exchange between artists, intellectuals, students, and the general public is enhanced.

Another cultural centre is the *Centre Cívic Barri Vell—Mercadal*, where different cultural events are made, such as performances, shows, courses, workshops and collective and individual exhibitions.

The *Teatre Municipal* has stable programming of shows for adults and family audiences. They also organise the *EscèNIT* street programming during the summer. They are open to all school-age boys and girls so that they can get to know and enjoy the world of the performing arts.

The *Cinema Truffaut* is also in the area of the old quarter and offers original version[9] films.

Besides from these site institutions, there are many other activities during the year which have the Barri Vell as the main location, as *Temporada Alta*, one of the best European performing arts festivals between the months of October and December, a meeting point for the display of theatre proposals and a platform for supporting the creation and production of shows. *Temps de Flors* is another visitor attraction. For 10 days Girona's historical buildings, courtyards and gardens are adorned with floral arrangements with a programme of cultural events in the Barri Vell and all over the city.

---

[8]CaixaForum belongs to *"La Caixa" Foundation*, a bank that has a network of cultural centres all over Spain; in total, 8 locations: Barcelona, Girona, Lleida, Madrid, Palma de Mallorca, Sevilla, Tarragona, and Zaragoza.

[9]The films are normally dubbed versions in Spain, only some cinemas offered the original language version.

Product

Among the diversity of heritage resources in the Barri Vell, the project's promotion team has been able to identify a theme that has a coherent historical and artistic basis, and that is likely, as has been shown, to become a powerful and solid cultural brand. The objectives of the project are:

1. The enhancement of the monumental episcopal heritage of Girona. As those in charge of the Tresor and the Cathedral have stated, the goal of the project is to generate value from a rigorous narrative that allows a better understanding of the episcopal heritage but, at the same time, also involves the revaluation of all the city's monumental heritage. Therefore, there is a broad view beyond the product itself which reinforces the value of all local historical heritage.
2. The creation of a discourse based on historical research that helps to break with the individuality of the three centres and relate them based on the concept of the episcopal neighbourhood. Above all, a discourse based on the fact that the three buildings developed a function in relation with the Bishop (liturgical, residential, or administrative). This corresponds to the plot of streets surrounding the three nodes and includes the whole series of buildings linked to the life and the daily functions of government and administration of the bishop, the curia, and the canonical chapters of the cathedral and Sant Feliu. European episcopal districts normally occupy the epicentre of power in medieval cities, along with the residence of kings, counts or delegated authorities, and municipal governments. In Girona, it is in a markedly symbolic place, the highest part of the hill overlooking the city, on the east bank of the Onyar river. It is an elevated location that makes visible the imposing facade of the cathedral, the facade and tower of the basilica of Sant Feliu, and even the Episcopal Palace from different parts of the city.
3. A schedule of joint educational or cultural activities such as visits, exhibitions, itineraries, audio-visuals, and research projects in the future. Until now, this programme has been limited to the visit to the three centres (2016–2021). However, as those responsible say, there is a firm intention to expand to the creation of thematic itineraries that include other emblematic buildings linked to episcopal and canonical functions as well as to educational activities (*Girona Episcopal. Memòria 2019,* 2020).
4. The joint ticket and the pricing policy includes offers and discounts to facilitate the accessibility to Girona's episcopal heritage to a larger group of people. One of the key aspects of the project was to generate a functional operation that would facilitate the definition of this unique ticket, which would result in a flow of audiences that would go from the cathedral (main attraction) and feed the Basilica of Sant Feliu and, above all, the md *Girona Episcopal. Memòria, 2016,* 2017).

The strategic lines discussed were:

1. The creation of a brand that defines and visualises the mission and vision of the project since its launch in 2016: Girona Episcopal.

Girona
Episcopal

Catedral de Girona
Basílica de Sant Feliu
Museu d'Art de Girona

Fig. 1 Logo of the Girona Episcopal project

2. The development of dissemination and promotional actions that are agreed upon and specified annually. Its materialisation falls on the md'A, which is endowed with a wider, multidisciplinary team than the Treasury staff.
3. The creation of a rigorous discourse on Girona Episcopal that will be the basis of the programming of joint activities.

Communication

The priority for the project development team was to define the brand. According to brand creation and management experts, naming is a key factor in brand consolidation. The term created—Girona Episcopal—fully reflects the mission and vision of the product and, most importantly, the content of the product generated.

Once the name was chosen, the initial promotional actions aimed to create the graphic image and promotional materials. The first step was the logo with the geometric and simplified profile of a mitre as a motif, one attribute of the bishop, accompanied by the name of the three institutions that make up the network (Fig. 1). The second action was selecting an aerial image that would help locate the three centres within the urban topography of the old town. The design of these two graphic elements is accompanied by a slogan: "Passeja't per més de mil anys d'art i història" (Stroll through over a thousand years of art and history), which emphasises and intensifies the experiential nature of the proposed visit and gives the keys of a profile of a cultural visitor who is characterised by looking for personal experiences that make them feel and live history and its imprints; this is a trend in tourism and consumer culture. In fact, the use of the verb "passejar" (to walk, to stroll, to wander) already suggests a kind of dynamic visit that connects, therefore, with an active tourist profile looking for a plurality of experiences through an itinerary.

These graphic and textual elements were applied to the dissemination materials created, often combined with three images, one from each site. They designed the entrance tickets, a quadriptych, signposts for the lobbies and entrance points of sale of the three sites and printed paper towels in 2016. They also prepared flyers in 2017 and banners and outdoor advertising boards the following year. Girona Episcopal took a step into the digital world by creating a website in 2018[10] that sought to visualise and position the project in the online cultural offer as well as to facilitate the

[10] https://gironaepiscopal.cat

sale of tickets. The new platform also contains a promotional video. At the level of social networks, Girona Episcopal does not have its own channels except for the hashtag Girona Episcopal, so its promotion is linked to the Facebook, Twitter, and Instagram accounts of md'A and with valuable support, for the scope of its impact, from the channels of the Agència Catalana de Patrimoni Cultural (Catalan Cultural Heritage Agency) of the Generalitat de Catalunya.

Periodically, the Girona Episcopal management team establishes promotional campaigns focused on urban promotion through the distribution of dissemination materials (flyers, quadriptychs), from printed towels to restaurants and bars in Girona, the advertising in the agenda of the md'A, the placement of banners in the city (2018) and advertising boards for the 2020 Christmas campaign (Fig. 2). Promotional campaigns have also been launched in Barcelona through the collocation of mupis at the subway and at Renfe train stations (July 2018 and 2020). At the digital level, regular campaigns are carried out on Facebook, Twitter, and Instagram. They also decided to launch various advertising spots in local and national newspapers and media.[11]

Another strategic focus for promotion has been the synergies generated with cultural institutions and tourist agents. Turisme Girona stands out in this sense, especially the work from the beginning of the project with the Patronat de Turisme de la Costa Brava (the Tourism Board).[12] In a post-pandemic position, Girona Episcopal participates in the cultural and tourist campaign promoted by the Agència Catalana de Turisme and the Agència Catalana de Patrimoni Cultural in the summer of 2020. This year, they are working on positioning and online sales through general leisure and tourism platforms.

5.2 Brand Orientation in Girona Episcopal

As it has been said, Evans et al. (2012) propose a brand orientation model based on two philosophical aspects *culture* and *compass*, and four behaviours (or capabilities) that we will apply to our study case: *distinctiveness*, *functionality*, *augmentation*, and *symbolism*.

In the case of Girona Episcopal, the brand's *distinctive* character is based on the name itself and the slogan that accompanies it, which suggests a unique experience. But also, the unique character of the buildings or the works of art that they show inside reinforces this value. The most eloquent case is the cathedral of Girona, an emblematic building of the Catalan Gothic architecture, with the widest nave in the world (22.98 m), which preserves in its treasure unique works such as the famous

[11] *Hora Nova* (2018), *Diari de Girona* (2019 and 2020), with a specific campaign on Christmas in *Diari Ara* and *El Punt Avui* (2020).

[12] Patronat de Turisme de la Costa Brava is an entity made up of the Catalan public administrations and the tourism sector that aims to develop tourism on the Costa Brava and the Girona Pyrenees.

Fig. 2 Advertising board of the Girona Episcopal project

Tapís de la Creació (Tapestry of Creation), the Beatus of Girona, or the casket of Hixam II among other objects. The basilica of Sant Feliu[13] and the Md'A also stand out for their architectural value and the outstanding collection of goods on display.

To analyse the *functionality* of the brand the lack of audience satisfaction surveys is a serious handicap that will have to be resolved in the future in order to increase the quality of the visit. However, despite this, the overall number of visitors reached since the project's birth shows the correct identification of the product for its public.

The *augmentation* is one of the strengths of the brand. As we have seen, the people in charge of Girona Episcopal have the express will to generate new content, the basis of a wider programming and, on the other hand, to implement aspects of communication that are now starting to be developed, such as relations with local agents, the positioning of the brand in browsers, or improvements to the website.

In the case of Girona Episcopal, *symbolism*, as augmentation, is one of the main strengths. It is linked to the type of experience created from a story with a marked experiential character that wants to help its consumers walk through 1000 years of art and history. The impact of this discourse on local visitors can result in a close, and emotional, connection with the heritage, its enhancement, and, therefore, the strengthening of a sense of belonging, which is often necessary in a context where the volume of tourism can have the effect of alienating the local population.

6 Discussion

The qualitative information extracted from the interviews and the quantitative data provided by the reports allow an X-ray of the brand management and brand orientation of Girona Episcopal, at least until before the pandemic.

We want to start the discussion based on visitor data as they are the only quantitative element, we have today to measure the impact of Girona Episcopal. Thus, although the pandemic outbreak has broken the upward logic of the line of audiences drawn from 2016 to 2019, the data provided by the reports show a very positive growth. It went from 6659 visitors in 2016 to 12,331 (2017), 13,268 (2018), and 14,395 (2019).[14] Of course, these figures hide different realities for each site, as the cathedral continues to have a significant lead in the number of visitors. However, Girona Episcopal has generated a positive effect in the other two attractions, bringing a significant flow of public to the md'A that oscillates between the 61.96% of visitors who acquired the Girona Episcopal ticket (in the Cathedral or in Sant Feliu) going later to md'A in 2018 and 69.96% in 2019.

[13]Declared BCIN, Bé Cultural d'Interès Nacional (Cultural Heritage of National Interest) in 1931, in the category of historic monument.

[14]The report includes the annual quantification of visitors to Girona Episcopal, the seasonality, the point of purchase of the ticket and the flow of public between the centres. However, there is no data on the profile and origin of the public. Satisfaction surveys were not conducted, and they would have provided highly meaningful information for subsequent marketing actions.

Along with the positive evolution of visitors, the identity and solidity of the brand and the clear definition of its content are also positive aspects. The brand is based on previous, intrinsic, and objective elements: the historical-artistic value of each of the three heritage sites. It is highly commendable that the three sites, endowed with a marked personality and accumulated prestige for decades, have been able to come together to create a common story that unites them and that the brand created expresses that perfectly. The naming, the graphic image, and the slogan sum up the essence of the content. In addition, as mentioned, the slogan fits the profile of the tourist public (internal or external) who choose to get to know the Barri Vell while looking for an atmosphere that makes them relive history. The "experiential" connotations of the slogan connect, as suggested in previous studies, with the typology of visitors to cities or monuments that base their experience on walking around (Galí, 2004; Alcalde, 2008).

Also noteworthy are the consistency and rigour of the content. Indeed, the current and future potential of the brand is based on a concept—Girona Episcopal—that has a historical and artistic basis. Interestingly, this obvious strength is yet to be exploited because, so far, the project has basically materialised in the joint ticket and some specific activity promoted by one of the institutions with the involvement of the others. Therefore, the creation of a powerful and attractive narrative that generates several guided tours or joint audio-guided tours that allow different audiences to visualise the link between the three sites and the rest of the buildings in the Episcopal district is still missing today. As stated in the 2020 report and as the director of md'A mentioned, they are working on this, thanks to the commissioning of some previously presented academic studies, which will be the touchstone for the construction of informative materials for the new programming (Fumanal, 2021a, 2021b; *Estudi arqueo-arquitectònic del Palau Episcopal*, n.d.).

In addition to the positive aspects, the interviewees reflected other aspects that show certain weaknesses. On the one hand, there is a dependence on an agreement with short-term renewals that can be stopped if one of the parties shows a significant change in the institutional strategies, or they are no longer interested in the project, or there is a lack of understanding. On the other hand, the lack of specific human resources dedicated to the programming and communication of the project has also been highlighted.

Regarding the public, the segment of visitors sought by the promoter team for this type of activity was initially the visiting tourist of Girona, who chooses the cathedral as the first option to visit. Therefore, they were also offered two other iconic spaces to visit. Now, however, the aim is to target local audiences who, due to their origins and interests, may be more sensitive to content linked to their own history, as has been demonstrated in studies of temporary exhibitions that have a more ethnocentric focus (Camarero et al., 2010). The nature of the content and its treatment can lead to a closer emotional and identity bond. The types of symbolic values that generate identification and intensify a close relationship with heritage are also crucial for consolidating the cultural brand. They also have a major social impact in the medium and long term because they cooperate in strengthening a local identity, deepening social cohesion, and also contributing to the enhancement of one's own heritage by creating a sense of belonging based on an emotional and intellectual bond (Camarero et al., 2010; Cerquetti, 2010).

However, other segments of the audience can be targeted. Audience studies in the Barri Vell have allowed Galí and Donaire to identify two groups of visitors—interested tourists and erudite tourists—both good consumers of cultural products and audiences who can easily be seduced by programming with relevant experiential components. However, they are a minority if we compare them to the global (Galí & Donaire, 2006b). On the other hand, as the director of md'A has pointed out, there are natural niches for growth, such as tourism that comes from the south of France and has easy access to the northern part of Catalonia, to which it usually travels for tourism, cultural, leisure, or commercial reasons. Costa Brava tourism, which takes part in the many performing arts festivals in the summer, as well as the Temporada Alta autumn festival, are also segments of the public to consider.

At a communication level, the importance of marketing and dissemination campaigns and institutional relations should be highlighted as positive. However, it is important to establish synergies and complicity in working with the associative fabric of the Barri Vell and with other local public or private cultural and tourist agents, including the city council itself and the other museums and heritage facilities with which it shares location and audiences. Beneficial alliances could be created that, without distorting their own brand, help to expand the number and scope of agents and institutions involved in the project and, as a result, increase their impact.

Another pending subject is the brand's positioning on a digital level, conditioned today by the too static typology of the website and the lack of social networks. This is a major front if we want to give Girona Episcopal a boost, at least in terms of the domestic public and foreign tourists who use browsers as a basic source for selecting cultural proposals and tourist destinations.

In order to summarise the discussion, we have carried out a SWOT analysis that allows us to visualise the strengths and weaknesses of the project and the opportunities and threats (Table 1).

7 Conclusions

The application of brand management and brand orientation models in Girona Episcopal has shown that, despite the youth of the project and the negative impact of COVID-19, the management and guidance of the brand are broadly positive and endowed with a high potential yet to be exploited.

The promotional team demonstrates a clear vision of the product, the goals to be achieved, and the values of the brand. And, most importantly, there is a strong commitment from those responsible for the product. As Evans et al. have pointed out, the philosophical aspects of the brand—culture and compass—are critical to determining its direction and involve its promoters in creating an internal debate about what the brand is and what its values and goals should be. These elements are even more important when the product generated is a network and, therefore, is born

Table 1 SWOT analysis of the project

Strengths	**Weaknesses**
• Relevance and distinctiveness of the three stakeholders	• Not joint activities developed at the moment
• The quality of the name, logo, slogan, and image of the brand	• Lack of specific human resources
• The strong link among the buildings around the bishop's historical role	• Static website and lack of social network
• The commitment of the team	• Need to develop the content
• Joint ticket and price	• No audience satisfaction surveys
• Communication strategy	• Temporariness of the agreement
	• No relationship with cultural agents of Barri Vell
	• Lack of relations with public administrations
Opportunities	**Threats**
• Prestige of Girona as an historical and cultural city	• Pandemic situation
• The visit to a monumental city as Girona is an experiential tour for the visitors	• The possible similar offer provided by other competitors
• Strategic location and good connections of the city	• Overtourism
• Increase of cultural visitors every year	

from the association of previous products and brands endowed with a strong personality of their own. Therefore, the harmony shown by those responsible for sharing is a key element in developing the product.

In relation to Cerquetti's and Evans et al.'s models, the definition of the product, the correct identification of a historical and attractive story for the profile of the public in tune with the given name (thematisation) and, on the other hand, the orientation towards communication are features that show the strength of the brand and show the leadership capacity—and cultural leadership—that the creators of the project have had. Their value also lies in the impact of enhancing the value of local heritage for both tourists and the local public itself. Nevertheless, there is still a long way to go to provide content that strengthens the brand and makes it more attractive and consistent to local audiences and external visitors.

However, it should be added that our study lacks the opinion of the public, as there are no satisfaction surveys and no information about the origin or socioeconomic profile of the visitors. We have neither a variety of opinions from external agents (except for an interview with Carme Saïs). Given the potential of the brand to be developed, we should wait for the recovery of post-pandemic tourism and therefore for the normalisation of the health situation and the launch of new joint activities between the three partners of the project, which can certainly strengthen the product for a further analysis of the evolution of the brand.

References

Alcalde, G. (2008). El uso de los museos y la visita turística a la ciudad de Girona: un análisis a partir del tique M5 del Girona museus. *Estudios Turísticos, 177*, 77–90.

Baumgarth, C. (2009). Brand orientation of museums: Model and empirical results. *International Journal of Arts Management, 11*(3), 30–45.

Baumgarth, C., & O'Reilly, D. (2014). Brands in the arts and culture sector. *Arts Marketing, 4*(1), 2–9.

Baumgarth, C., Kaluza, M., & Lohrisch, N. (2016). Brand audit for cultural institutions (BAC): A validated and holistic brand controlling tool. *International Journal of Arts Management, 19*(1), 54–68.

Blasco, D., Palou, S., & Poch, M. (2021). Emergence and spread of critical discourses opposing tourism: The case of Girona, and urban satellite destination of Barcelona. *Tourism Planning & Development*. https://doi.org/10.1080/21568316.2021.1873840

Bridson, K., & Evans, J. (2004). The secret to a fashion advantage is brand orientation. *International Journal of Retail & Distribution Management, 32*(8), 403–411.

Caldwell, N. G. (2000). The emergence of museum brands. *International Journal of Arts Management, 2*(3), 28–34.

Caldwell, N. G. (2003). Do museums have 'brand identities' that are recognised by visitors? In *7th International Conference on Arts and Cultural Managements*, s.p.

Caldwell, N., & Coshall, J. (2002). Measuring brand associations for museums and galleries using repertory grid analysis. *Management Decision, 40*(4), 383–392.

Camarero, C., Garrido, M. J., & Vicente, E. (2010). Components of art exhibition brand equity for internal and external visitors. *Tourism Management, 31*(4), 495–504.

Camarero, C., Garrido-Samaniego, M. J., & Vicente, E. (2012). Determinants of brand equity in cultural organisations: The case of an art exhibition. *The Service Industries Journal, 32*(9), 1527–1549.

Camprubí, R., & Planas, C. (2020). El Storytelling en la marca de destinos turísticos: el caso de Girona. *Cuadernos de Turismo, 46*, 269–289.

Cap de premsa i comunicació del Museu d'Art de Girona [Entrevista a Àngels Miralles] 26 de novembre 2021.

Cap de servei de Museus, Patrimoni Cultural i arts visuals [Entrevista a Carme Sais]. Ajuntament de Girona. Girona 29 de novembre 2021.

Catedral de Girona. Memòria. (2019).

Cerquetti, M. (2010). Strategie di branding del cultural heritage nella prospettiva esperienziale (Branding strategies of cultural heritage according to the experiential perspective). *Sinergie, 82*, 123–142.

Cerquetti, M., & Montella, M. M. (2015). Museum networks and sustainable tourism management. The case study of Marche region's museums (Italy). *Enlightening Tourism. A Pathmaking Journal, 5*(1), 100–125.

Chamorro, M. A. (2004). *La construcció de Sant Feliu al segle XIV*. Tesi doctoral 2004. Universitat de Girona, Departament de Geografia, Història i Història de l'Art.

Clara, J., & Marquès, J. M. (1992). *Sant Feliu de Girona. Col·lecció Sant Feliu.* Edita Parròquia de Sant Feliu.

Comas, J., Aulet, S., Bouallala, Y., & Bernardó, V. (2013). *Estudi d'impacte econòmic de Girona Temps de Flors 2012*. Ajuntament de Girona.

Crous-Costa, N. (2010). *La gestió turística en els museus. El cas del Museu d'Art de Girona.* Treball de Pràctiques del Màster en Planificació i Direcció del Turisme. Girona.

Crous-Costa, N., & Vidal-Casellas, D. (2013). Urban regeneration and cultural development of Girona. *Cultural Tourism, 178*, 178–189.

Director del Tresor de la Catedral de Girona i Gustau Torres, tècnic del conjunt catedralici [Entrevista a Joan Piña]. Girona 17 de setembre 2021.

Directora del Museu d'Art de Girona [Entrevista a Carme Clusellas]. Girona, 17 de setembre 2021.

Evans, J., Bridson, K., & Rentschler, R. (2012). Manifestations of brand orientation in museums: An exploratory study. In *AMA Summer Educators Conference 2011*. Delivering value in turbulent times (Vol. 22, pp. 435–442).
Ferrer, J., Sureda, M., Nolla, J. M., Gironella, A., Puig, J., & Freixas, P. (2020). *Girona: catedrals, esglésies i convents (I): del naixement de l'església local a l'època moderna.* Ajuntament de Girona, Servei de Gestió Documental, Arxius i Publicacions SGDAP.
Font, L. (1952). *Gerona. La Catedral y el Museo Diocesano*. Editorial Carlomagno.
Freixas, P. (2016). *La Basílica de Sant Feliu: primera catedral de Girona*. Ajuntament de Girona.
Fumanal, M. A. (2021a). *Estudi dels escuts i les llindes del Palau Episcopal de Girona.*
Fumanal, M. A. (2021b). *Continguts per a una visita de 180° al Palau Episcopal de Girona.*
Fumanal, M. A. *Estudi arqueo-arquitectònic del Palau Episcopal de Girona* (n.d.). Institut de Recerca Històrica de l'Universitat de Girona. In progress.
Galí, N. (2004). *Mirades turístiques a la ciutat. Anàlisi del comportament dels visitants del Barri Vell de Girona*. Tesi doctoral. Universitat de Girona.
Galí, N., & Donaire, J. A. (2003). La imagen a priori de los destinos turísticos monumentales: el caso de Girona. *Papers de turisme, 34*, 78–97.
Galí, N., & Donaire, J. A. (2005). *Itineraris turístics a la ciutat de Girona.* Treballs de Patrimoni Cultural. Universitat de Girona.
Galí, N., & Donaire, J. A. (2006a). La història del turisme a la ciutat de Girona. *Revista de Girona, 239*, 34–41.
Galí, N., & Donaire, J. A. (2006b). Visitors' behavior in heritage cities: The case of Girona. *Journal of Travel Research, 44*(4), 442–448.
Girona Cultura. (2021). Àrea de Cultura de l'Ajuntament de Girona.
Girona Episcopal. Memòria 2016. (2017). Catedral de Girona, Basílica de Sant Feliu, Museu d'Art de Girona.
Girona Episcopal. Memòria 2017. (2018). Catedral de Girona, Basílica de Sant Feliu, Museu d'Art de Girona.
Girona Episcopal. Memòria 2018. (2019). Catedral de Girona, Basílica de Sant Feliu, Museu d'Art de Girona.
Girona Episcopal. Memòria 2019. (2020). Catedral de Girona, Basílica de Sant Feliu, Museu d'Art de Girona.
Girona Episcopal. Memòria 2020. (2021). Catedral de Girona, Basílica de Sant Feliu, Museu d'Art de Girona.
Hankinson, P. (2001a). Brand orientation in the top 500 fundraising charities in the UK. *Journal of Product & Brand Management, 10*(6), 346–360.
Hankinson, P. (2001b). Brand orientation in the charity sector: A framework for discussion and research. *International Journal of Nonprofit and Voluntary Sector Marketing, 6*(3), 231–242.
Iglesias, L. (2013). Territori de grans festivals. *Revista de Girona, 280*, 82–88.
Izquierdo, C. C., Samaniego, M. J. G., & Hernández, E. V. (2008). Valor de marca de las entidades culturales: el caso de la exposición Las Edades del Hombre. In *XX Encuentro de Profesores Universitarios de Marketing [Recurso electrónico]: Universidad de Las Palmas de Gran Canaria, 17-18-19 de septiembre de 2008* (p. 39). Escuela Superior de Gestión Comercial y Marketing, ESIC.
Jiménez Morales, M., & San Eugenio Vela, J. D. (2009). La Organización de eventos como estrategia identitaria y evocadora de imagen turística. Estudio de caso: Girona. Temps de flors. *Pasos, Revista de turismo y patrimonio cultural, 7*(1), 73–84.
Kammerer, N. C. (2019). Festa i festival. Una història sobre dos models de cultura a la ciutat de Girona. *Revista d'etnologia de Catalunya, 44*, 192–198.
Kladou, S., & Kehagias, J. (2014). Developing a structural brand equity model for cultural destinations. *Journal of Place Management and Development, 7*(2), 112–125. https://doi.org/10.1108/JPMD-03-2013-0007
Laborde, A. (entre 1806 i 1820). *Voyage pittoresque et historique de l'Espagne.* De l'Imprimerie de Pierre Didot l'ainé.

Llorens, J. M. (2011). *Sant Pere de Galligants. Un monestir al llarg del temps*. Museu d'Arqueologia de Catalunya-Girona.

Mapa Cultural. (2021). Àrea de Cultura de l'Ajuntament de Girona.

Memòria 2019. (2020). Patronat Call de Girona.

Nadal, J., Freixas, P., Jiménez, M., Nolla, J. M., & Roura, G. (2002). *La catedral de Girona. Una interpretació*. Lunwerg.

Nolla, J. M. (2020). Els Orígens de la basílica de Sant Feliu de Girona. In J. Ferrer i Godoy, M. Sureda Jubany, J. M. Nolla, A. Gironella i Delgà, J. Puig i Oliver, & P. Freixas (Eds.), *Girona: catedrals, esglésies i convents (I) : del naixement de l'església local a l'època moderna* (pp. 55–69). Ajuntament de Girona, Servei de Gestió Documental, Arxius i Publicacions SGDAP.

Nolla, J. M., & Palahí, L. (2011). *Els Orígens de la basílica de Sant Feliu de Girona : dades documentals i arqueològiques : estat de la qüestió*. Universitat de Girona, Institut de Recerca Històrica, Laboratori d'Arqueologia i Prehistòria.

Phillips, M., & O'Reilly, D. (2007). Major case study: Rethinking Tate Modern as an art museum 'brand'. In R. Rentschler & A. M. Hede (Eds.), *Museum marketing* (pp. 186–191). Routledge.

Pla Cargol, J. (1956). *Los museos de Gerona*. Ed. Dalmau Carles.

Pla Cargol, J. (1965). La vieja Revista de Gerona, exponente de un afán de cultura. *Revista de Girona, 32*, 10–19.

Pla Especial de Rehabilitació Interior per al Barri Vell. (1982). Ajuntament de Girona.

Pla estratègic de Turisme de Girona. (2018). UTE conformada per l'Institut d'Estudis Turístics (INSETUR) de la Universitat de Girona i les consultories Joan Manuel Ribera i Advanced Leisure Services.

Pusa, S., & Uusitalo, L. (2014). Creating brand identity in art museums: A case study. *International Journal of Arts Management, 17*(1), 18–30.

Recuero Virto, N., García de Madariaga, J. G., & Blasco, M. F. (2017). The influence of museum's brand image and satisfaction on visitors' loyalty. In *Conference: 16th International Marketing Trends Conference*, Madrid, s.p.

Reixach, F. (1982). *Farmàcia antiga de l'Hospital de Santa Caterina*. Centre de Conservació i Catalogació de Monuments de la Generalitat de Catalunya i la Diputació de Girona.

Scott, C. (2000). Branding: Positioning museums in the 21st century. *International Journal of Arts Management, 2*(3), 35–39.

Servitja, L., & Burch, J. (2014). Els Museus de Girona. Un passeig per la seva història. *Annals de l'Institut d'Estudis Gironins,* LV, 531–544.

Soler, N. (1984). El Museu d'Art de Girona. *Revista de Girona any XXX, 106*, 1–10.

Soler, N. (1988). El Museu d'Art de Girona. In A. Pladevall (Ed.), *Catalunya Romànica* (Vol. 23, pp. 13–15).

Sureda, M. (2004). La Catedral de Girona, matèria històrica. Historiografia a l'entorn de la seu (ss. XVII-XXI). *Annals de l'Institut d'Estudis Gironins, 45*, 69–109.

Vidal-Casellas, D., Crous-Costa, N., & Oliveras-Schwarz, M. (2013). Case study 8: Reflection on the constituent elements of cultural tourism: Theatre festivals. A case study of Temporada Alta (Girona, Spain). In R. Raj, K. Griffin, & N. Morpeth (Eds.), *Cultural tourism* (pp. 190–203). Cabi. https://doi.org/10.1079/9781845939236.0190

Wallace, M. (2006). *Museum branding. How to create and maintain image, loyalty and support*. AltaMira.

Wong, H., & Merrilees, B. (2008). The performance benefits of being brand-orientated. *Journal of Product & Brand Management, 17*(6), 372–383.

Websites

https://cultura.gencat.cat/ca/temes/museus/dades/visitants/museus-registrats/
https://gironaepiscopal.cat
http://ptop.gencat.net/rpucportal/AppJava/cercaExpedient.do?reqCode=veureDocument&codintExp=211863&fromPage=

Multidimensional Value Creation Through Cultural Programmes? Challenges for Veszprém-Balaton ECoC 2023

Mara Cerquetti, Katalin Lőrincz, and Ágnes Raffay-Danyi

Abstract This research aims to enrich the international debate on cultural programmes by investigating the challenges the European Capital of Culture (ECoC) faces when promoting the long-term cultural and social development of cities and the innovation of their cultural offer. Focusing on the case of Veszprém-Balaton, ECoC in 2023, the field research investigates: the main weaknesses of the urban cultural policy and management; if and how the cultural sector is currently contributing to the quality of life and sustainable tourism development; expected innovations that ECoC could and should trigger to bring culture to people and widen its understanding over the long term. Taking into account the increasing scientific literature devoted to value creation, the paper focuses on the role of community engagement and networking in the cultural and tourism sector for promoting long-term development. The literature review allowed us to build a conceptual framework that guided and structured the analysis of the case study. The research findings reveal the need to improve the levels of involvement and networking and highlight the role of *process innovations* in addition to *product innovations*, for ensuring long-term sustainable development. The results also provide policy and managerial implications for cities entering into cultural programmes.

1 Introduction

In a recent article, Sacco et al. (2018) classify EU countries into six classes by matching their performances both in innovation and cultural practice. In the final ranking, Hungary is in the last but one category, together with Cyprus, Portugal,

M. Cerquetti (✉)
Department of Education, Cultural Heritage and Tourism, University of Macerata, Macerata, Italy
e-mail: mara.cerquetti@unimc.it

K. Lőrincz · Á. Raffay-Danyi
Faculty of Business and Economics, University of Pannonia, Veszprém, Hungary
e-mail: lorincz.katalin@gtk.uni-pannon.hu; raffay.agnes@gtk.uni-pannon.hu

E. Borin et al. (eds.), *Cultural Leadership in Transition Tourism*, Contributions to Management Science, https://doi.org/10.1007/978-3-031-14121-8_11

Greece and Italy. This group is characterised by average performance in innovation and bottom performance in cultural practice.

In this scenario, the present research focuses on the case of Veszprém-Balaton, European Capital of Culture (ECoC) in 2023, to investigate if and how cultural institutions in Veszprém and the Balaton region are currently facing the need to innovate their cultural offer, expand audiences and increase active participation. Given that Lake Balaton is a well-known tourist destination, and also the project partner of Veszprém for the 2023 ECoC year, the analysis investigated the contribution of cultural institutions to the quality of life and the role of cultural heritage and cultural activities in promoting a sustainable tourism offering. The pandemic situation will have an impact on the city and the project but it cannot be estimated yet.

The research is based on two premises: first, if cultural institutions have to contribute to the quality of life, people and local organisations (e.g. citizens, tourism industry and civic organisations) have to *use* them; second, the promotion of sustainable tourism development demands a high level of *cooperation* among institutions in the culture and tourism field.

The field activity aimed to answer the following research questions:

1. The main weaknesses of the urban cultural policy and management.
2. If and how the cultural sector is currently contributing to the quality of life and sustainable development of the city and its surrounding area.
3. Expected innovations that ECoC could and should trigger to bring culture to people and widen its understanding over the long term.

Specifically, the level of engagement (if and how cultural institutions in Veszprém and the Balaton region involve local communities) and the level of cooperation (if and how these institutions cooperate with other cultural and tourism organisations) were investigated, in order to identify the strengths and weaknesses of the current approach. Given that Veszprém in partnership with the Balaton region will be a European Capital of Culture in 2023, the case study also examined the needs and opportunities for innovation.

From a theoretical point of view, the research aims to contribute to the debate on ECoCs to understand how cultural programmes can help sustainable urban and tourism development and ensure long-term benefits for all local stakeholders.

The chapter is structured as follows. First, we discuss the main scientific contributions to the socio-economic impacts of culture and public value creation, with a focus on the role of community engagement and networking in the cultural and tourism sector for promoting sustainable development. This section also describes the role of ECoCs in promoting the long-term cultural and social development of cities, not only by attracting tourists from other European countries, but also by activating the participation of the residents of the city and its surrounding area. This literature review allowed us to identify the dimensions and indicators investigated in the field research. Then, after presenting research methodology, we discuss research results and implications for cities entering into cultural programmes.

2 Theoretical Background

2.1 The Multidimensional Impact of Culture

Over the past 15 years, cultural value and the value of culture to individuals and society have become buzzwords in the public policy debate: there has been growing interest in the impact of culture—specifically cultural initiatives and organisations—not only on the economy and society, but also on quality of life, health, and well-being, i.e. mental health, older people, marginalised people, learning disabilities and rehabilitation (O'Brien, 2010, 2015; Bollo, 2013; Scott, 2013; Tuck & Dickinson, 2014; Brown, 2019).

Scholars have identified different instrumental benefits of the arts (i.e. cognitive, attitudinal and behavioural, health, community-level social and economic benefits) (McCarthy et al., 2004), and investigated the ability of arts and cultural engagement to improve social cohesion, civic engagement, health and well-being and contribute to economic development (Lee, 2013; Crossick & Kaszynska, 2016; OECD, ICOM, 2018).

Broadly speaking, the concept of "cultural welfare" has begun to make headway in its definition as the appropriate and effective inclusion of cultural production and dissemination processes within a welfare system; and also as an integral part of the social welfare and health services that guarantee citizens the care and guidance needed to overcome critical issues related to health, ageing, disabilities, social integration and all the problems associated with a duty of social protection. In this vein, recent studies have explored the problematic relationship between cultural participation and well-being (Oakley et al., 2013; Bodo & Sacco, 2017; Cicerchia, 2017a, 2017b). For instance, a study by Grossi et al. (2011, 9–10) conducted on a large sample of Italian residents to understand the impact of health status and cultural access on psychological well-being revealed that "cultural access clearly stands out as the second variable in terms of absolute effect size on perceived well-being, right after health condition" and that it "acts as a sort of modulator of the disease-associated psychological burden of the perceived loss of well-being".

2.2 Towards a Holistic Approach

From the point of view of cultural management and policy, the results of current research on cultural welfare point to the need to measure and evaluate the real capacity of cultural activities to make a meaningful impact (Vermeulen & Maas, 2021). Far from being simple, measuring and assessing the impact of culture is still an open challenge today, for at least two reasons. First, there is a lack of adequate methodologies capable of accounting for the direct and/or indirect tangible (e.g. employment, wages and business development) and intangible (e.g. cultural capital, social cohesion and quality of life) impact of cultural activities (Whelan,

2015). Second, it is still difficult to establish a cause-and-effect link between the actions taken and the social benefits achieved (Viganò & Lombardo, 2018, 333).

When focusing on cultural institutions, at least two relevant issues should be addressed, regarding value creation and its measurement. On the one hand, an organisation's culture and practices should be innovated towards value co-production, by involving its stakeholders in all the different phases of the cultural production process (Spence et al., 2013). Co-creation is the central premise of the service logic (Bollwerk, 2015; Williams et al., 2020). Sharing this perspective, the scientific debate on the impact of culture is moving away from audience development towards community engagement and empowerment. Indeed, value co-creation not only involves customers (Boorsma, 2006; Antón et al., 2018; Ross, 2020)—from "passive audiences" to "active players"—but all different categories of people engaged in or affected by a cultural institution's activity—from users to employees to local communities. On the other hand, if a multi-stakeholder approach is adopted, measurement systems should be holistic, by including different perspectives (i.e. the organisation, users and internal stakeholders) and integrating different indicators (Cerquetti, 2017). As already argued, effectiveness is a multidimensional concept in cultural organisations and "financial measurements by themselves are inadequate to evaluate performance" (Boorsma & Chiaravalloti, 2010, 306). In addition, measurement should not merely be an institutional obligation in the context of evaluation systems, but should aim to detect the socio-economic impact generated.

In short, understanding the contribution of cultural institutions to value creation over the long-term (e.g. quality of life and sustainable tourism development) calls for a multi-level framework (Cerquetti, 2017), which defines *recipients* (categories and sub-categories of stakeholders to address), balances *objectives* (goals to achieve for each category) and adopts *indicators* (measures for assessing value creation in each category).

Sharing the need for a multidimensional framework that includes all stakeholders, we can draw five pillars of value creation from previous research on cultural value co-creation and measurement in cultural institutions (Bollo, 2013; Scott, 2013; Tuck & Dickinson, 2014; Cerquetti, 2017; Carlucci, 2018) (Fig. 1): (1) the contribution made by cultural institutions to knowledge building through their facilities and services; (2) user attraction and satisfaction, that is, audience development, engagement and empowerment; (3) community participation and involvement, including local schools, associations, groups and volunteers; (4) employment, staff well-being and staff learning; (5) networking within the cultural and tourism sector.

2.3 *From Theory to Practice: Culture-Led Urban Regeneration Through ECoCs*

Since the end of the 1990s, economic and management studies have devoted increasing attention to the contribution of culture and cultural programmes to

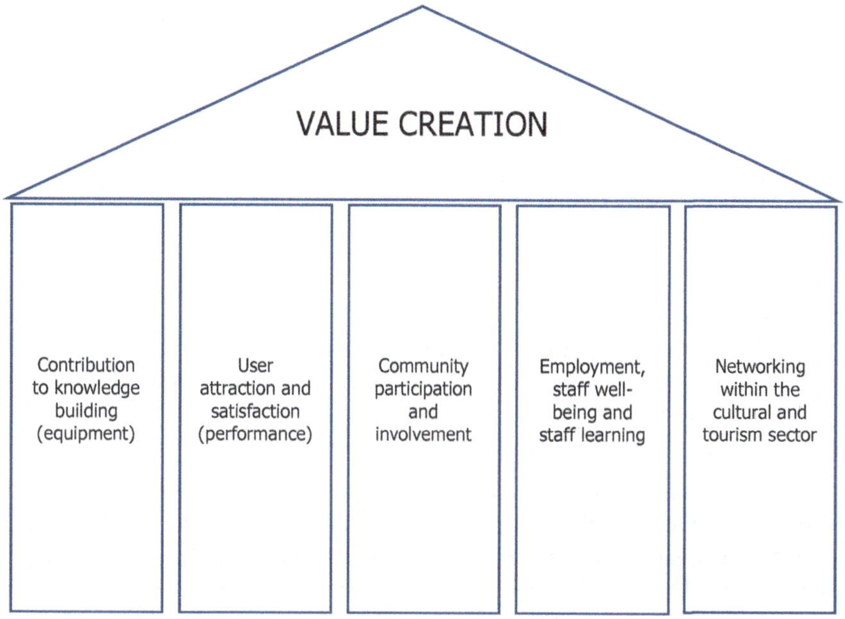

Fig. 1 The five pillars of multidimensional value creation in cultural organisations (own elaboration)

urban regeneration, quality of life and sustainable local development (Evans, 2005; Babkova, 2006). Among culture-led initiatives aimed at promoting sustainable growth, it is worth mentioning the European Capital of Culture (ECoC) (Schneider & Jacobsen, 2019). Launched in 1983 as "European City of Culture", beginning in Athens in 1985, and renamed as "European Capital of Culture" in 1999, the programme currently aims:

> (a) To safeguard and promote the diversity of cultures in Europe and to highlight the common features they share as well as to increase citizens' sense of belonging to a common cultural area; (b) to foster the contribution of culture to the long-term development of cities in accordance with their respective strategies and priorities (European Parliament and Council of the European Union, 2014, Art. 1).

On the one hand, the programme points to the richness of European cultures and the need to boost greater mutual understanding between European citizens. On the other hand, it aims to contribute to a European city's sustainable development, by promoting local citizens' participation, in addition to attracting tourists through cultural events.

In the early stages, the title rewarded established cultural capital cities every year for being "bearers of the quintessential expressions of European high culture" (Sassatelli, 2008, 235): Athens (1985), Florence (1986), Amsterdam (1987), West Berlin (1988) and Paris (1989) "already enjoyed the interest of visitors from all over Europe and beyond and experienced an even greater boost in tourism" (Carrijn,

2019, 27). In 1990, a gradual move towards the rhetoric of culture-led urban regeneration began with the designation of Glasgow, a post-industrial city with weighty socio-economic problems (García, 2005). As summarised by Sassatelli (2008, 236), this marked a significant shift from cities that already proved to *be* European, towards cities that were *becoming* European.

More widely, when tracing the evolution of the ECoC initiative, Lamour and Durand (2019, 41) identified four series of transformations since the beginning of the programme:

1. A progressive downscaling from major international urban centres to more modest urban areas.
2. Increasing use of the initiative as a driver of urban regeneration.
3. A strongly perceived need and will to broaden the ECoC's audience beyond its formerly elite basis.
4. A growing number of ECoCs planned in the European borderlands.

After the 2004 enlargement of the EU, an East-West development divide emerged, in addition to those already existing between North and South, rich and poor, thus helping to reshape the traditional core-periphery distinction and idea (Tölle, 2016). Designated cities in the Eastern European hinterland were able to put themselves in the spotlight (Turşie, 2015, 73) and had an opportunity to "broaden the notion on Europe and European cultural identity by narrating the socialist history, heritage and experience as a part of Europeanness" (Lähdesmäki, 2014b, 491).

In this new comprehensive scenario, *Decision No 445/2014/EU of the European Parliament and of the Council of 16 April 2014 establishing a Union action for the European Capitals of Culture for the years 2020 to 2033* has finally widened the criteria for the assessment of applications, including the following categories: "contribution to the long-term strategy", "European dimension", "cultural and artistic content", "capacity to deliver", "outreach" and "management" (Art. 5).

Taking into account the criteria identified by the European Parliament and the Council of the European Union, we can assume that long-term cultural, social and economic impact, including urban development, depends on: (a) the involvement of local cultural organisations in the implementation of the cultural programme; (b) the development of long-term links between the cultural, economic and social sectors in the candidate city; (c) the involvement of the local population and civil society in terms of audience development, engagement and empowerment; (d) the capacity to enhance local cultural heritage, including by combining it with innovative and experimental cultural expressions.

When analysing the cultural, social and economic impact of previous experiences, most case studies find a positive effect of the ECoC title on tourism flows in the year of the event (Rátz, 2014), but a negative effect on the well-being of the regional population (Steiner et al., 2015); the long-term tourism effects are less clear-cut (Falk & Hagsten, 2017, 2176). Even though some scholars find long-term effects on GDP per capita to be significant, effects on output and employment in accommodation and food services and their related industries are insignificant (Gomes &

Librero-Cano, 2018). Similarly, in the ex post evaluation of the 2012 Maribor ECoC, Srakar and Vecco (2017) find no impact on local employment, despite positive tourism effects for the city in the year of the event.

Another drawback could be the increasing emphasis on creative concepts in ECoC programmes. The debate on local development and urban regeneration is overladen with talk of creativity, which has become "extraordinarily banal" because of "a dominant culture of uncritical acceptance" (Schlesinger, 2006, 4). Thus, when discussing the contribution of creative sectors to economic development, like Oakley (2004, 72), we can argue that "if the notion of the creative industries is problematic, even more so perhaps is the sense that these are sectors that can be replicated and developed pretty much anywhere, without regard for the specifics of place". Similarly,

> when considering the context in which ECOC mega-events take place, the same strategy is unlikely to deliver the same results in different conditions. The conditions of foremost cultural centres of European significance (Florence, Paris) are entirely different from smaller towns with a historical and cultural heritage (Avignon, Pátra, Salamanca) or post-industrial cities looking for an entirely new identity when creating an appropriate culture-led strategy (Essen, Glasgow, Košice) (Urbančíková, 2018, 47).

For this reason, an emphasis on local idiosyncrasies, characteristics and "already existing local and grass-root level cultural phenomena" (Lähdesmäki, 2014b, 493–494) should become a priority in the promotion of European diversity. In the ECoC debate, it is not just the need to ensure a balance between "ephemeral" and "permanent" activities (Liu, 2019), but also the relevance of the temporary events to the cultural framework of the place.

Promoting diversity should also be addressed as a process of diversification that embraces different voices and boosts their active participation and collaboration (Sassatelli, 2008, 239). This calls for multi-level governance—through a decision-making process involving different actors both vertically and horizontally—and a "bottom-up approach in the entire phase of planning through the systematisation of a network of relationships involving community representatives, artists, local stakeholders and research centres" (Tubaldi, 2014b, 1097).

Finally, several other further issues should be tackled: on the one hand, the idea of Europe as being about good quality and high standards (Lähdesmäki, 2014b, 490); on the other, the development of a new relationship between the designated city and its region (Ivona et al., 2019, 10; Jacobsen, 2019, 94).

As our focus in this research is on Hungary, the experience of the city of Pécs, ECoC in 2010 (Lähdesmäki, 2014a), emerges as a typical case and a point of reference for Eastern Europe, whose dark sides are a warning to future experiences (Tubaldi, 2014a). The low economic impact and the lack of long-term vision "to make tourism a strategic issue for the development of the city" (Tubaldi, 2014b, 1093) are confirmed by several factors, facts and figures: the one-year delay in completing the reconstruction of the Zsolnay Cultural Quarter; low-quality materials used for renovation; urban restoration imitating other regions without considering the social and economic characteristics of Pécs; the 35.6% decrease in tourist arrivals between 2001 and 2012 and a similar trend for "nights spent". It has already been argued that the central government played a crucial role in this process because it never saw the initiative as a project to be implemented on a local basis (Tubaldi, 2014b, 1095). Many opportunities for developing a cultural space for citizens were missed: the implementation of a bottom-up approach involving the citizens, local artists, public officials of Pécs and research centres; spotlighting the city's uniqueness; the development of sustainable cultural programmes including through a professional management task force (Song & Lee, 2017). This experience stands out as a lesson for future ECoCs in Hungary and beyond.

3 Research Methodology

The explorative research discussed in this chapter adopts a case study approach, which is coherent with the aim of conducting and in-depth investigation into a specific phenomenon (Eisenhardt, 1989; Stake, 2008; Yin, 2018). Case study research is a qualitative method that allows to examine a single setting through a variety of interpretive lenses, thus revealing the multiple and sometimes contradictory facets of the phenomenon under investigation.

Focusing on the case of Veszprém-Balaton ECoC 2023, the analysis was carried out with a double objective. First, it aimed to investigate if and how cultural institutions in Veszprém and the Balaton region are tackling the issues of stakeholder engagement, value co-creation and measurement, that is, how they address their different stakeholders, how they co-create value with and for them and if and how they measure it. Four categories of stakeholders were considered, namely users, staff, the local community and other organisations in the cultural and tourism sector. Second, it explored opportunities arising from the ECoC initiative.

As a first step, secondary data on the cultural and tourism offer and demand, policy documents, reports, websites etc. were examined, including preparatory and planning documents related to Veszprém-Balaton ECoC 2023.

Then, in September and October 2019, field research was conducted involving a sample of nine cultural organisations (one cultural centre, one library, three theatres, three museums, one art gallery), which are the most important cultural institutions in Veszprém and the Balaton region.[1] Except for the Balaton Museum in Keszthely, they correspond to the cultural organisations already involved in the ECoC process. We also visited the cultural institutions involved in the study.

Following the five-pillar model presented in Sect. 2.2 (Fig. 1), the semi-structured interview investigated the following dimensions:

[1] Agora Cultural Centre (https://agoraveszprem.hu/english-site/), Eötvös Károly County Library (https://www.ekmk.hu/index.php/201-angol/664-angol), Petőfi Theater (https://www.petofiszinhaz.hu/), Hangvilla (https://www.hangvilla.com/), Kabóca Puppet Theatre (https://www.kabocababszinhaz.hu/en), Laczkó Dezső Museum (https://www.ldm.hu/en), Salezianum (https://szalezianum.hu/), Balaton Museum in Keszthely (https://balatonimuzeum.hu/), House of Arts (https://www.muveszetekhaza.hu/)

1. *The contribution of cultural institutions to knowledge building (equipment)*, i.e. opening hours; shows/plays; cultural spaces for laboratories, teaching activities, conferences, temporary exhibitions; on-site and online services.
2. *User attraction and satisfaction (performance)*, i.e. number and trend of visitors/users/audience members, audience composition (places of origin and age); strategies and activities for attracting different audiences; visitor/user/customer satisfaction measurement.
3. *Community participation and involvement*, i.e. partnerships with and programmes for local schools/associations/groups; outdoor activities; volunteers; community and volunteer satisfaction measurement;
4. *Employment, staff well-being and staff learning*, i.e. employees, contract type, length of employment; career opportunity; staff training; staff satisfaction measurement.
5. *Networking within the cultural and tourism sector*, i.e. participation in networks, partnerships with organisations in the cultural and tourism industry; involvement in local policies and city planning.

The interview ended with a focus on expectations for the innovation of the cultural offer arising from Veszprém-Balaton ECoC 2023, i.e. innovation projects and further opportunities.

Two destination management organisations were also interviewed for triangulation purposes (Bryman, 2016) to understand the critical issues that need to be addressed and the potentialities that should be valued in order to ensure a good balance between quality of life and tourism development. A semi-structured interview was adopted to analyse the characteristics of DMOs (i.e. their mission, organisation, strategic objectives and activities) and the tourism system (i.e. tourism flows, main innovations implemented in the last few years, opportunities and threats, the role of culture and cultural heritage, level and quality of networks and partnerships, top-down and bottom-up processes).

Finally, a project manager involved in the ECoC programme was interviewed to get more information about the ECoC process, the main difficulties and challenges faced and changes already in progress.

Face-to-face semi-structured interviews included closed and open questions. The interviews were recorded, transcribed and sent to the interviewees for a final review before being manually analysed.

4 Research Results

4.1 Veszprém-Balaton: A Changing Context

With a population of nearly 60,000, Veszprém lies where the Bakony Hills meet the Balaton Uplands, 110 km southwest of Budapest, 15 km north of Lake Balaton. Built on seven hills, it is also called "the Rome of Transdanubia". It is one of the

Table 1 Tourism system in Veszprém: a SWOT analysis (Veszprém DMO)

Strengths	Weaknesses
• Unique natural and environmental conditions. • Rich built cultural heritage. • Internationally renowned programmes. • Good accessibility; closeness to Balaton and Bakony. • Direct link to Lake Balaton Bike Trail. • Image enhancement power of handball team. • A human-scale city. • Strong local patriotism, civic associations. • Family-friendly attractions, programmes. • Major developments (Zoo, Salesianum, city Centre). • Exemplary TDM organisation. • Strong international relations. • Advanced infrastructure.	• Difficult car transport. • Limited number of parking spaces in the city centre. • Few good quality restaurants in the city centre. • Emptying city Centre, lowering standard of remaining stores. • Limited (timewise) attendance at the exhibition grounds in the castle. • Local decrees to prevent "sitting down" terraces. • Fragmentation of cultural resources. • Low duration of tourist stays. • Lack of high-capacity hotel. • University students are not integrated into the city's sociocultural life.
Opportunities	**Threats**
• ECoC project resources and opportunities. • More attractive opportunities for young people to stay in place. • Boosting domestic tourism. • Chosen destinations based on easy and fast access—More frequent and shorter trips are typical.	• Further ageing of the local community, migration of skilled young people. • Expected economic crisis (shrinking demand in car manufacturing). • Intensified competition among municipalities for equity. • Increasing competition between destinations: Destinations with more and more competitive offerings worldwide.

oldest cities in Hungary, boasting a long history, time-honoured traditions, and a strong local identity, as well as internationally renowned programmes (Table 1). Residents and visitors enjoy its proximity to the largest freshwater lake in Central Europe and the countless leisure activities offered by the lake and the towns along the shore. The hiking and educational trails in the forests of the Bakony Hills and the sites of the Balaton Uplands National Park can also be easily combined with sightseeing in Veszprém. Veszprém primarily focuses on cultural tourism within the urban area, and, thanks to its developed economy, business tourism is significant.

The city's two most visited attractions are the historic Veszprém Castle and Veszprém Zoo, which is located in a green setting. The "city of queens"[2] is a city of festivals. It offers major events in every season: the Festival of Dance, the Gisela Days, the Beer Festival, Kabóciádé Family Festival, the Street Music Festival, the VeszprémFest, Rosé, Riesling and Jazz Days, Saint Michael's Feast Day, the Bakony Expo and the Fragrant Advent are the main events in Veszprém's festival calendar. The listed Baroque complex of Veszprém Castle stands on a 40-metre high dolomite rock. There are museums, art galleries (Csikász Gallery, Modern Art

[2]For centuries, Hungary's queens were crowned by the bishop of Veszprém.

Gallery—Vass László Collection, the Irokéz Collection exhibited in Dubniczay Palace, Castle Gallery) in Castle Street (Vár utca), as well as mainly renovated listed buildings housing various institutions. The gem of Holy Trinity Square is the Archbishop's Palace, an outstanding example of Hungarian Baroque architecture. The Castle hosts Hungary's first bishop's cathedral, Saint Michael's Cathedral, and the Byzantine style Gisela Chapel displaying thirteenth-century frescoes of the Apostles, which also attracts many visitors. The Salesianum Archbishopric's Visitor Centre was also opened in the Castle and offers interactive exhibitions, a violin-making workshop, a glass-floored room with an herb exhibition and many events. Those who wish to view the most prominent sights from above can do so from the 48-m high Fire Tower or from the lookout terrace of the Heroes' Gate (Castle Gate) Visitor Centre.

Even though it is not far from Lake Balaton, the Balaton tourist destination did not involve the city in its policies and activities until 2017. In October 2017, Veszprém, which is the second most common reason after the lake for visiting the Balaton region, was finally included in the Balaton tourism development area (National Tourism Development Strategy 2030 by MTÜ—Hungarian Tourism Agency). The Veszprém DMO is currently working on its strategic plan for the 2020–2025 period. The purpose is to turn the city into the most liveable cultural and tourist centre of the Balaton destination, leading in quality and sustainability within 5 years.

An analysis of the tourism system reveals certain drawbacks. First, Veszprém has rather limited four-star hotel capacity and lacks enticing spa and wellness facilities. In the highest season (summer), demand exceeds supply, while spending is concentrated by the Lake Balaton (with unreasonably priced accommodation and services). Moreover, tourists only stay for short periods (two to two and a half nights per tourist). Finally, the city's uniqueness is poorly communicated. The "city of queens" slogan has, unfortunately, no real content, owing to a lack of historical evidence and artefacts that would make it more appealing and understandable to visitors. The slogan should be reformulated and the city should reinvent itself as a cultural city.

The main external threats are the ageing population, the migration of skilled young people and increasing competition with similar destinations. An economic crisis is also expected due to shrinking demand in the car manufacturing industry, which several of the city's large firms serve.

In 2018, the European Year of Cultural Heritage, Veszprém and the Lake Balaton region won the title of ECoC 2023. Thanks to the ECoC project, Veszprém could become the most important cultural site in the north of the Balaton region, even though Hungarians still associate the city with the Castle and the Veszprém Zoo. The event could help provide attractive opportunities for young people and boost domestic tourism. Several infrastructure projects for sustainable transport (i.e. bicycle lanes and buses) are being implemented, along with optimisation of train/boat changes. Every year, investments are made not only in the zoo, but also in sports tourism (e.g. the new swimming pool near the Veszprém arena) and in enhancing the attractiveness of the historic city centre (e.g. the renovation of historic buildings).

Table 2 Veszprém 2023 ECoC's vision (Veszprém Candidate City 2023 European Capital of Culture, 2018, 10)

Vision	
A competitive place to live in—a high-quality cultural "offer" that attracts people to live here and businesses to invest	*Smart heritage protection*—cultural preservation and creative arts and industries are working in close synergy
A quality tourism economy—More integrated regional tourism development and a major shift towards quality, year-round cultural tourism	*Culture as a tool to build communities*—Strengthened neighbourhood cohesion and cooperation of communities
A stronger base for creative industry—Supporting backgrounds for creative enterprises to flourish and grow	*Audience development and artistic education*—Educational framework for future artists and a new approach to cultural outreach both in Veszprém and the Balaton region

The keyword of Veszprém ECoC 2023 is *Beyond*, and one of the first objectives is to reach out to new audiences (Lőrincz & Raffay, 2019). This means bringing culture to people and widening the understanding of culture: mindful of the Pécs experience, by creating content rather new infrastructures. The project presented in the bid book (Veszprém Candidate City 2023 European Capital of Culture, 2018) is based on the vision set out below (Table 2).

Looking forward to 2023, it is worth analysing how cultural institutions are approaching this vision, what they are currently doing, what is still missing and how ECoC could trigger innovation processes to increase value creation for different categories of stakeholders, thus promoting quality of life and sustainable tourism development.

4.2 *Cultural Institutions in Veszprém and the Balaton Region*

In this context, the research results show that the contribution of cultural institutions to quality of life and the sustainable tourism development of the area is increasing and can find significant opportunities in Veszprém-Balaton ECoC 2023.

4.2.1 Contribution to Knowledge Building

Cultural institutions in Veszprém and the Balaton region regularly provide their services and performances. Even though buildings need renovation, the infrastructure and facilities are adequate for their core activities and are also equipped with spaces for other cultural activities.[3] Only one institution which functions as a

[3]The Agora Cultural Centre for Veszprém's residents, for example has a theatre room (300 seats), more than ten presentation rooms (hosting from 20 to 100 people), language laboratories, computer rooms, one room/open space for temporary exhibitions, two handcrafts rooms, three meetings

"travelling theatre" (Kabóca Puppet Theatre) has just two rooms and has to rent the main hall, the stage and the technicians. The Petőfi Theater presents plays every season for people with disabilities in the region, and offers sign language and subtitles for the deaf and narration for the blind.

There are concerns regarding the opening hours of museums and art galleries, which are closed on Sundays, like the Salezianum in the winter season and some temporary exhibitions at the Laczkó Dezső Museum and the House of Arts.

Other weaknesses relate to online services. Some websites provide information in Hungarian only (Petőfi Theater, Hangvilla, House of Arts), while others are partially translated into English (Agora Cultural Centre, Salezianum). The Kabóca Puppet Theatre and the Laczkó Dezső Museum have a new website in English and Hungarian, while the Balaton Museum has recently restyled its website, but at the moment it is only in Hungarian. One exception is the Eötvös Károly County Library, whose website is translated—even if only partially—in English, German, French and Finnish.

At the moment, partly because of a lack of dedicated staff, these institutions only have a light presence and activity on social media, mainly for posting events or exhibitions.

4.2.2 User Attraction and Satisfaction

The main users of cultural institutions in Veszprém and the Balaton region, like the Agora Cultural Centre and the Eötvös Károly County Library, are local residents who mostly come from the city and the county. Except for the Kabóca Puppet Theatre, which organises events specifically for children and families, the concerts, performances, shows and other events organised by theatres in Veszprém mainly attract people aged above 45 years. Approximately 40% of users of the Agora Cultural Centre are over 65 years. Museums attract 25,000–30,000 visitors per year, while the House of Arts receives fewer than 20,000 visitors per year. Visitors come mostly from Hungary (70% at the Balaton Museum) and strategies and activities to attract audiences mainly target schools and teachers. The Balaton Museum and Laczkó Dezső Museum go to schools in the surrounding area to present their offer. For the Balaton Museum, it is difficult to draw other clusters of visitors in a city where the main attraction for tourists is the Castle. Visitors to the Salezianum come from all over the country, but events and book presentations mostly attract people from Veszprém.

Cultural offerings are promoted online, using Facebook and other online channels, or through programmes/brochures sent to different institutions, e.g. theatres,

rooms for the public and associations, one small library. The Eötvös Károly County Library has one event hall for up to 100 people, one meeting room for up 25 people, with fixed tables, and a community space for up to 50 people.

libraries, museums, shops, and schools. Newspaper advertisements and outdoor signage around the city are also employed.

Unfortunately, as clearly stated by some cultural institutions (Laczkó Dezső Museum, House of Arts and Hangvilla), collecting and analysing data about audience composition is not very common. In general, customer satisfaction is measured sporadically, and mainly for temporary exhibitions or EU programmes. In the past, some institutions, such as the Eötvös Károly County Library, measured customer satisfaction occasionally, but this activity never became systematic.

4.2.3 Community Participation and Involvement

All cultural institutions in Veszprém and the Balaton region have close relationships with schools in the city and the surrounding area, less so with the local university. In some cases, such as the Agora Cultural Centre and the Eötvös Károly County Library, there are partnerships with primary and high schools and with nearby villages. Partnerships with other cultural institutions are also well established. With the exception of Salezianum, all the cultural institutions organise outdoor events and activities at other venues and, in some cases, in nearby villages.

Even though it is not measured, volunteer participation is quite high for all cultural organisations, except for Hangvilla, which works with staff within its own network to cover the various tasks. Volunteers are mainly students, within the framework of compulsory community service for secondary school students. Nevertheless, adult volunteers (mostly pensioners) are also involved organising the cultural offer.

4.2.4 Employment, Staff Well-being and Staff Learning

The cultural institutions investigated are mainly public and are funded by the city, the county and the government; only Salezianum is a Church institution. They have between 4 (Hangvilla) and 102 employees (Petőfi Theater). In general, there is no financial motivation to work in the cultural sector: people love their jobs and work environment, even though there is not much opportunity to move up to higher positions. In some cases, employees are not satisfied with their salary level (Agora Cultural Centre and Petőfi Theater). Not all institutions provide staff training and staff satisfaction is not measured.

4.2.5 Networking Within the Heritage and Tourism Sector

Only some cultural institutions in Veszprém and the Balaton region are part of the networks. These networks are sectorial—professional or scientific. The Eötvös Károly County Library belongs to the Hungarian Library Association, the Balaton Museum to the Association of Hungarian Museums, the Laczkó Dezső Museum and

Table 3 Main critical points and areas for improvement for each research dimension (own elaboration)

Research dimensions	Main critical points and areas for improvement
Contribution to knowledge building	• Some buildings need renovation. • Some museums and art galleries are closed on Sunday during the winter season. • Institutional websites provide information only in Hungarian or partially translated into English. • Cultural institutions have a light presence and activity on social media.
User attraction and satisfaction	• Users are mainly local residents. • The cultural offer mainly attracts people aged over 45 years. • Collecting and analysing data about audience composition is not a common activity. • Customer satisfaction is measured sporadically, and mainly for temporary exhibitions or EU programmes.
Community participation and involvement	• Local university students are not involved in the city cultural offer yet. • Volunteer participation is not measured.
Employment, staff well-being and staff learning	• There is no financial motivation to work in the cultural sector. • In some cases, employees are not satisfied with their salary level. • Not all institutions provide staff training. • Staff satisfaction is not measured.
Networking within the heritage and tourism sector	• Only some cultural institutions in Veszprém and the Balaton region are part of networks, i.e. professional or scientific networks. • At a local level, cooperation is limited to tourism promotion and the organisation of special events.

the Balaton Museum to the network of rural museums, the Kabóca Puppet Theatre to international networks of small cities for the organisation of festivals. Moreover, the House of Arts cooperates with museums and galleries at national and international levels.

At a local level, cooperation is limited to tourism promotion and the organisation of special events.

The main critical points and areas for improvement are summarised in Table 3.

4.2.6 Expectations and Opportunities for the Innovation of the Cultural Offer Arising from Veszprém-Balaton ECoC 2023

Except for the Balaton Museum, which is located in Keszthely, all the cultural institutions investigated in the research have already been involved in the organisation of Veszprém-Balaton ECoC 2023 and have high expectations of the event. The main hopes are for: the renovation or refurbishment of the building so that more people from the district can be hosted (Agora Cultural Centre); a more comfortable venue (Eötvös Károly County Library); better access for people with disabilities and the installation of modern and interactive permanent exhibitions (Laczkó Dezső Museum); the ability to organise cultural events to attract younger generations (Petőfi Theater).

The House of Arts is going to create a future lab from one of its derelict buildings, whereas the Kabóca Puppet Theatre is planning to provide a place to reopen the "Small House" and the Laczkó Dezső Museum would like to go to small villages to do archaeological research with other museums and exhibit the finds to the public. Hangvilla also wishes to involve new places/living areas in its activities and the Eötvös Károly County Library wants to open more street libraries (glass cabinets containing books accessible to anyone who wants to borrow them) in several spots around the city and provide venues for various ECoC events.

As stated by the House of Arts, the question is how to engage with the slogan *Beyond*, and provide opportunities for foreign artists to come and for local artists to go abroad and make connections through computer techniques and art media.

The Kabóca Puppet Theatre recognises that audience development is a general mandate that calls for the organisation of workshops and conferences on cultural marketing aimed at raising the professional skills of staff members. Finally, the cultural institutions would like to collaborate on "tandem" projects.

5 Discussion

The research results show that the available unused spaces can be repurposed and old buildings can be maintained and renovated instead of being replaced by newly constructed cultural spaces. The cultural institutions are trying to reach out to new audiences and strengthen their relationship with their communities. However, the route to making a significant cultural, social and economic impact requires some social innovations. First, the cultural organisations should improve their service culture, which acts as the *software* that ensures the smooth running of the renovated buildings (*hardware*). If online and on-site information is not provided in English, access to local services can be difficult for immigrant communities and international students. Moreover, if the city aims to be the most important cultural site in the northern Balaton region, museums and temporary exhibitions should be open on Sundays to attract national and international tourists.

A new 2.0, or even 3.0, approach to audience development is needed (Sacco et al., 2018), one which considers marketing a strategy, not merely a tool (Boorsma, 2006), and customer satisfaction measurement a means for accomplishing the core mission, rather than an obligation only for European projects. Even though the cultural institutions promote many activities for schools and children, the same cannot be said for their relationship with university students. More attention could be devoted to strategies to engage and empower them in the cultural offering and the life of the city. As argued by the Bakony-Balaton DMO tourism manager, Veszprém's cultural sites should be promoted and presented so that people are drawn to them, rather than in a boring way.

At a local level, the low degree of cooperation and coordination between public and private institutions should be highlighted, particularly between museums and heritage sites located in the Castle area, the city's historical centre. Among the main weaknesses, it is worth mentioning the management of museums that belong to the Church. There is a partnership between the Veszprém DMO and Church cultural institutions (e.g. tickets to visit religious buildings and their museums), but it is difficult to achieve good results because of an apparent conflict of interests (heritage maintenance vs growing market). As argued by the director of the Veszprém DMO, "in order to generate meaningful outcomes, we need genuine openness on the part of the Church". Even though the public and Church institutions have a good relationship and coexist amiably, there is no real cooperation. However, ownership should not be an obstacle and public and private institutions and tourist attractions should work together.

In this context, the ECoC programme could be a great opportunity to invest in people, by bringing the different stakeholders together and connecting the ECoC project to the region's local population (Farmaki et al., 2019). Cooperation could attract more tourism through culture. As argued by one of the managers of the ECoC programme, it is also about community and capacity building. The cultural institutions are involved and are asked to define a strategy and make a connection with the city. The challenge is to innovate by integrating the project into the life and routines of the city, rather than as a separate entity.

Moreover, education plays a crucial role. In Veszprém the change is in progress. The ECoC programme is organising English courses and working to create an English website for the city's cultural institutions, as well as opening a volunteering platform to improve local participation.

Finally, we can argue that the ECoC programme should become a means to trigger *process innovations* besides *product innovations*. In order to ensure long-term benefits, cultural organisations should be supported in understanding their needs and priorities, by learning and implementing a customer-centric approach and building new relationships with their internal and external stakeholders (Fig. 2).

6 Concluding Remarks, Limitations and Further Research

This research aimed to understand the main weaknesses of the urban cultural policy and management in Veszprém and the Balaton region (*RQ1*), the current contribution of the cultural sector to the quality of life and sustainable tourism development of the city and its surrounding areas (*RQ2*), and the expected innovations that ECoC could and should trigger to bring culture to people and widen its understanding over the long term (*RQ3*). To this aim, the study proposed a model based on five pillars for investigating multidimensional value creation (Fig. 1).

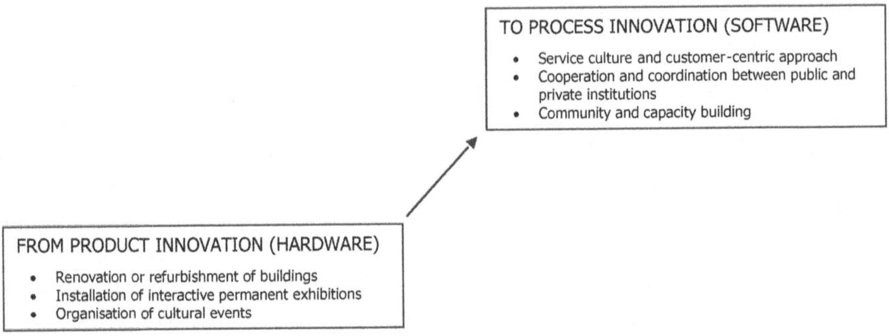

Fig. 2 Value creation through cultural programmes: from *product innovations* to *process innovations* (own elaboration)

The analysis shows different levels of involvement and networking that could benefit from improvements and highlights the need for incremental innovations—sociocultural rather than technological, organisational rather than structural (Chun, 2020)—to promote long-term sustainable development.

The research results contribute to research on ECoCs, showing the need to improve the current management and organisation of the cultural offer (Aquilino et al., 2019), by rethinking the way value creation mechanisms can be operated and innovated to achieve sustainability (Carlucci, 2018). Following Schiuma and Lerro (2017, 3), we can argue that "arts and cultural organisations need to undertake a development journey towards the capacity of managing and innovating the ways how they operate, interact with stakeholders, define their value objectives, acquire and deploy resources, and support economically their artistic, cultural and social programmes". In this context, the role of business models for innovating arts and cultural organisations could be further investigated.

Although the case study methodology does not allow the generalisability of results, the current experience of the cultural organisations in Veszprém and the Balaton region and the process started within the framework of ECoC 2023 provide policy and managerial suggestions for similar cases, thus contributing to the scientific debate on the sustainability of cultural mega-events and the opportunities and threats they can generate in historical cities (Ponzini, 2022). In addition, the model proposed in this research (Fig. 1) can be applied to other cultural organisations to evaluate their capability to create multidimensional value for different stakeholders. Further research should include longitudinal analysis by monitoring, measuring and evaluating the process triggered by ECoC in Veszprém and the Balaton region and should also analyse cultural events in addition to cultural institutions.

Even during and after the COVID-19 pandemic, the connection with local communities and proximity to tourists is emerging as an aspect that can be improved (Flew & Kirkwood, 2020; Romagosa, 2020). As argued by Pratt (2020, 1), "how the cultural economy will get rebuilt is an open question: it provides scholars with an (unlooked for, and unwanted) opportunity to witness the 'cultural heart attack' at first hand, it also provides a unique opportunity to address the systemic problems of precarity and injustice in the urban cultural economy and (hopefully) to change it".

Acknowledgements This research has been supported by the European Union and Hungary and co-financed by the European Social Fund through the project EFOP-3.6.2-16-2017-00017, titled "Sustainable, intelligent and inclusive regional and city models" (Research topic: "Aspects of the quality of life by the Lake Balaton as a living space"; Research field: II.2).

References

Anton, C., Camarero, C., & Garrido, M.-J. (2018). Exploring the experience value of museum visitors as a co-creation process. *Current Issues in Tourism, 21*(12), 1406–1425.

Aquilino, L., Armenski, T., & Wise, N. (2019). Assessing the competitiveness of Matera and the Basilicata region (Italy) ahead of the 2019 European Capital of Culture. *Tourism and Hospitality Research, 19*(4), 503–517.

Babkova, S. (2006). Cultural programs for economic development: A comparison of European and Russian models. *The Journal of Arts Management, Law, and Society, 36*(3), 197–212.

Bodo, C., & Sacco, P. L. (Eds.). (2017). Cultura, salute, benessere / Culture, health, wellbeing. *Economia della Cultura, 27*(2).

Bollo, A. (2013). *Measuring museum impacts*. Report 3. LEM – The Learning Museum.

Bollwerk, E. (2015). Co-creation's role in digital public archaeology. *Advances in Archaeological Practice, 3*, 223–234. https://doi.org/10.7183/2326-3768.3.3.223

Boorsma, M. (2006). A strategic logic for arts marketing: Integrating customer value and artistic objectives. *International Journal of Cultural Policy, 12*(1), 73–92.

Boorsma, M., & Chiaravalloti, F. (2010). Arts marketing performance: An artistic-mission-led approach to evaluation. *The Journal of Arts Management, Law, and Society, 40*(4), 297–317.

Bryman, A. (2016). *Social research methods*. Oxford University Press.

Brown, K. (2019). Museums and local development. An introduction to museums, sustainability and well-being. *Museum International, 71*(3–4), 1–13. https://doi.org/10.1080/13500775.2019.1702257

Carlucci, D. (2018). Fostering excellence in business model management in arts and cultural organisations: Insights from focus group research across Europe. *Measuring Business Excellence, 22*(1), 14–30.

Carrijn, F. (2019). Universities in Europe – Europe in the universities. In W. Schneider & K. Jacobsen (Eds.), *Transforming cities. Paradigms and potentials of urban development within the "European Capital of Culture"* (pp. 25–33). OLMS.

Cerquetti, M. (2017). From value assessment to public value creation and measurement in the museum sector. Theoretical approaches and critical issues in a changing world. In M. Cerquetti (Ed.), *Bridging theories, strategies and practices in valuing cultural heritage* (pp. 121–147). Eum.

Chun, L. M. (2020). Cultural product innovation strategies adopted by the performing arts industry. *Review of Managerial Science, 15*, 1139. https://doi.org/10.1007/s11846-020-00393-1

Cicerchia, A. (2017a). Assessing the impact of culture on Well-being: From indicators to narratives, and back. In M. Cerquetti (Ed.), *Bridging theories, strategies and practices in valuing cultural heritage* (pp. 181–195). Eum.

Cicerchia, A. (2017b). *Culture and wellbeing. Theory, methodology, stories and other challenges: An itinerary*. Reflection paper n. 4. Culture action Europe.

Crossick, G., & Kaszynska, P. (2016). *Understanding the value of arts & culture. The AHRC cultural value project*. Arts and Humanities Research Council.

Eisenhardt, K. M. (1989). Building theories from case study research. *Academy of Management Review, 14*(4), 532–550.

European Parliament, & Council of the European Union. (2014). *Decision No 445/2014/Eu of the European Parliament and of the Council of 16 April 2014 establishing a Union action for the European Capitals of Culture for the years 2020 to 2033 and repealing Decision No 1622/2006/EC.* Accessed June 13, 2022, from https://eur-lex.europa.eu/legal-content/EN/TXT/PDF/?uri=CELEX:32014D0445&from=EN

Evans, G. (2005). Measure for measure: Evaluating the evidence of culture's contribution to regeneration. *Urban Studies, 42*(5/6), 959–983.

Falk, M., & Hagsten, E. (2017). Measuring the impact of the European Capital of Culture programme on overnight stays: Evidence for the last two decades. *European Planning Studies, 25*(12), 2175–2191. https://doi.org/10.1080/09654313.2017.1349738

Farmaki, A., Christou, P., Saveriades, A., & Spanou-Tripinioti, E. (2019). Perceptions of Pafos as European Capital of Culture: Tourism stakeholder and resident perspectives. *International Journal of Tourism Research, 21*(2), 234–244.

Flew, T., & Kirkwood, K. (2020). The impact of COVID-19 on cultural tourism: Art, culture and communication in four regional sites of Queensland, Australia. *Media International Australia*, 1–5. https://doi.org/10.1177/1329878X20952529

García, B. (2005). Deconstructing the City of Culture: The long-term cultural legacies of Glasgow 1990. *Urban Studies, 42*(5/6), 841–868.

Gomes, P., & Librero-Cano, A. (2018). Evaluating three decades of the European Capital of Culture programme: A difference-in-differences approach. *Journal of Cultural Economics, 42*, 57–73. https://doi.org/10.1007/s10824-016-9281-x

Grossi, E., Tavano Blessi, G., Sacco, P. L., & Buscema, M. (2011). The interaction between culture, health and psychological well-being: Data mining from the Italian culture and well-being project. *Journal of Happiness Studies*, March. https://doi.org/10.1007/s10902-011-9254-x

Ivona, A., Rinella, A., & Rinella, F. (2019). Glocal tourism and resilient cities: The case of Matera 'European Capital of Culture 2019'. *Sustainability, 11*(4118), 1–12. https://doi.org/10.3390/su11154118

Jacobsen, K. (2019). Out into the region? In W. Schneider & K. Jacobsen (Eds.), *Transforming cities. Paradigms and potentials of urban development within the "European Capital of Culture"* (pp. 93–97). OLMS.

Lähdesmäki, T. (2014a). Discourses of Europeanness in the reception of the European Capital of Culture events: The case of Pécs 2010. *European Urban and Regional Studies, 21*(2), 191–205.

Lähdesmäki, T. (2014b). European capital of culture designation as an initiator of urban transformation in the post-socialist countries. *European Planning Studies, 22*(3), 481–497. https://doi.org/10.1080/09654313.2012.752438

Lamour, C., & Durand, F. (2019). European capitals of culture across state borders. In W. Schneider & K. Jacobsen (Eds.), *Transforming cities. Paradigms and potentials of urban development within the "European Capital of Culture"* (pp. 41–50). OLMS.

Lee, D. (2013). How the arts generate social capital to Foster intergroup social cohesion. *The Journal of Arts Management, Law, and Society, 43*(1), 4–17. https://doi.org/10.1080/10632921.2012.761167

Liu, Y.-D. (2019). Event and sustainable culture-led regeneration: Lessons from the 2008 European Capital of Culture, Liverpool. *Sustainability, 11*(1869), 1–18. https://doi.org/10.3390/su11071869

Lőrincz, K., & Raffay, Á. (2019). 'Beyond, azaz túllépni saját magunkon' – a turizmus szerepe a Veszprém2023 Európai Kulturális Főváros projektben. *Turisztikai És Vidékfejlesztési Tanulmányok, 4*(2), 18–38.

McCarthy, K. F., Ondaatje, E. H., Zakaras, L., & Brooks, A. (2004). *Gifts of the muse. Reframing the debate about the benefits of the arts*. RAND.

Oakley, K. (2004). Not so cool Britannia: The role of the creative industries in economic development. *International Journal of Cultural Studies, 7*(1), 67–77.

Oakley, K., O'Brien, D., & Lee, D. (2013). Happy now? Well-being and cultural policy. *Philosophy & Public Policy Quarterly, 31*(2), 18–26.

O'Brien, D. (2010). *Measuring the value of culture: A report to the Department for Culture Media and Sport*. Department for Culture, Media and Sport.

O'Brien, D. (2015). Cultural value, measurement and policy making. *Arts & Humanities in Higher Education, 14*(1), 79–94. https://doi.org/10.1177/1474022214533892

OECD, & ICOM. (2018). *Culture and local development: Maximising the impact*. Guide for Local Governments, Communities and Museums.

Ponzini, D. (2022). Introduction to the special issue – Cultural mega-events and heritage: Challenges for European cities. *European Planning Studies, 30*(3), 427–436. https://doi.org/10.1080/09654313.2021.1959719

Pratt, A. (2020). COVID – 19 impacts cities, cultures and societies. *City, Culture and Society, 21*, 1–2. https://doi.org/10.1016/j.ccs.2020.100341

Rátz, T. (2014). Az EKF-turizmus mint lehetséges niche termék. A kulturális turizmus egy különleges területének vizsgálata. *Turizmus Bulletin, 16*(2), 24–32.

Romagosa, F. (2020). The COVID-19 crisis: Opportunities for sustainable and proximity tourism. *Tourism Geographies, 22*(3), 690–694. https://doi.org/10.1080/14616688.2020.1763447

Ross, D. (2020). Towards meaningful co-creation: A study of creative heritage tourism in Alentejo, Portugal. *Current Issues in Tourism, 23*(22), 2811–2824.

Sacco, P. L., Ferilli, G., & Tavano Blessi, G. (2018). From culture 1.0 to culture 3.0: Three socio-technical regimes of social and economic value creation through culture, and their impact on European cohesion policies. *Sustainability, 10*, 3923, 1–23. https://doi.org/10.3390/su10113923

Sassatelli, M. (2008). European cultural space in the European cities of culture. *European Societies, 10*(2), 225–245.

Schiuma, G., & Lerro, A. (2017). The business model prism: Managing and innovating business models of arts and cultural organisations. *Journal of Open Innovation: Technology, Market, and Complexity, 3*(3), 1. https://doi.org/10.1186/s40852-017-0066-z

Schlesinger, P. (2006). Creativity: From discourse to doctrine? *E-Compós. Revista da Associação Nacional de Programas de Pós-graduação em Comunicação*, 1–23.

Schneider, W., & Jacobsen, K. (Eds.). (2019). *Transforming cities. Paradigms and potentials of urban development within the "European Capital of Culture"*. OLMS.

Scott, C. A. (Ed.). (2013). *Museums and public value. Creating sustainable futures*. Routledge.

Song, E. J., & Lee., J. O. (2017). European capital of culture program in Hungary: Focused on Pécs in terms of culture-led urban regeneration. *Advanced Science Letters, 23*, 1595–1598. https://doi.org/10.1166/asl.2017.8600

Spence, D., Wareham, T., Bressey, C., Bam-Hutchison, J., & Day, A. (2013). The public as co-producers: Making the London, sugar and slavery gallery, Museum of London Docklands. In C. A. Scott (Ed.), *Museums and public value. Creating sustainable futures* (pp. 95–109). Routledge.

Srakar, A., & Vecco, M. (2017). Ex-ante versus ex-post: Comparison of the effects of the European Capital of Culture Maribor 2012 on tourism and employment. *Journal of Cultural Economics, 41*(2), 197–214. https://doi.org/10.1007/s10824-017-9294-0

Stake, R. E. (2008). Qualitative case studies. In N. K. Denzin & Y. S. Lincoln (Eds.), *Strategies of qualitative inquiry* (3rd ed., pp. 119–149). Sage.

Steiner, L., Frey, B., & Hotz, S. (2015). European capitals of culture and life satisfaction. *Urban Studies, 52*(2), 374–394.

Tölle, A. (2016). Transnationalisation of development strategies in east central European cities: A survey of the shortlisted polish European Capital of Culture candidate cities. *European Urban and Regional Studies, 23*(3), 374–388.

Tubaldi, M. (2014a). *An attempt at socio-economic regeneration through culture in a central European city: The case of Pécs*. Discussion papers, 94. Institute for Regional Studies Centre for Economic and Regional Studies Hungarian Academy of Sciences.

Tubaldi, M. (2014b). The ECoC title for a central European City: The case of Pécs 2010. In *5th Central European Conference in Regional Science – CERS* (pp. 1087–1099).

Tuck, F., & Dickinson, S. (2014). *The economic impact of museums in England*, prepared by TBR's Creative & Cultural Team in partnership with Pomegranate LLP and Scott Dickinson & Partners Ltd. Accessed May 10, 2021, from https://www.artscouncil.org.uk/sites/default/files/download-file/Economic_Impact_of_Museums_in_England_report.pdf

Turşie, C. (2015). Re-inventing the Centre-periphery relation by the European capitals of culture. Case-studies: Marseille-Provence 2013 and pecs 2010. *Eurolimes, 19*, 71–84.

Urbančíková, N. (2018). European capitals of culture: What are their individualities? *Theoretical and Empirical Researches in Urban Management, 13*(4), 43–55.

Veszprém Candidate City 2023 European Capital of Culture. (2018). Bid Book. Veszprém-Balaton 2023 Ltd.

Vermeulen, M., & Maas, K. (2021). Building legitimacy and learning lessons: A framework for cultural organizations to manage and measure the social impact of their activities. *The Journal of Arts Management, Law, and Society, 51*(2), 97–112. https://doi.org/10.1080/10632921.2020.1851839

Viganò, F., & Lombardo, G. (2018). L'impatto sociale generato dai musei. L'applicazione della metodologia SROI. In A. Luigini & C. Panciroli (Eds.), *Ambienti digitali per l'educazione all'arte e al patrimonio* (pp. 332–350). FrancoAngeli.

Whelan, G. (2015). Understanding the social value and well-being benefits created by museums: A case for social return on investment methodology. *Arts & Health. An International Journal for Research, Policy and Practice, 7*(3), 216–230. https://doi.org/10.1080/17533015.2015.1065574

Williams, M., Biggemann, S., & Tóth, Z. (2020). Value creation in art galleries: A service logic analysis. *Australasian Marketing Journal, 28*, 47–56. https://doi.org/10.1016/j.ausmj.2019.08.002

Yin, R. (2018). *Case study research and applications. Design and methods* (6th ed.). Sage.

Eliminating Overtourism in UNESCO Destinations: A Case Study from Slovakia

Diana Kvasnová and Matúš Marciš

Abstract UNESCO heritage sites represent one of the most important resources on which travel and tourism are based. Focusing on quantitative growth causes problems with environmental sustainability but also with the quality of life of local communities. In this context, the term overtourism has begun to be used in recent literature. Overtourism in cultural heritage destinations led to the search for new approaches in cultural destination management based on sustainability, taking into consideration the sociocultural dimensions rather than quantitative indicators. Implementation of various innovations in cultural destinations leads to sustainable development. Therefore, the chapter builds on these statements and presents a new approach to the management of UNESCO destinations. The research methodology is based on the exploratory case study approach to investigate the methods and strategies used to minimise overtourism in UNESCO destinations. The study adopts qualitative and quantitative research techniques. The study presents a conceptualisation of the phenomenon of overtourism in UNESCO destinations and offers the theoretical framework of a more sustainable approach to destination management based on innovations. The framework is tested in the Central European UNESCO destination Banská Štiavnica, Slovakia.

> Good ideas are not adopted automatically. They must be driven into practice with courageous patience. (Admiral H. G. Rickover)

D. Kvasnová (✉) · M. Marciš
Matej Bel University in Banská Bystrica/Univerzita Mateja Bela v Banskej Bystrici, Banská Bystrica, Slovakia
e-mail: diana.kvasnova@umb.sk; matus.marcis@umb.sk

E. Borin et al. (eds.), *Cultural Leadership in Transition Tourism*, Contributions to Management Science, https://doi.org/10.1007/978-3-031-14121-8_12

1 Introduction

Cultural tourism has changed dramatically since the late 1970s, when it was recognised as a distinct product category. Currently, tourism marketers and tourism researchers realised that some tourists travel especially to better understand the cultural heritage (Du Cros & McKercher, 2015). Tourism growth has been accompanied by seasonality of demand, concentration of tourism flows, leading to oversaturation, exceeding carrying capacity and environmental sustainability (Capocchi et al., 2019). As Croce (2018) predicts, although shocks and crises will continue to negatively affect tourism, the long-term trend over the next decade is tourism growth. Now, sustainable tourism presents a major paradigm in tourism research and industry practises, based on the balance among economic, sociocultural and environmental sustainability (Mihalic, 2020). UNESCO sites are among the most popular tourist destinations and increasingly attract tourists from all over the world. Cities like Venice, Rome, Dubrovnik and Amsterdam are victims of their own tourism success (Séraphin et al., 2018). UNESCO destinations must face overtourism and tourism saturation. The need to manage unsustainable tourism flows is today strongly recognised as a priority (UN, 2015; UNWTO, 2018a).

Overtourism can be characterised as the acceleration of tourism supply and demand in the destination. It is often accompanied by negative impacts on social and economic development and the environment. A tipping point has been reached in 2015, when numerous media reports on tourism growth, overtourism, and tourism phobia in various destinations. These issues have also led to a strong focus on sustainable development in UNESCO destinations. The challenge is to manage tourism sustainably for residents, tourists and businesses (Goodwin, 2017). However, overtourism and its related issues are not yet fully defined academically and remain subject to various interpretations (Koens et al., 2018; Dodds & Butler, 2019; Gretzel, 2019; Mihalic, 2020) from different disciplines. Systematic research on the monitoring and evaluation of practices aimed at abating and avoiding overtourism and tourists saturation is still scarce (Peeters et al., 2018; Scuttari et al., 2020). Due to the growing need to design a sustainable tourism development model, the present chapter considers the research gap in the literature and the knowledge around the topic of overtourism in UNESCO destinations.

The main purpose of the chapter is to conceptualise the phenomenon of overtourism in UNESCO destinations and to offer the theoretical framework towards a more sustainable approach to destination management based on innovations in tourism destinations (Du Cros & McKercher, 2015; Goodwin, 2017; Postma & Schmuecker, 2017; UNWTO, 2018a; Peeters et al., 2018; Bertocchi & Visentin, 2019; Milan, 2019; Capocchi et al., 2019; Mihalic, 2020; Pechlaner et al., 2020). The research methodology is based on the exploratory case study approach to investigate the methods and strategies used to eliminate overtourism in the UNESCO town Banská Štiavnica, Slovakia. What the current literature does not explore is how innovations resulting from activities of destination management organisations can decrease overtourism and disperse tourist flows. Our chapter is as follows: (1) The conceptualisation of the most significant approaches to cultural tourism, overtourism and innovations of destination management in UNESCO destinations. (2) The methodology part. (3) The results are based on the explorative case study of the UNESCO destination Banská Štiavnica, Slovakia and discussion and (4), the conclusion about the future perspectives.

2 Literature Review

Due to increased prosperity, digitalisation and technological advancements, cultural tourism has boomed over recent decades. Cultural tourism is one of the oldest forms of tourism. People have been traveling for cultural reasons since the days of the ancient Romans visiting Greece and Egypt. There is still much debate about what cultural tourism is and this fundamental question is difficult to answer. Over the years, several definitions have been proposed that contain some elements of cultural tourism. Early definitions place cultural tourism within a broader framework of tourism or as a form of special interest (Du Cros & McKercher, 2015). The second group understands cultural tourism as the need to learn about cultural heritage and as a major travel motive (Whyte et al., 2012). The third group of definitions adopts an experiential approach that argues that motivation alone does not fully encapsulate the extent of this sector. According to this, cultural tourism involves a unique experience and the special character of the destination (Blackwell, 1997; Schweitzer, 1999). The last group of definitions adopts a more marketing-oriented approach. These definitions explain cultural tourism as a product class defined by four factors: tourism, the use of cultural heritage, the consumption of experience and products, and the tourist. As can be seen, there is no consensus on understanding cultural tourism. The fundamental question of what cultural tourism is has proved to be difficult to answer definitively, as there are as many variations of definitions as there are tourists. The authors tend to a marketing-oriented approach to defining cultural tourism as a form of tourism that relies on the assets of the cultural heritage of a destination and transforms them into products consumed by tourists (Du Cross & McKercher, 2020). The preservation of cultural heritage makes cultures and civilisations better known, improving local conditions and the experience of tourism.

UNESCO destinations have always attracted many tourists, generating both benefits and costs. Despite the original objective of the UNESCO—to identify and protect sites of outstanding universal value, the UNESCO label has generally been used as a marketing tool to attract more tourists (Vecco & Caust, 2020). Therefore, the UNESCO listing presents a clear dilemma as the relationship between tourism development and cultural heritage preservation is dichotomous (McKercher et al., 2005). A higher number of tourists generates income, but, on the other hand, could lead to the degradation of cultural assets. Many UNESCO destinations suffer from the consequences of overtourism, resulting in a loss of cultural identity, degradation of cultural assets and the quality of life of local communities (De Luca et al., 2020). UNESCO destinations have limited capacity to adapt to overtourism due to their complex and multidimensional structure. Overtourism is not only related to the economic status quo, but defines the occurrence of far too many visitors for a particular destination to absorb over a given period. In fact, "too many" is a subjective term, and it is not possible to define the specific number for different UNESCO destinations (Séraphin et al., 2018). Especially in these destinations, tourists and residents share the same services and places (Bertocchi & Visentin, 2019).

The term overtourism has been in use only for a few years (Dickinson, 2018), however, its roots have been widely discussed in tourism literature since the mid-1960s (Mihalic, 2020). In 1973, Kaspar (1973) suggested a new field of tourism, tourism ecology. Almost a decade later, Swiss ecologist Kripendorf (1987) called for a new form of tourism, one that will bring benefits to all stakeholders, tourists, residents and businesses, without causing environmental and sociological damage. Since then, many researchers (e.g. Swarbrooke, 1999; Goodwin, 2017) and international institutions (UNWTO, 2018b) have done the same, and the sustainable paradigm is now the main school of tourism thought.

Overtourism can be described as the phenomenon of a popular destination becoming overrun with tourists in an unsustainable way (Fontanari & Berger-Risthaus, 2020). But there is more behind this statement. It refers not only to the physical destruction of the natural and cultural heritage, but also to the decrease in the quality of life of the residents and the quality of the experience of the tourists. The researchers pointed out a wide range of causes of overtourism, including the neoliberal capitalism tourism business model (Mihalic, 2020). Marketing, changes in tourism behaviour, lack of facilities, seasonality and the form of tourism in destinations can also contribute to overtourism. How the designation is managed has a direct impact on resilience to overtourism (Goodwin, 2017). Koens et al. (2018: 4384) define overtourism as:

> accumulation of different impacts and perceptions that relate to both tourist behaviour and actions by and encounters with stakeholders, as well as changes in social, economic and physical environment.

The UNWTO (2018a) argues that overtourism is not itself a threat to UNESCO destinations, it depends on how site managers control tourist flows and manage their spatial and time dispersion. Cultural tourism becomes sustainable when it can expand its dominion beyond the economic realm and embrace human development by incorporating concepts such as cultural values, cultural diversity and cultural significance of destinations (Milan, 2019). Cultural attractions and UNESCO destinations are, in fact, increasingly important in modern societies. In essence, the core is to find the balance between development of tourism industry, while still conserving the authenticity of cultural heritage, promoting and celebrating its educational, historic and cultural values (Carbone et al., 2020). Defying overtourism is therefore a very complex issue and requires a proactive attitude and participation of all stakeholders in the destination, including visitors, residents and the destination management organisation (DMO). Furthermore, numerous scholars emphasise the implementation of new procedures and innovations in managing destinations responsibly (Du Cros & McKercher, 2015; Goodwin, 2017; Postma & Schmuecker, 2017; Peeters et al., 2018; Bertocchi & Visentin, 2019; Milan, 2019; Capocchi et al., 2019; Mihalic, 2020; Pechlaner et al., 2020). Current approaches determine these elements as the effective management system: (1) a thorough shared understanding of the property by all stakeholders; (2) a cycle of planning, implementation, monitoring, evaluation and feedback; (3) development of a mechanism for the involvement and coordination of the various activities between stakeholders and (4) an allocation of necessary resources. However, due to the significant differences between UNESCO destinations, there is no "one-size-fits-all solution" to overcome overtourism (Goodwin, 2017; Koens et al., 2018).

Indeed, according to experts, it is quite difficult to measure the real effects (positive and negative as well) of the UNESCO World Heritage List inscription (Prigent, 2011; D'Eramo, 2014; Cuccia et al., 2016; Parga Dans & Alonso González, 2019; Dosquet et al., 2020). The leitmotif of UNESCO is to protect the sites of outstanding universal value, the unique testament and the exemplary character of sites. For example, in 2009, an agreement was signed between UNESCO and the TripAdvisor rating site to increase awareness of cultural heritage sites (Marcotte et al., 2017). In this context of massification of visits, it is necessary to ask the following question: It is not paradoxical to want to protect and at the same time encourage promotion and visits to cultural heritage sites? UNESCO has realised this paradox and has offered as a solution the concept of sustainable tourism that would unite these two dichotomous elements, the protection of cultural heritage and tourism (Dosquet & Lorey, 2017; UNESCO, 1993). This approach is linked to a new strategy oriented towards multidimensional effort through the reduction of seasonality, careful planning that respects carrying capacity, product diversification and the participation of the local community, entities from both the public and private sectors (Capocchi et al., 2019). Different UNESCO destinations have implemented different approaches to deal with the negative consequences of overtourism. Some destinations have focused on redirecting tourist flows to other areas not visited by many tourists or suggesting alternative attractions. As Benner (2019) argues, a multidimensional effort must be achieved to overcome this phenomenon, including careful planning and destination management that respects the limits of destination capacity, product diversification, implementation of modern technologies, reduction of seasonality, development of alternative cultural routes, establishment of DMOs as partnerships of the public and private sectors, and education of visitors about overtourism and its negative impacts. All these different activities are the results of the implementation of innovations. Various categories of innovations have different effects on the sustainability of destinations (Jafari et al., 2003; Hall & Williams, 2008; Hjalager, 2010; Gajdošíková et al., 2018).

Throughout history, tourism has been a phenomenon characterised by immense innovations. However, in recent decades, there has been an increasing focus on the topic of tourism innovations (Hall & Williams, 2008; Hjalager, 2010). Innovation can be characterised as a combination of production factors that affect business activity (Schumpeter, 1987). In the context of destinations, innovations represent changes, focus on the introduction of new and better products or services and the implementation of new processes for destination management. The main reasons for the implementation of innovations are usually constant changes in visitor needs, to increase market share and the impact on the destination life cycle. According to Nordhorn (2015), through innovations, destinations gain competitive advantages and ensure sustainable development. The definitions of innovation vary in the literature, as well as the approaches towards its categories. The Schumpeterian approach to innovation categories has been extended in tourism research. The study by Hall and Williams (2008) complies with the four categories of innovations of the OECD (2005), which are product, process, marketing and organisational innovations. Mayer (2009) focuses on organisational, institutional and customer innovations.

Product, process, managerial, marketing and institutional innovations presented by Hjalager (2010) are the most relevant classification of innovation in tourism (Maráková & Kvasnová, 2016). Innovations in products or services refer to changes directly observed by the customer and regarded as something new. Innovations can represent altered or completely new products or services, new features that are evident to customers with a substantial effect on consumer behaviour. Process innovations are aimed at escalating efficiency, productivity and flow. Technology investments are the anchor of mainstream process innovations (Hjalager, 2010). Managerial innovations deal with new ways of organising internal cooperation, motivating and empowering staff (Ottenbacher et al., 2005). Marketing innovations refer to the regulation of information and the building of a relationship between service providers and customers. Institutional innovations include the establishment of different collaborative structures that coordinate tourism development in the destination (e.g. partnerships, networks, clusters and DMOs). These entities are considered crucial for fostering the innovation process in destinations.

Implementing innovations requires the presence of the DMO and the participation of local authorities, who must pay attention to management strategies that improve the quality of life of residents, mitigate the negative impacts of overtourism and lead to sustainable tourism development in UNESCO destinations.

3 Methodology

An exploratory case study approach was used to investigate methods and strategies used to minimise tourist oversaturation in the UNESCO destination town Banská Štiavnica (Slovakia). This method has been often used in research on tourist destinations (see, e.g. Goldberg et al., 2018), including those facing overtourism (e.g. Atzori, 2020; Oklevik et al., 2019). To explore the methods and strategies used to minimise overtourism in the UNESCO town of Banská Štiavnica, this study applied a combination of qualitative and quantitative research.

Qualitative research involved a semi-structured interview with a managing director of the DMO Region Banská Štiavnica. DMO was established in 2012 as one of the first destination management organisations in Slovakia. In the beginning, the DMO had 50 members, of which 13 were municipalities, 33 businesses and 4 non-businesses. By December 2021, the number of members increased to 70, of which 57 are businesses and 13 municipalities. The interviewee has worked as the managing director of DMO since its establishment and has more than 30 years of experience in the field of tourism. He also works as an advisor to several institutions in the development of tourism in Slovakia and abroad (e.g. The Association of Towns and Communities of Slovakia).

Following the guidelines of Patton (2014), the interview protocol was created. The questions were contextualised based on the literature review to relate to the study objective and reflect the problem of overtourism in the UNESCO destination town of Banská Štiavnica:

1. How do you perceive the social, economic and environmental impacts of tourism development in the town of Banská Štiavnica?
2. In your opinion, are there any signs of overtourism in Banská Štiavnica?
3. Which destination management strategies should be adopted to develop tourism sustainably in Banská Štiavnica?
4. Which innovative activities should be adopted to develop tourism sustainable in Banská Štiavnica?

First, an interview protocol was sent to the DMO manager. After agreeing to participate, a face-to-face interview with the managing director of DMO was conducted on August 13, 2020. The interview lasted approximately 2 h and included answers to the protocol questions and open dialogue. The responses of the managing director were recorded manually for analysis. To analyse recorded data, content analysis was adopted on the basis of the Braun and Clarke (2006) guide. The procedure contained three steps: initial reading of recorded text, repeated readings and identifying important thoughts and exploring results.

To further explore the context of the case study and the development of tourism in the town of Banská Štiavnica, secondary data from the Statistical Office of the Slovak Republic (SO SR, 2021) and internal materials from the DMO Region Banská Štiavnica in the last 10 years were also used.

4 Case Study Context and Results

Banská Štiavnica is a historic town located in the central part of Slovakia. It is the oldest mining town in Slovakia, established in the thirteenth century. The historical Banská Štiavnica is a unique compact town that is harmonious with the landscape. The mining wealth of the town supported its rapid development of crafts and fine arts, the results of which are visible in the town and its architecture. Banská Štiavnica represents an extraordinary example of advanced urban design from the sixteenth

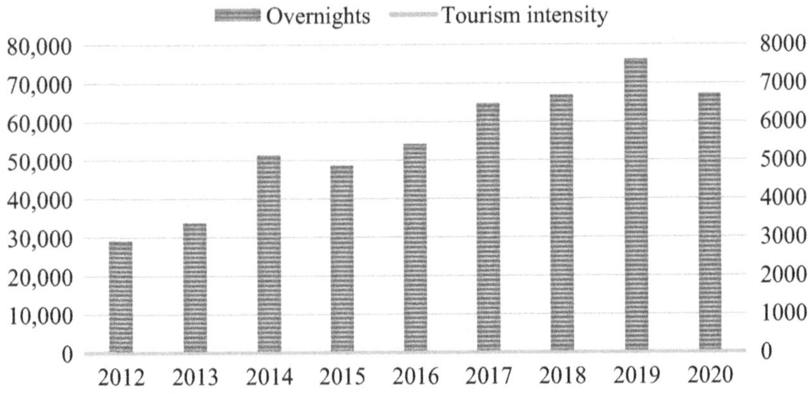

Fig. 1 Overnights and tourism density in Banská Štiavnica (SO SR, 2021)

century to the present day. Its exploitation and processing of precious metals had a profound influence on European developments in this field (World Heritage List no. 618 rev. 1991, p. 42). The town progressively became the biggest mining centre in the Habsburg monarchy in the eighteenth century. The establishment of the first Mining and Forestry Academy in Europe in 1762 demonstrates the importance of this town as a centre for the education of mining experts. Owing to the most important technical and scientific discoveries in the field of mineral processing and metallurgy, Banská Štiavnica was also a European centre for mining education for several centuries (Timčák et al., 2010). However, the development of the town reached a deadlock at the end of the nineteenth century when mining fell into decay. This trend also continued in the twentieth century. The last mine in Štiavnica was closed down in 2001. The following years were accompanied by higher unemployment, stagnation of economic development and the labour outflow.

Despite the challenges in economic development, the architecture in and around the town has been preserved. In 1993, the historic town and technical monuments of the area were inscribed in the UNESCO World Heritage List based on the criteria iv and v (UNESCO, 1993: 46):

> Criterion (iv): The urban and industrial complex of Banská Štiavnica and the Technical Monuments in its Vicinity is an outstanding example of a mediaeval mining centre of great economic importance that continued into the modern period and assumed a characteristic and distinctive form. Criterion (v): Banská Štiavnica and its surrounding areas are outstanding mining areas which have become vulnerable to the potential erosion of their character and urban fabric, following the cessation of mining activities as well as the removal of the Mining Academy.

In the following years, the economic and social transformation of the town followed.

Slowly, the economically lagging town, also known as the "ghost town" (Darulová, 2011), has become a centre of cultural and tourism importance. With a population of around 10,500, tourism has become an important part of the development of the town and more and more stakeholders have become aware of its benefits.

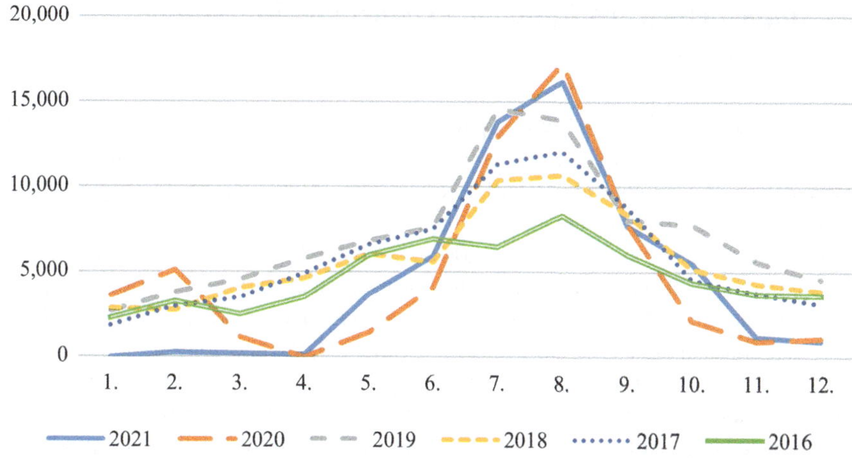

Fig. 2 Seasonality of overnights in Banská Štiavnica (SO SR, 2021)

In 2012, the DMO Region Banská Štiavnica was established. Since 2012, the number of overnights has increased by more than 150%, reaching 80,000 in 2019 (Fig. 1). This contribution of tourism led to the overall development, the gradual renovation of historical buildings, an increase in the income of business entities and the confidence of investors. The growing importance of tourism is also expressed in the comment of the DMO manager:

> Banská Štiavnica has become the most popular city destination in Slovakia. Annually, 500 thousand tourists visit the town (including same-day visitors) and spend 12 million euros in this region. Nowadays, there is no doubt that tourism supports the development of the town, but there are also its negative consequences. In summer, the Štiavnica is at the limit of its capacity.

With the increasing popularity of tourism destinations, the first signs of overtourism occur (Fig. 2).

> In our country, the government does not deal with tourism regulation, even where it would be appropriate, for example, the Liptov and High Tatras regions. The responsibility for tourism regulation is devolving to local stakeholders, and it is limited by finances, capacities and laws. It is not easy for locals to ensure a proper collection of accommodation tax or to regulate a worldwide giant such as Airbnb. Prices are increasing in cafes and restaurants, as well as in real estates, which have reached record levels among comparable towns in Slovakia. Small traditional shops for locals are disappearing from the city centre as their owners can make more money selling souvenirs. The same happens to residents when Airbnb incomes far exceed the prices of long-term rentals. (Banská Štiavnica DMO manager 2020)

With this negative impact of tourism development in Banská Štiavnica, tourists are singled out as agents of destruction of cultural heritage (Lansing & De Vries, 2007; Marcotte & Bourdeau, 2010), contributing to seasonality of demand, rising prices of goods and services, pollution, the degradation of areas, as overcrowding.

5 Implemented Institutional, Product and Marketing Innovations

It is no longer a question of giving up tourism development in Banská Štiavnica and other UNESCO destinations, but finding the balance between tourism, heritage, exploitation and exploration (Charlet & Dosquet, 2020). By applying innovative strategies of destination management based on innovation and sustainability, it is possible to achieve long-term growth (Mihalache & Mihalache, 2016; Smith, 2017; Séraphin et al., 2018). On this theoretical basis, radical solutions are not needed in Banská Štiavnica, but innovations are essential to reduce the negative consequences of overtourism in the UNESCO destinations. In 2019, the new long-term orientation in Banská Štiavnica has been applied, based on selected pillars, which are (a) understanding of the negative impacts of overtourism and applying new strategies; (b) creating adding value through investments to infrastructure and products; (c) monitoring and evaluating of strategies and action plans and (d) implementation of innovations. In Banská Štiavnica, the DMO has modified the theoretical approaches of sustainable tourism development in compliance with the uniqueness of the destination. According to the DMO manager, the basis for future tourism development was the understanding of overtourism and the willingness of stakeholders to modify the strategy of tourism development.

> The current stage of life cycle of tourism development in Banská Štiavnica needs a new qualitative approach to the participation and cooperation of stakeholders and the management of the destination. This recent development is the result of a unilateral profit orientation and brings with it a steady loss of the authenticity and identity of the town. There is no universal panacea for dealing with overtourism. However, some UNESCO destinations have adopted various measures, which can be assumed to be an example of best practices. Crucial is to reduce tourist flows in the city centre, to diverse visitors in time and space by building a year-round tourism infrastructure in the suburbs and also in the adjacent villages. (Banská Štiavnica DMO manager 2020)

Therefore, mainly institutional, product and marketing innovations could be the tool to reverse overtourism to sustainability. The implementation of innovations in destination management activities has been shown to be a successful strategy without abandoning tourism in UNESCO destinations (Benner, 2019). The new strategic orientation of Banská Štiavnica is to implement a range of actions to improve the spatial distribution of visitors throughout destinations, a control of visitor numbers throughout peak/off-peak seasons and the time of day, data collection and evaluation, precisely segmentation of visitors experience, promotion of less known attractions in destinations, and collaborative effort stakeholders to change visitor behaviour patterns (Postma & Schmuecker, 2017; Kolb, 2017; Capocchi et al., 2019; du Cross & McKercher, 2020; De Almeida et al., 2020; De Luca et al., 2020; Demartini et al., 2021; Pásková et al., 2021).

Institutional innovations are represented by the destination management organisation (DMO) Región Banská Štiavnica. The DMO was established with the aim of replacing the mining industry with tourism as the basis for future economic and social development. The increasing tourism indicators confirm the growth, as a result of the collaborative effort of the DMO members. However, the massive development of tourism and the seasonality of demand are not defensible.

> Tourism in Banská Štiavnica is in the stage of saturation, the need for qualitative change in the destination management approach, and the ability of stakeholders to cooperate is essential. The DMO cannot be the only responsible institution for the transformation from overtourism to sustainable development. The success of the tourist destination is directly proportional to the ability of the municipality, the tourism business, the residents, and other stakeholders to cooperate in favour of the destination. (Banská Štiavnica DMO manager 2020)

To ensure sustainable development, from 2019, DMO modified its approach to destination management and strategic priorities as follows: to educate all stakeholders about overtourism and its negative impact, to prolong the peak season, to redirect more tourist flows from the city centre to the rural areas, to create new/to adapt existing product according to visitor expectations, and to implement more modern technologies as an important source of data to monitor visitor flows. The second, but no less significant institutional innovation was establishment of Regional DMO Banská Bystrica Tourism in 2018. This regional DMO has created a platform for systematic coordination of tourism development. For Banská Štiavnica, the most significant result of this cooperation is the financial support—providing two employees for marketing activities of Banská Štiavnica, co-financing of new thematic products and a chatbot (the winemaker from the Banská Štiavnica), who can accompany visitors across Banská Štiavnica and its surroundings. Thereby, the effort of regional DMO Banská Bystrica Tourism is to eliminate overtourism in Banská Štiavnica and to support stakeholders in the region to cooperate more in creating innovative experiences for visitors in suburbs, with the aim of their spatial and timing dispersion.

Innovations in product and marketing result from the strategic priorities of DMO mentioned above. While it is not possible to create more space in a destination, it may be possible to act on the distribution of visitors within existing places (Charlet & Dosquet, 2020). With this in mind, DMO, with the cooperation of museums and other key stakeholders, has implemented several innovations to improve the spatial distribution of visitors throughout the destination. To control tourist flows, segmentation can be done with the implementation of modern technologies. DMO developed the smartphone application, "the smart guide of Banská Štiavnica" or the outdoor team game, to disperse visitors in space. Museums also innovate their product with the use of virtual reality and holograms during exhibitions. However, the DMO strategy is to offer visitors more than the mining history of the town. The history of Banská Štiavnica is inseparable from famous historical personalities of Slovak history (poets, inventors, geologists). With the cooperation of the private sector, DMO has created several products related to these personalities and their lives. The aim of this effort is not only to spread the concentration of visitors to the less visited attractions of the city, but also to restore the interest of visitors in the history of Banská Štiavnica.

At the beginning of tourism development, Banská Štiavnica was an enigmatic destination that attracted curious and open-minded tourists. Today, most of the same-day visitors are not very interested in the cultural heritage and history. They prefer coffees, restaurants, parties and night-life activities. As a result of overtourism and the popularity of Banská Štiavnica as a destination, the superficiality of visitors' experiences has grown. (Banská Štiavnica DMO manager 2020)

For managing visitor flow, it is important to control three parameters: the number of visitors across high and low seasons, across the time of day tourists are attracted to visit (peak/off-peak times in the morning, at midday, in the evening), and visit duration (Carbone et al., 2020). In recent years, DMO has devoted much effort to innovation in marketing communication of Banská Štiavnica, to encourage the return of visitors and rediscovery of the town in the middle season. Visitors can find online real-time information about the attendance of the most visited museum in the destination, the open-air mining museum. Extension of opening hours can also be the solution to eliminate tourist oversaturation, as cultural events take place in the evening or in the low season.

The example of a solution to cope with seasonality is to organise more cultural events in the suburbs and neighbouring rural areas. In 2019, DMO in 2019 collaborated on the realisation of 16 cultural events of local, regional, national and international importance. Crucial is to promote the events, to create innovative, mostly on-line campaigns. Events organised in the off-peak season have a positive impact on the occupancy rate of accommodation facilities, as well as the rates of restaurants. To prolong the season and to appeal to new segments of visitors, DMO has organised experience gastronomy events or site-specific theatre (any type of theatrical production that is performed in a unique, specially adapted location other than a standard theatre) festival. (Banská Štiavnica DMO manager 2020)

These DMO activities are only the beginning. As important product and marketing innovations are preparing a new design and functions of the official destination website for better data collection, promotion videos of suburbs (villages Svätý Anton, Vyhne, Hodruša-Hámre), a new promotion video spot will be prepared in cooperation with UNESCO and National Geographic, and creating new social media campaigns to promote this UNESCO destination in a new, sustainable way. For a better spatial distribution of visitors, it is necessary to build new or revitalise existing cycle tracks and other tourist infrastructure in rural areas near the city. Moreover, step by step, to change visitors from mass tourism consumers to mature visitors. Innovative solutions are an efficient way to avoid the effects of overtourism without limiting the welcome of new visitors and to protect the cultural heritage. And to restore the balance among three pillars of sustainable tourism development in destination: economic, social and environmental.

In Banská Štiavnica, we realise the need for changes in a way to develop tourism sustainably. The theoretical approaches how eliminate overtourism and its negative impact is, to put it mildly, hard to transform into practise. Luckily, the stakeholders in our destination are aware that present model of tourism development orientated into unsustainable development, could be devastating in several years not only for tourism businesses, but visitors could lose the motivation to visit Banská Štiavnica irretrievably.

6 Conclusion

In the past decade, the rapid growth of international and domestic tourism has negatively affected UNESCO destinations. While the global pandemic has reduced foreign tourism demand, domestic tourism has increased in these destinations. According to recent studies (McKinsey et al., 2017; Vecco & Caust, 2020), the challenges associated with overtourism are related to alienated residents, limited or degraded tourist experience, overcrowded infrastructure and damage to nature and cultural heritage. The impact of overtourism and uncontrolled development can be dramatic in the coming years. Researchers and supranational institutions (UNESCO, 2013; UNWTO, 2018a; European Commission, 2020) are calling for the development of distinctive approaches to better manage tourism and eliminate overtourism.

The case study of UNESCO destination Banská Štiavnica, Slovakia, is interested in the views of the issues proposed in our chapter. Mostly, we illustrate the problem between heritage and tourism: too much tourist activity at the UNESCO site can put it in danger. In contrast, the lack of visitors means vanishing the cultural heritage and human memory (Dosquet et al., 2020). The case studies of other popular UNESCO destinations (e.g. Dubrovnik, Athens, Roma, Budapest and Lisbon) have proved the fragile balance between tourism development and cultural heritage. When the destination will reach the capacity of attendance limit, DMOs must change the development of uncontrolled tourism to sustainable tourism (Postma & Schmuecker, 2017; Mihalic, 2020; Pásková et al., 2021). Based on the results of the case study, we were able to point out the innovative management of destinations, which is economically, socially and environmentally sustainable. Inventions and their transformation into innovations are crucial for all sectors of the global economy, and the tourism sector cannot be excluded.

The purpose of this chapter is to conceptualise the phenomenon of overtourism in UNESCO destinations and to offer a theoretical framework towards a more sustainable model of cultural tourism based on innovations in tourism destinations. The UNESCO destination Banská Štiavnica has to cope with the same consequences of overtourism as the more famous UNESCO destinations worldwide. In the destination has been applied new long-term strategy, based on awareness of negative consequences of overtourism as well as the common obligation of all stakeholders in a destination to develop tourism sustainably. For elimination of overtourism in the Banská Štiavnica, creating adding value to visitors' experience is needed, through implementation of several innovations. Despite of very unknown Word Heritage site situated in Central Europe, this implementation of theoretical approaches for more sustainable destination management based on innovations, adjusted to the uniqueness of each UNESCO destination, can be contributive to international community. In compliance with literature review and results of case study, the author's contribution is in the conceptualisation of the framework, how to manage the tourism destination in a sustainable way, with an emphasis on innovation, to eliminate overtourism in destinations (Fig. 3).

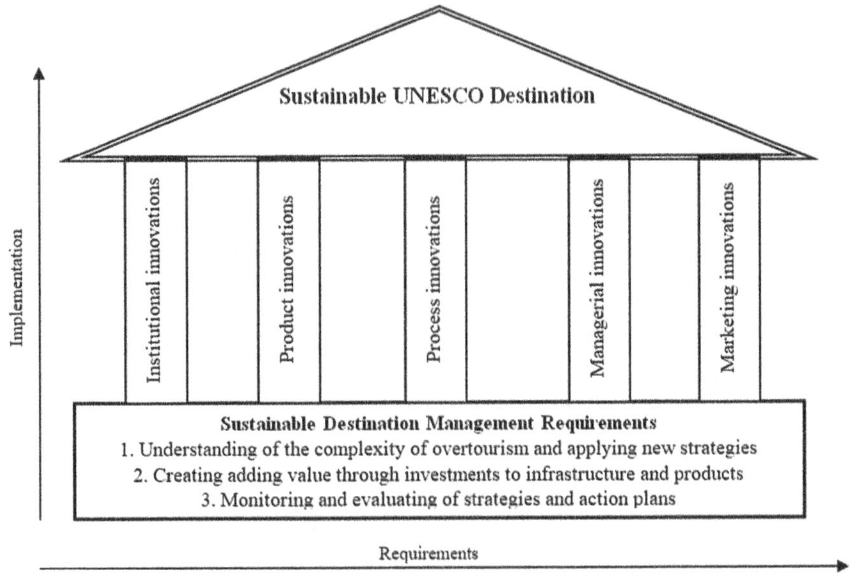

Fig. 3 Proposed conceptual framework for destination management to eliminate overtourism (Authors' results 2021)

Going beyond this initial case study, we have formulated the framework as the basis for sustainable tourism development in UNESCO destinations. The conceptual framework for destination management is based on a deep understanding of the complexity of the phenomenon of overtourism by all stakeholders in tourism destinations. The understanding is the precondition for the common effort to decrease negative attendant effects on overtourism, resulting in formulating new strategies based on offering the visitor an alternative to the less sustainable options offered previously. The key is to create added value through experiences, and investments in services in infrastructure and new technologies, resulting in various innovation processes, which can lead to sustainable UNESCO destinations. Implication of innovations can lead to eliminating overtourism as well to secure sufficient number of visitors and beneficial economic impact. Many UNESCO destinations around the world face common problems (Oikonomopoulou et al., 2017; Benner, 2019; Bertocchi & Visentin, 2019; Carbone et al., 2020). Due to the local specification and differences between UNESCO destinations in the legal, institutional and managerial system of destinations (Milan, 2019), this theoretical framework can be the inspiration for changes in destination management.

> The core of destination management lies in the opportunities for cooperation among different stakeholders in destination. The negative consequences of overtourism and COVID-19 should be seen as an opportunity to deepen the collaboration within the destination. The people responsible for finding solutions are all stakeholders, not only the DMO, municipalities or tourism companies. In the tourist destination, there is no "we and them", it is just "we—together". (Banská Štiavnica DMO manager 2020)

Management of World Heritage sites means managing part of the common heritage of humanity. According to Milan (2019), UNESCO has launched and pushed forward a number of normative tools and actions important in terms of culture, tourism and development. However, the role in its implementation and the implementation of a new sustainable tourism development belongs to the local stakeholders in the UNESCO destinations, their cooperative effort to adopt an innovative approach to traditional destination management strategies focused primarily on the economic effects of tourism.

The study is an attempt to add to academic schools of thought on tourism in destinations by proposing a conceptualisation of overtourism in UNESCO destinations, based on relevant theoretical approaches, supported by the results of the exploratory case study approach and strategies used to minimise tourist oversaturation in UNESCO destination town Banská Štiavnica. The contribution of the paper is to conceptualise overtourism in UNESCO destinations through the novelty of a theoretical framework towards a more sustainable approach to destination management based on innovations in tourism destinations. As early as 1991, Canestrelli and Costa (1991) warned policymakers and public authorities about the possible negative impact of uncontrolled tourism development in UNESCO destinations. However, so far, no destination governance system has been developed capable of making tourism development compatible with the economic and social well-being of inhabitants and sustainably with the environment (Bertocchi & Visentin, 2019).

The limitation of the study was its approach. It was based on the review of the literature, an exploratory case study approach to investigate methods and strategies used to minimise tourist oversaturation in the UNESCO destination town Banská Štiavnica (Slovakia). The study applied a combination of qualitative and quantitative research. Despite the fact that Banská Štiavnica is located in central Europe, this UNESCO destination is facing the negative consequences of overtourism as the other, more famous, UNESCO destinations. The geographical area of UNESCO destinations in Central and Easter Europe is under researched, but, due to its unique offer, it is viewed as a future playground for the European tourism market. As a limitation, the inclusion of a UNESCO destination in a country with specific conditions for the development of tourism can be considered. Therefore, the authors suggest providing a quantitative analysis between innovation and the elimination of overtourism in other UNESCO destinations for a longer period of time.

Future debates and research will show whether the new approach to destination management results in the implementation of innovations that reduce overtourism and its negative effects on UNESCO destinations.

References

Atzori, R. (2020). Destination stakeholders' perceptions of overtourism impacts, causes, and responses: The case of Big Sur. California. *Journal of Destination Marketing and Management, 17*, 100440.

Benner, M. (2019). From overtourism to sustainability: A research agenda for qualitative tourism development in the Adriatic. *Zeitschrift für Wirtschaftsgeographie, 2*(64), 74–87.

Bertocchi, D., & Visentin, F. (2019). "The overwhelmed city": Physical and social overcapacities of global tourism in Venice. *Sustainability, 11*(24), 6937.

Braun, V., & Clarke, V. (2006). Using thematic analysis in psychology. *Qualitative Research in Psychology, 3*(2), 77–101.

Blackwell, C. (1997). Tourism and cultural tourism: Some basic facts. *Preservation Issues 7*(3). Retrieved December 10, 2020, from https://www.umsl.edu.services/library/blackstudies/culttour.html

Canestrelli, E., & Costa, P. (1991). Tourist carrying capacity: A fuzzy approach. *Annals of Tourism Research, 18*(2), 295–311.

Capocchi, A., Vallone, C., Amaduzzi, A., & Pierotti, M. (2019). Is 'overtourism' a new issue in tourism development or just a new term for an already known phenomenon? *Current Issues in Tourism, 23*(18), 2235–2239.

Carbone, F., Oosterbeek, L., Costa, C., & Ferreira, A. M. (2020). Extending and adapting the concept of quality management for museums and cultural heritage attractions: A comparative study of southern European cultural heritage managers' perceptions. *Tourism Management Perspectives, 35*, 100698.

Charlet, N., & Dosquet, F. (2020). Case study 4: Overtourism – The case of the palace of Versailles. In H. Séraphin, T. Gladkikh, & V. T. Thanh (Eds.), *Overtourism. Causes, implications and solutions.* Palgrave Macmillan, Springer.

Croce, V. (2018). With growth comes accountability: Could a leisure activity turn into a driver for sustainable growth*? Journal of Tourism Futures, 4*(3), 218–232.

Cuccia, T., Guccio, C., & Rizzo, I. (2016). The effects of UNESCO World Heritage list inscription on tourism destinations performance in Italian regions. *Economic Modelling, 53*, 494–508.

D'Eramo, M. (2014). *UNESCO-cide. New left review*. 88, July–August 2014. Retrieved March 20, 2022, from https://newleftreview.org/II/88/marco-d-eramo-unescocide

Darulová, J. (2011). The influence of the registration of the locality into the list of UNESCO on tourism (as a case of the town Banská Štiavnica). In V. Klímová & V. Žítek (Eds.), *14th international colloquium on regional sciences* (pp. 286–294). Brno: June 2011.

De Almeida, C. R., Quintano, A., Simancas, M., Huete, R., & Breda, Z. (2020). *Handbook of research on the impacts, challenges, and policy responses to overtourism.* IGI Global.

De Luca, G., Shirvani Dastgerdi, A., Francini, C., & Liberatore, G. (2020). Sustainable cultural heritage planning and management of overtourism in art cities: Lessons from atlas world heritage. *Sustainability, 12*(9), 3929.

Demartini, P., Marchegiani, L., Marchiori, M., & Schiuma, G. (2021). *Cultural initiatives for sustainable development*. Springer International Publishing.

Dickinson, G. (Ed.). (2018). *Overtourism. New word suggestion*. Collins Dictionary.

Dodds, R., & Butler, R. W. (2019). Introduction. In R. Dodds & R. W. Butler (Eds.), *Overtourism: Issues, realities and solutions*. Walter de Gruyter GmbH.

Dosquet, F., & Lorey, T. (2017). La nouvelle gouvernance des Chemins de ST Jacques de Compostelle peut-elle construire à terme un contrat de destination touristique? In S. Bede & J. Spinder (Eds.), *Les contrats de destinations*. L'Harmattan.

Dosquet, F., Lorey, T., Bourliataux-Lajoinie, S., & del Olmo Arriaga, J. L. (2020). Case study 5: A paradox of the UNESCO "world heritage" label? The case of the way of St James of Compostela in France. In H. Séraphin, T. Gladkikh, & V. T. Thanh (Eds.), *Overtourism. Causes, implications and solutions.* Palgrave Macmillan, Springer.

Du Cros, H., & McKercher, B. (2015). *Cultural tourism* (2nd ed.). Routledge.

Du Cros, H., & Mckercher, B. (2020). *Cultural tourism* (3rd ed.). Routledge.
European Commission. (2020). *Call: Innovative approaches to urban and regional development through cultural tourism*. Retrieved January 15, 2021, from https://www.euro-access.eu/calls/
Fontanari, M., & Berger-Risthaus, B. (2020). Problem and solution awareness in overtourism. A Delphi study. In H. Pechlaner, E. Innerhofer, & G. Erschbamer (Eds.), *Overtourism. Tourism management and solutions*. Routledge.
Gajdošíková, Z., Gajdošík, T., & Maráková, V. (2018). Innovation process in mountain destination: Does sustainability matter? The high Tatras case study. In A. Dias, B. Salmelin, D. Pereira, & M. S. Dias (Eds.), *Modeling innovation sustainability and technologies*. Springer International Publishing.
Goldberg, J., Birtles, A., Marshall, N., Curnock, M., Case, P., & Beeden, R. (2018). The role of great barrier reef tourism operators in addressing climate change through strategic communication and direct action. *Journal of Sustainable Tourism, 26*(2), 238–256.
Goodwin, H. (2017). *The challenge of overtourism, responsible tourism partnership*. Working paper 4, October 2017. Retrieved April 15, 2021, from http://haroldgoodwin.info/pubs/RTP\T1\textquoter ightWP4Overtourism01\T1\textquoteright2017.pdf
Gretzel, U. (2019). The role of social media in creating and addressing overtourism. In R. Dodds & R. W. Butler (Eds.), *Overtourism: Issues, realities and solutions*. Walter de Gruyter GmbH.
Hall, C. M., & Williams, A. M. (2008). *Tourism and innovations*. Routledge.
Hjalager, A. M. (2010). A review of innovation research in tourism. *Tourism Management, 31*(1), 1–12.
Jafari, J., et al. (2003). *Encyclopaedia of tourism*. Routledge.
Kaspar, C. (1973). Fremderverkehrsoekologie - eine neue Dimension der Fremdenverkehslehre. In E. Walter, W. Hunziker, & H. P. Schmidhauser (Eds.), *Festscrift zur Vollendung des 65. Lebensjahres von P. Bernecker*. Institut für Fremdenverkehrsforschung.
Koens, K., Postma, A., & Papp, B. (2018). Is over tourism overused? Understanding the impact of tourism in a city context. *Tourism Review, 10*, 4384.
Kolb, B. (2017). *Tourism marketing for cities and towns. Using social media and branding to attract tourists* (2nd ed.). Routledge.
Kripendorf, J. (1987). *Die Ferienmenschen*. Zürich Orell Füssli.
Lansing, P., & De Vries, P. (2007). Sustainable tourism: Ethical alternative or marketing ploy? *Journal of Business Ethics, 72*, 77–85.
Maráková, V., & Kvasnová, D. (2016). Cooperation as a driving force of innovations in destination marketing management. *Forum Scientiae Oeconomia, 4*(3), 67–80.
Marcotte, P., & Bourdeau, L. (2010). La promotion des sites du Patrimoine mondial de l'UNESCO: Compatible avec le développement durable? *Management & Avenir, 34*, 270–288.
Marcotte, P., Bourdeau, L., & Sarrasin, B. (2017). Tourisme et développement durable. Un exercice d'adaptation, d'intégration et de conciliation. *Téoros, 36*(1).
Mayer, M. (2009). Innovations as a successful factor in tourism: Evidence from Western Austrians cable-car companies. *Erdkunde, 63*(2), 123–139.
McKinsey, Company and the World Travel & Tourism Council. (2017). *Coping with success: Managing overcrowding in tourism destinations*. WTTC Report.
McKercher, B., Ho, P. S. Y., & Du Cros, H. (2005). Relationship between tourism and cultural heritage management: Evidence from Hon Kong. *Tourism Management, 26*, 539–548.
Mihalic, T. (2020). Conceptualising overtourism: A sustainability approach. *Annals of Tourism Research, 84*, 103025.
Mihalache, M., & Mihalache, O. R. (2016). Organizational ambidexterity and sustained performance in the tourism industry. *Annals of Tourism Research, 56*, 128–163.
Milan, S. B. (2019). Word heritage management plans as opportunities to Foster cultural sustainable development: Fact or myth? In Stankov, U., Boemi, S. N. Attla, S., Kostopoulpu, S., & Mohareb, K. (Eds.), *Cultural sustainable tourism. A selection of research papers from IEREK conference on cultural sustainable tourism (CST)*. Springer International Publishing.

Nordhorn, C. (2015). *Innovation and product development*. Lecture at the Faculty of Economics at Matej Bel University in Banská Bystrica.

OECD. (2005). *Oslo manual: Guidelines for collecting and interpreting innovation data*. OECD.

Oikonomopoulou, E., Delegou, E. T., Sayas, J., & Moropoulou, A. (2017). An innovative approach to the protection of cultural heritage: The case of cultural routes in Chios Island, Greece. *Journal of Archaeological Science: Reports, 14*, 742–757.

Oklevik, O., Gössling, S., Hall, M. C., Steen Jacobsen, K. J., Grøtte, I. P., & McCabe, S. (2019). Overtourism, optimisation, and destination performance indicators: A case study of activities in Fjord Norway. *Journal of Sustainable Tourism, 27*(12).

Ottenbacher, M., Shaw, V., & Lockwood, A. (2005). An investigation of the factors affecting innovation performance in chain and independent hotels. In M. Peters & B. Pikkemaat (Eds.), *Innovation in hospitality tourism*. Routledge.

Parga Dans, E., & Alonso González, P. (2019). Sustainable tourism and social value at World Heritage Sites: Towards a conservation plan for Altamira, Spain. *Annals of Tourism Research, 74*, 68–80.

Pásková, M., Wall, G., Zejda, D., & Zelenka, J. (2021). Tourism carrying capacity reconceptualization: Modelling and management of destinations. *Journal of Destination Marketing & Management, 21*, 100638.

Patton, M. Q. (2014). *Qualitative research & evaluation methods: Integrating theory and practise* (4th ed.) Sage.

Peeters, P., Gössling S., Klijs, J., Milano, C., Novelli, M., Dijkmans, C., Eijgelaar, E., Hartman, S., Heslinga, J., Isaac, R., Mitas, O., Moretti, S., Nawijn, J., Papp, B., & Postma, A. (2018). *Research for TRAN committee - Overtourism: Impact and possible policy responses*. European Parliament.

Pechlaner, H., Innerhofer, E., & Erschbamer, G. (2020). *Overtourism: Tourism management and solutions*. Routledge.

Postma, A., & Schmuecker, D. (2017). Understanding and overcoming negative impacts of tourism in city destinations: Conceptual model and strategic framework. *Journal of Tourism Futures, 3*(2), 144–156.

Prigent, L. (2011). Le patrimoine mondial est-il un mirage économique? Les enjeux contrastés du développement touristique. *Téoros, 30*(2), 6–16.

Scuttari, A., Isetti, G., & Habicher, D. (2020). Visitor management in world heritage sites: Does overtourism-driven traffic management affect tourist targets, behaviour and satisfaction? The case of the Dolomites UNESCO World Heritage Site, Italy. In H. Pechlaner, E. Innerhofer, & G. Erschbamer (Eds.), *Overtourism. Tourism management and Solutions* (pp. 187–205). Routledge.

Schumpeter, J. A. (1987). *Teória hospodárskeho vývoja*. Pravda.

Schweitzer, C. (1999). Cultural tourism: The hot ticket to cool meetings. *Association Management, 51*(8), 121–130.

Séraphin, H., Sheeran, P., & Pilato, M. (2018). Over-tourism and the fall of Venice as a destination. *Journal of Destination Marketing and Management, 9*, 374–376.

Smith, S. (2017). Organisational ambitexterity: Welcome to paradox city. *Human Resources Management International Digest, 25*(1), 1–3.

Statistical office of the Slovak Republic. (2021). Retrieved August 13, 2021, from https://slovak.statistics.sk

Swarbrooke, J. (1999). *Sustainable tourism management*. CABI.

Timčák, G. M., Rybár, P., & Jablonská, J. (2010). Geotourism site development in Slovakia. In M. Colin & L. Jolliffe (Eds.), *Mining heritage and tourism*. Routledge.

UN. (2015). *The 2030 Agenda for Sustainable Development*. Retrieved January 3, 2022, from https://sdgs.un.org/goals

UNESCO. (2013). *Managing Cultural World Heritage*. Retrieved January 9, 2022, from https://whc.unesco.org/en/managing-cultural-world-heritage/

UNESCO World Heritage List. (1993). *Advisory Body Evaluation (ICOMOS)*. Retrieved January 3, 2022, from https://whc.unesco.org/en/list/618/documents/

UNWTO. (2018a). *Overtourism? Understanding and managing urban tourism growth beyond perceptions*. UNWTO.

UNWTO. (2018b). *Tourism and the sustainable development goals – Journey to 2030*. UNWTO.

Vecco, M., & Caust, J. (2020). UNESCO, cultural heritage sites and tourism. A paradoxical relationship. In H. Pechlaner, E. Innerhofer, & G. Erschbamer (Eds.), *Overtourism. Tourism management and solutions* (pp. 67–77). Routledge.

Whyte, B., Hood, T., & White, B. (Eds.). (2012). *Cultural and heritage tourism: A handbook for community champions*. Federal provincial Territorial Ministers of Culture and Heritage. Retrieved January 15, 2021, from https://tpr.alberta.ca/tourism/tourismdevelopment/pdf/CulturalTourismHeritageHandbook.pdf

Sustainability and Commodification? The Role of Cultural Assets in the Development of New Paths for Tourism in Catalonia

Nicolás Barbieri and Camila del Marmol

Abstract Recognising and even producing cultural assets has been one of the emerging strategies in the development of new potential paths for sustainable tourism. This type of process has integrated into the same frame practices and concepts such as cultural rights and economic competitiveness, cultural diversity and the economy of identity or cultural heritage and the intangible economy. Even if significant research has been developed on these issues, some questions remain still unexplored. Can we see a kind of paradox in these strategies, i.e. an opportunity for transforming development processes (making them more sustainable) and at the same time for reproducing economic and cultural inequalities? Can we see an opportunity to ensure equity in public decision-making about cultural tourism initiatives and at the same time the risk of deepening social and political exclusion? This chapter will address these questions through the analysis of a historical view of cultural policies in Catalonia and, specifically, a study case on the relation between the promotion of heritage and the development of tourism in the Catalan Pyrenees. Both the theoretical framework and methodology of this contribution are based on an interdisciplinary approach, including knowledge from the fields of political science, social anthropology and cultural studies.

1 Introduction

Recognising and even producing cultural assets has been one of the emerging strategies in the development of new potential paths for cultural tourism (Urry, 1995; Kirschenblatt-Gimblett, 1998; Stylianou-Lambert & Christodoulou-Yerali, 2014; De Frantz, 2005). Cultural planning on this subject, such as the creation of

N. Barbieri (✉)
Universitat Internacional de Catalunya, Barcelona, Spain
e-mail: nbarbieri@uic.cat

C. del Marmol
Universitat de Barcelona, Barcelona, Spain
e-mail: cdelmarmol@ub.edu

E. Borin et al. (eds.), *Cultural Leadership in Transition Tourism*, Contributions to Management Science, https://doi.org/10.1007/978-3-031-14121-8_13

new public and private institutions or the promotion of cultural, natural and intangible heritage (Labadi, 2011; Winter, 2011) are some of the initiatives coming from the cultural tourism sector. This type of process has integrated in the same frame practices and concepts such as cultural rights and economic competitiveness (Baltà & Dragićevic, 2017), cultural diversity and the economy of identity (Barbieri, 2015) or cultural heritage and the intangible economy (Del Mármol & Santamarina, 2019).

Even if significant research has been developed on these issues (Throsby, 2001; Pratt, 2008; Anheier & Isar, 2008; Bendix et al., 2012; Scott, 2014), some questions remain still unexplored. Can we see a kind of paradox in these strategies, i.e. an opportunity for transforming development processes (making them more sustainable) and at the same time for reproducing economic and cultural inequalities? Can we see an opportunity to ensure equity in public decision-making about cultural tourism initiatives and at the same time the risk of deepening social and political exclusion? In the specific field of cultural heritage, for example several studies have shown how the heritagisation of cultural phenomena have resulted both in processes of commodification and normativisation, whilst at the same time opening breaches towards contra-hegemonic dynamics that can or cannot end up into new developmental paths (Herzfeld, 1991; Hafstein, 2014; Cortés-Vázquez et al., 2017).

This chapter addresses these questions through a case study that includes an analysis of a historical view of cultural policies in Catalonia and, specifically, the relation between the promotion of heritage and the development of tourism in the Catalan Pyrenees. The research combined, on the one hand, the methodology of frame analysis (Goffman, 1975; Verloo, 2005) of debates in the Parliament of Catalonia with content analysis of historical documents (cultural policy plans, laws, governmental budgets etc.) and semi-structured interviews with 18 informants (including policymakers from the Catalan and local governments, third sector organisations, private companies and cultural experts). We focused on these interview profiles because our research wanted to analyse policymaking and formulation processes. And these informants directly participated in those processes or have expert knowledge about them. On the other hand, this chapter includes knowledge from a long-term ethnographic approach based on participant observation for 14 months between 2006 and 2012 together with qualitative methodologies of analysis. Both the theoretical framework and methodology of this contribution are based on an interdisciplinary approach, including knowledge from the fields of political science, social anthropology and cultural studies.

2 Theoretical Framework

This theoretical framework brings together two fields of study: the analysis of cultural policy and heritage studies. In particular, it will define and interrogate the relationship between cultural policy, heritage and regional and economic development strategies where tourism plays an important role.

2.1 Cultural Policies: Legitimacy, Instrumentalisation and Opportunities for Sustainable Development

Cultural policy refers to the institutional support structures that direct both aesthetic creativity and our collective lifestyles: it is (or means to be) a bridge between aesthetic and anthropological registers (Miller & Yúdice, 2004). Occupying this liminal space has posed important challenges for cultural policies in regard to their public legitimacy, in essence, the legitimacy of public institutions intervening in the field of culture. The process of institutionalisation for European cultural policies is thus carried out with a significant deficit in perceived legitimacy, implying a consequent limitation of recognition and resources, especially if we compare this situation with that of the Welfare State's central public policies.

During the second half of the twentieth century, cultural policies built their legitimacy by overlapping different developmental models. Firstly, came that of the democratisation of culture, specifically the promotion of access to those things which are considered the key references of Western culture (Mulcahy, 2006). Hence, the England Arts Council's first slogan: *Great art and culture for everyone*. Secondly, in the face of criticisms levied against this conceptualisation of culture, cultural policies sought to rebuild their legitimacy using a model labelled *cultural democracy*: the government's responsibility is to provide equal opportunities for citizens to be culturally active on their own terms (Urfalino, 1996). However, these strategies require mediation, and this role is often realised by the so-called *cultural sector*, a deliberately imprecise concept that includes (but also excludes) certain public and private actors.

Support for this sector, and ultimately for all cultural (and therefore creative) industries tend to become not only a facilitating tool, but an objective. This sectorial logic is then compounded with the logic that conceives of culture as a right, perceived as just as, if not more legitimatised. The tension between these two logical frameworks has shaped the history of cultural policy (Barbieri & Fina, 2020). As an underlying issue then, inseparable from the tension previously described, cultural policies have a legitimacy problem; a constant need to find solid elements that explain their importance and direction. For this reason, on this path towards institutionalisation, we see attempts to make compatible very diverse objectives such as the construction of national identity, the promotion of artistic excellence, the impulse towards economic development, the safeguarding of heritage and access to culture.

One of the logical frameworks for the justification or legitimisation of the most established cultural policies advocates a State as the principal sponsor of cultural and creative industries (Wu, 2003). The underlying idea is that powerful industries will simultaneously ensure a competitive national economy, artistic quality and cultural diversity (in a globalised context). During this process, conceptual frames of reference are adopted such as the economy of identity or economy of experience. What kind of result has this type of policy had? In many cases, it has been the creation of cultural industries that are not only unequal from country to country (or with transnational agents that monopolise the production of content), but also internally unequal (inequality of class, origin etc.) (O'Brien, 2019).

In the particular case of regional and local development, cultural policies have attempted to integrate complex and contentious concepts and practices such as the economy of knowledge and urban regeneration with social cohesion, governance and sustainability. Frequently, the balance of these objectives has tended to focus on the economic impact of culture, implementing cultural policies and the idea of territory or *cultural proximity* (Rius-Ulldemolins & Klein, 2020). This proximity, which ought to have served to incorporate the unique and heterogenous character of the territory into cultural policy, instead became above all a lever for economic growth, wherein an ensemble of private-sector actors controlled the policymaking process. This was the case, amongst others, with the cultural policies of New Labour in the United Kingdom which were very interested in the instrumental capacity of arts and culture. With these processes in mind, it was crucial for culture professionals to have an increasing level of visibility in the political arena and above all, that this visibility was accompanied by a growing capacity for the sector to exploit and take advantage of economic resources associated with the budgets of other political sectors such as tourism or urban development (Belfiore, 2012).

However, it is necessary to emphasise that cultural policies of this kind have also created opportunities for wider democratic participation in their elaboration. Various agents (not only culture professionals and governmental bodies) have been able to influence the cultural policy agenda, introducing new issues like cultural sustainability. Culture is considered the fourth pillar (together with economic, social and environmental factors) of the process of sustainable development (Hawkes, 2001; Pascual, 2018). The discourse surrounding cultural rights and community-based culture also fits into this process (Baltà & Dragićevic, 2017).

A large part of the tensions, contradictions and even paradoxes present in this methodology for cultural policy are captured in the field of heritage politics and its relation with the cultural tourism sector. The origin of heritage policies must be sought in the philosophies of conservation indebted to nineteenth-century thought with its first legal definition appearing in the French national legislation of the nineteenth century (Poulot, 2006). This marked the beginning of a series of theoretical and managerial transformations, which would lead us through various changes of direction to the contemporary configuration of heritage (see Santamarina, 2013). The following section of this chapter will be dedicated to succinctly exploring these debates.

2.2 Heritage, Cultural Policies and Tourism

Heritage politics has developed as a crucial dimension of cultural politics over the past decades. UNESCO implantation of a complex corpus of conventions and regulations was fundamental in the growing presence of heritage in the international scene (Smith, 2006; Di Giovine, 2009). The deployment of this refined heritage regime (Bendix et al., 2012) instituted not only a governmental sphere but also a powerful system of thought including elaborated dichotomies of legitimacy in relation to the conservation of the past (Lowenthal, 1998; Herzfeld, 2010; Meskell & Brumann, 2015). Built as a legitimate representation of the past, cultural heritage constitutes a powerful field of political action. Heritage thus became a doctrine, a mandatory field imposing specific Western ideas of conservation, restoration, monumentalisation and beauty (Lowenthal, 1998; Smith, 2006). But the latest twist in heritage regimes has been the adoption of a new conceptualisation: *Intangible Heritage*, coined within UNESCO discussions to enlarge the restrictive frame of the 1972 World Heritage Convention (Hottin, 2011). The concept became known worldwide in a vertiginous and successful process starting with the approval of the Convention for the Safeguarding of Intangible Cultural Heritage (UNESCO) in 2003 (Aikawa-Faure, 2009). As observed before, the adoption of the Convention prompted a cascade effect on national legislation worldwide, proliferating the local protection, identification and safeguarding of what became known through UNESCO definitions as intangible heritage. To grasp the complexity of the new concept, we need more than the official definition given by UNESCO:

> the practices, representations, expressions, knowledge, skills—as well as the instruments, objects, artefacts and cultural spaces associated therewith—that communities, groups and, in some cases, individuals recognize as part of their cultural heritage (UNESCO, 2003, Art. 2).

A refurbishment of former disputed and criticised fields of knowledge such as folklore (Bortolotto, 2011), intangible heritage represented as well a deep effort to prompt more inclusive notions of heritage in tune with ideas of social inclusion and participation (Sánchez-Carretero et al., 2019). The extent to which these efforts are successful is still being strongly discussed (Smith & Akagawa, 2009; Hafstein, 2009, 2014; Santamarina, 2013, amongst many others).

Beyond the well-intended attempts of consolidating alternative definitions of heritage, the rise of intangible heritage came about together with the rise of intangible assets in the economic domain (see Del Mármol & Santamarina, 2019). The emergence of the intangible economy has been widely celebrated since the early 2000s, in reference to different phenomena such as knowledge, social relations, ideas, brands, R+D, product design or human capital and its role in the economic system (Bloomberg, 2013[1]; Haskel & Westlake, 2017). In tune with the expansion of definitions of new economic regimes in relation to the encroaching of the cultural domain, such as a new *cultural economy* or *cognitive economy* (Throsby, 2001; Anheier & Isar, 2008; Scott, 2008), the rise of intangible assets is being widely celebrated. The tricky relationship between heritage and economic dynamics has long been a controversial topic of debate (see Kirschenblatt-Gimblett, 2006; Bendix et al., 2012), pointing to the risky potential of turning heritage into a cultural and economic asset. Nonetheless, tourism development has largely seized cultural, intangible and natural heritage in its thirst for shaping attractive destinations

[1] https://www.bloomberg.com/news/articles/2013-07-18/the-rise-of-the-intangible-economy-u-dot-s-dot-gdp-counts-r-and-d-artistic-creation

(MacCannell, 1976; Korstanje, 2012, 2019; Fyall & Rakic, 2006; Timothy, 2018). This aligns with recent statements of the United Nations World Tourism Organization (UNWTO) praising cultural tourism as a key market niche in the next years (UNWTO, 2018; see Cousin, 2008). In recent years, many authors have identified UNESCO with a brand (Meskell, 2015; Grätzer et al., 2015; Lai & Can-Seng, 2015), posing that the mandate to protect heritage through the inscription of elements into UNESCO Lists has been replaced by the mobilisation of the brand in the hope to achieve development, in many cases conflated with tourism.

In this vein, cultural and intangible heritage has been deployed as a strategy to reshape local and regional markets in consonance with global directives emanating from developmental agencies or supranational bodies such as the UNESCO or the UE, for the cases that we are dealing with in this text. If the intangible investment is being identified as a well-known source of economic development from an economic perspective (Haskel & Westlake, 2017) it is crucial to identify the role of heritage politics (whether cultural, natural or intangible) in the shaping of specific regions as tourist's destinations. The fact is that states and local governments, as well as NGOs, and international agencies are increasingly promoting development based on the exploitation of intangible heritage (Labadi, 2011; Winter, 2011; Collins, 2008). Already in 2009 Bendix has argued that the economic potential of heritage has grown to be the primary incitement in the development of heritage projects. But the risk of this situation is also denounced by UNESCO itself, that has intensified its warnings against the *freezing* or *folklorisation* of heritage, opposing *market value* to *cultural value*. But this translation of cultural value into a market value is not a contradiction as such, and many authors have identified it as a usual strategy of current neoliberal economies (Miller, 2008; Franquesa, 2013). As Tsing (2015) has recently and eloquently shown us, the connection of different regimes of value is a key neoliberal strategy for reaching the slippery promise of local development. It is important, though, to identify how cultural, natural, and intangible heritage are being incentivised and deployed, and how they are translated into economic assets. A critical analysis must strive to identify which sectors and social classes are able to display, develop and benefit from the new resources growing under the shade of heritage politics (see Brumann & Berliner, 2016; Alonso, 2014). Since the promises of development wrapping heritage politics are deeply intertwined with tourist economies, the risks of unequal development are frequently found in the form of transnational actors and capital seizing the benefits (Brumann & Berliner, 2016). Furthermore, the evidence shows an increase in restrictions on the rights of access and use of what is declared as heritage in different local contexts (see Breglia, 2006; Del Mármol, 2012; Copertino, 2014).

Nonetheless, we must not ignore that the expansion of notions of heritage referred above has also set up notions of common inheritances that can in some cases brought about paths of resistance and political struggles for new democratic claims. As Coombe and Weiss reminded us, important shifts from liberal state-based regimes of protection of patrimony to multiscalar and multisectoral assemblages of governmental regulation are taking place (2016: 43). Within this, cultural heritage is promoted as a resource, enshrined mainly as an economic asset in marketised relationships by governmental and developmental agencies, but seized as well as a token of empowerment and legitimacy by some subaltern groups. As many authors have shown, the increasing of regulatory regimes such as environmental or cultural and intangible protection schemes cannot be solely understood in its dominative capacity but must also be analysed as a contradictory process providing both opportunities and constraints (e.g. Herzfeld, 1991; Leblon, 2013; Coombe & Weiss, 2016; Kiddey, 2017). The extent to which each of these options is deployed in a particular setting requires a rooted and multiscalar approach in order to apprehend the specific configuration of local power enabling the final outcome. The institutional and legitimised language of cultural and intangible heritage can become not just a hegemonic idiom of oppression (Herzfeld, 2010), but also a vocabulary for reclaiming alternative rights, unique property rights and raise up the voice for nonconventional futures. We are witnessing thus new fields of confrontation and legitimation bringing about the possibility of paradoxical configurations.

3 Case Study: From the Historical Development of Cultural Policies in Catalonia to the Heritage Promotion in the Pyrenees

3.1 Cultural Policies in Catalonia: From Normalisation to Instrumentalisation

The case study of cultural policy in Catalunya illustrates a good many of the theoretical debates that have been previously presented here. We have referred to the structural weakness of cultural policies and the necessity of continuing to find discourses and practices that reinforce their legitimacy for intervention in the field of culture. Next, we will focus on understanding the historical evolution of these policies since their implementation at the beginning of the 1980s, up to the present day.

With the return of democracy in Spain, the government of Catalonia (Generalitat de Catalunya) and local governments (led by the Barcelona City Council and the Barcelona Provincial Council) developed cultural policies practically on a parallel timeframe with more conflicts than points of convergence. The Generalitat based its culture policy discourse on the idea of cultural "normalisation" or standardisation. As the Catalan Minister of Culture Joan Guitart (1990) pointed out, to standardise Catalan culture was to define it as a market (where goods are produced) and at the same time as an expression of national identity. Catalan culture had to be as *normal* as any other national culture. However, this framework only takes shape in the (re)construction of the great national cultural facilities (museums, theatres, auditoriums etc.).

Consequently, heritage policies (museums, heritage and archives) maintained a significant role during the 1990s, even exceeding remaining more important than those areas which we would call direct cultural promotion: performing arts and other sectors (visual arts, cinema, music and literature). However, these policies refer to the most part to the development of laws for the preservation of traditional cultural heritage. Ultimately, heritage was not recognised as a resource for regional development until well into the course of the 1990s.

For their part, local governments placed emphasis on policies that enabled access to culture, the recuperation and (re)invention of traditions and the construction of facilities used for cultural dissemination. The city councils took over from anti-Franco civil society and thus brought to light a significant phenomenon: that participation was constituted as a basic component of society. This shifted towards a process of cultural democratisation by means of the construction of facilities for cultural and artistic dissemination. However, the turning point for this stage of the process was the great cultural events staged during 1992 in Spain: the designation of Madrid as the European Capital of Culture, the Seville Expo'92 and the Barcelona Cultural Olympiad.[2] These events functioned as significant festive occasions but from the point of view of cultural policies, did not leave much behind.

Towards the end of the 1990s, a combination of factors led to the consolidation of the cultural industries paradigm: the failure of democratisation policies, very little development on policies for cultural democracy, and important conflicts with the professional culture sector (more organised each time). It became necessary to relegitimise cultural policies. The underlying idea is that public policy should act as the catalyst for the culture sector and its professional agents. At the level of the Generalitat, the Catalan Institute of Cultural Industries was created and, at the city level in Barcelona, the strategic plan for the culture sector (1999) defended the "strategic character of culture and cultural policies for urban economic competitiveness and the configuration of Barcelona as a centre for knowledge and services of an international nature" (Ajuntament de Barcelona, 1999). Furthermore, facing the centrality of this new model, heritage policies and institutions (museums, heritage and archives) lost relative influence during the decade of the 2000s.

From this point onwards, the budget and the organisational structures dedicated to the cultural industries grew significantly. Boosting the cultural industry and its competitiveness, it was argued, allowed the achievement of quality cultural production, access to culture for the population and the protection and reproduction of cultural identity in the long run. In the words of a Catalan Minister of Culture (Jordi Vilajoana), Catalan culture should take part in the initiatives that constitute contemporary culture from a worldwide perspective (Parlament de Catalunya, 2000). The accounts that support the cultural policies of the Generalitat in this period define culture as a productive process and attempts were made to construct the legitimacy of cultural policies based on concepts such as the economy of experience

[2]The Cultural Olympiad was a cultural programme of the Organising Committee for the Barcelona 1992 Olympic Games which offered cultural activities during the four previous years of the games. See Moragas (2008) for more details.

(Mascarell,[3] 2005) and the economy of identity (Tresserras,[4] 2008). This would imply a shift from a society based on industrialisation to one where services, leisure and free time are fundamental to the economy. This change was combined with a representation of Catalunya as an excellent, entrepreneurial and innovative society capable of global competition. The issue of Catalan identity was put at stake less during political negotiations with the central state and more so in economic relations within and outside those limits. Identity is considered both a condition of existence and a tool of competence. Therefore, the conceptual framework of cultural policies, which had until now linked culture and identity, expanded to incorporate notions such as competitiveness or economic attraction with differences depending on the body in question.

However, this change in the orientation of cultural policies left room for the emergence of other strategic directions. On the one hand, important policies for cultural facilities were developed aimed at the local level. The discourse of cultural rights was also incorporated and not only secondarily. However, access policies were often limited to the construction of infrastructure, leaving by the wayside the implications (social, educational, political) that their use may have for citizens. On the other hand, the Generalitat developed a policy of support for creativity, aimed at individual creators and cultural entities, with less structure and resources than that which was developed for the cultural industries, but in partnership with organisations that acquired additional relevance. Finally, the National Council for Culture and the Arts (CONCA) was created, with the initial objective of democratising the development of cultural policies by introducing the principle of *arm's length.*[5] However, this objective of improving governance did not come to fruition and CONCA has since reduced its advocacy capacity, again guided by a corporate logic.

The 2010s and the early 2020s were marked by multiple crises. Firstly, the economic crisis brought important cuts to the cultural budgets of many governments, but especially of the Generalitat. However, cultural policies also faced a new political crisis: a crisis that questions the legitimacy of their intervention in the field of culture. As an example of this crisis, we can cite the Generalitat de Catalunya's *Strategic Plan for Culture 2011–2021*. The development of the plan was initiated but remains unfinished.

And so, throughout the entire decade of the 2010s, attempts were made to search for new economic resources: a tax on ADSL companies was promoted (which was later declared unconstitutional), and the Fundación Catalunya Cultura and the Catalan Agency for Cultural Heritage were created. In the latter case, the functions of this body go beyond the search for new financing, but nevertheless the economic

[3]Ferran Mascarell was Catalan Minister of Culture (conseller de cultura), 2006 and 2010–2016.

[4]Joan Manuel Tresserras was Catalan Minister of Culture (conseller de cultura), 2006–2010.

[5]Even if this principle can be considered in different ways, following Per Mangset (2009), it is usually supposed to imply that a relatively independent and artistically competent *arts council* (or similar body) is established to take care of the allocation of public subsidies to the arts community.

concern is at the foundation of its creation. This process is better understood by noting that the cultural industries paradigm is gaining ground, also incorporating the concept of creative industries. In the search for renewed legitimacy, the remit of cultural policies is widening. The concept of *creative industries* is the conceptual umbrella for this expansion that allows the inclusion of fields such as design, video games, gastronomy or the digital industries (Garnham, 2005). Beyond this expansion of the definition, the logic for the "justification" of cultural policies becomes even more important, based on measuring their impact, both socially and especially economically.

Finally, in opposition to this dominant model, at the same time as the sphere of cultural policy expands, the door opens to new directions in this area or, indeed, to the recovery of previously displaced debates. In this way, policies are being developed, especially through the promotion of local governments, which connect culture and sustainability; the cultural system and the educational system; or the role of culture in community development policies. Many city councils in Catalonia adopt and develop the implementation of Agenda 21 for Culture, a directive that promotes the interconnectedness of the relationship between citizenship, culture, and sustainable development. For example, the Barcelona City Council recognises the need to orient the city's tourist activity towards sustainable development, with a responsible tourism model based on regional balance and the pillars of Agenda 21 for culture (Comissió de Cultura de CGLU, 2018; see UNWTO, 2013).

In the next section, we will analyse in detail how these challenges for cultural policy development are reflected in a region of the Catalan Pyrenees.

3.2 *Catalan Pyrenees*

The case study presented here, focusing on the Catalan Pyrenees, will allow us to deepen into the process of tourism promotion and its relationship with the development of heritage politics. This area covers two tourist brands coined by the Catalan Agency of Tourism: *Pirineus* and *Val d'Aran*, which are nonetheless far away from the two main tourism destinations of the country: the coastal zone and the city of Barcelona. The Catalan Pyrenees cover a vast area of 9652 km^2 (38% of the Catalan population) concentrating just 214,547 inhabitants (2.8% of the Catalan population). This area went through a substantial change in the past decades, in terms of economic orientation and the productive base of local development. Nowadays, this region shows one of the highest dependence rates on tourism revenues in the whole country (Duro & Rodrıguez, 2011) and has left behind its former emphasis on the primary sector together with the energetic and forest harnessing of natural resources that went on during the most part of the twentieth century (Campillo et al., 1992). In the framework of the EU structural funds, tourism has been identified as a strategic development sector for rural and mountainous areas of Europe with otherwise weak economic and social structures and rather low figures of development and growth. Even though tourism was not completely new to the area, with the first steps to be found in the nineteenth-century elite's hiking clubs or the first sky resorts within the Spanish tourist boom of the 1960s (Jiménez & Prats, 2006), the progressive abandonment and reorientation of the agricultural and livestock activities until the final blow given by the entry of Spain in the ECC in 1986, made it one of the pillars of local development.

Therefore, tourism in the region has been one of the most heavily promoted sectors in recent decades (López-i-Gelat et al., 2008). This is reflected in the body of legislation passed with this respect, as well as in the different projects undertaken in the area. A series of ideas began to take shape in the 1980s, oriented to the preservation of natural mountain areas and the identification of new values from where to draw on possible resources. Rural areas were urged to foster new modes of development based on *providing recreation and leisure for the city-dwellers* (European Commission, 1988: 32), identifying cultural and intangible heritage (by then still called folklore) as key areas of development. Heritage in its varied array of expressions was presented as a basic strategy for the production of a tourist destination.

Many initiatives were prompted into this new direction, dragging the area into the new cultural economy fostering alternative productive sectors such as service industries, tourism and leisure activities, as well as neo-artisanal forms of diversified production (food, clothes etc.) (Throsby, 2001; Pratt, 2008; Scott, 2008). The EC Leader Programme (*Liaison entre Activités de Développement de l'Économie Rurale*) operating since 1991, worked as a crucial funding channel aimed to promote the revitalisation of rural areas through productive diversification, a focus on tourism and the boosting of endogenous development (Luzón & Pi, 1999; Alonso & Macías, 2014). By means of the consolidation of public/private partnerships, local elites and regional governments fuelled new forms of neoliberal corporativism consolidating specific forms of modernisation and local development (Alonso & Macías, 2014). Whilst many investments are geared towards the promotion of rural tourism, some important resources were directed to what we referred here as the production of heritage, such as the restoration of Romanesque churches, the creation of local ethnographic museums (see Del Mármol, 2012) or the promotion of intangible heritage. Rural development in the form of tourism promotion was imposed as the right path towards modernisation, in line with the extended new uses and consumption of rurality (Halfacree, 1993), within what Sivaramakrishnan and Vaccaro (2007) called the new post-industrial landscapes.

For the sake of the argument, it is important to clarify that heritage in its varied and recent labeling (cultural, intangible and natural) is not considered here as a natural domain, as something given, following current trends of critical perspectives on heritage studies (see Smith, 2006, Harrison, 2013 amongst many others). On the contrary, heritage is taken as a complex concept involving the production of a political category, *a mode of cultural production in the present that has a recourse to the past* (Kirshenblatt-Gimblett, 1998: 7), and that involves a transvaluation of what has become obsolete (Kirshenblatt-Gimblett, 1998; Lowenthal, 1998; Herzfeld, 1991). In the Catalan Pyrenees, concrete actions included the declaration of Natural Parks, a set of urban planning regulations towards the protection of certain architectural traditional styles, the recovery of mountain paths and trails, and the promotion of new museums mostly focused on old local trades and activities turned into ethnological heritage. As we have seen, several documents, policies and plans both locally and from supranational instances such as the EU are clearly in favour of promoting tourism and heritage as fundamental development strategies (Bell & Jayne, 2010), presented as panaceas for rural abandonment and depopulation. A clear longing for what is considered as more natural and authentic, such as ecological food, antique fairs, and local rituals, underpinned from an ideological backdrop this reconceptualisation of the region.

Even though heritage is often viewed and presented as an economic asset, accompanying developmentalist official discourses and narratives, it has not always fulfilled its promises of cultural and social cohesion (Selwyn, 2007). The *demiurgic capacity* (Prats, 2003) of heritage in its pairing with tourist policies, has in many cases failed to bring profit to broader social sectors. As have been highlighted by many analysts, the differences between access and power amongst actors are unequal (see Porter & Salazar 2005). The imposition of heritage labels onto previous social uses such as local celebrations can turned into a resignification of popular culture, including massification or exhibition. Natural conservation has been largely denounced as new regulations restricting access and varied uses to local inhabitants (Frigolé, 2007; Vaccaro & Beltran, 2010), and urban regulations in some areas have induced an emphasis on decorative and aesthetic dimensions promoting the construction of residential areas and second homes (Vlès, 2014; Del Mármol, 2017). Whilst on the one hand many of the protective regulations on natural heritage and urban development based on the conservation of traditional styles (namely intangible heritage) helped avoid extensive rural gentrification and the intensive ski tourism model experience in some areas of the region, on the other hand, many voices were raised against the production of isolated landscapes that limited the uses of the land (see Tulla, 1994). The risk of a model of development fostering speculative dynamics that give priority to tourism sectors such as building of second residences or rural lodges can interfere with the aspiration of young inhabitants that are unable to find a place to live. During the recent sanitary confinement measures imposed due to the COVID-19 pandemic, the requests from urban dwellers to move to this mountain area have drastically escalated.[6] But the main problem identified by several civil servants working locally was the lack of available housing, since residents must compete in a highly gentrified rural property market. In a similar vein, protective actions within natural parks or protected areas have a direct impact on agricultural and livestock farming practices, unable to compete for scarce land in mountain areas. It goes without saying that this has a direct impact on the migration rates of the region, with limited possibilities of employment and important problem in the housing market.

Whilst our research in the Catalan Pyrenees has delivered enough evidence of the restrictive and coercive facet of heritage politics widely conceived (including its

[6]Interview August 2021. Local government agent.

footprint as a hegemonic idiom stretching the classical domain of cultural policies), it is relevant to bring up the Janus-face of heritagisation impinging on local contexts. One of the most boosted cultural policies from the 2000s on in the Catalan Pyrenees was the opening of ethnographic museums (Abella et al., 2012), celebrating traditional trades that were on the other hand relentlessly exterminated by the capitalisation of agriculture and husbandry during the twentieth century. As an example, we can cite the Turpentine Women Museum, a local exhibition in a highly depopulated valley of the Alt Urgell. With less than 100 people living in the central town of the valley in the early 2000s, the memory of the Turpentine women celebrating the local tradition of wandering peasants selling off mountain herbs and turpentine around Catalunya, allow the local council helped by the regional government (Consell Comarcal) to promote a new development path. The language of heritage enabled the town to seize local funds in order to inaugurate the museum and promote academic research (see Frigolé, 2005), prompting the neglected valley to reconfigure a local identity by organizing an array of activities to boost the local economy based mainly on handicrafts and a few of rural inns. This example is in tune with the definition Di Giovine (2009: 9) gives us of the field of heritage production and the field of tourism production:

> a multi-layered, global social structures wherein individuals struggle and negotiate to create, define, and promote formative encounters with place

4 Conclusion

This study enables us to test several well-established theories on the analysis of cultural policies, tourism and sustainability developed in different contexts. As has been widely analysed, cultural policies are conditioned by the lack of legitimacy which forced policymakers to integrate into the same framework the discourse on democratisation of culture and economic development through the promotion of cultural industries. This process results in the instrumentalisation of cultural policies and heritage in order to promote and enhance the territorial chances of economic development. Furthermore, the cultural policy process can become monopolised by a few agents strategically positioned in the cultural sector further enhancing corporativism and political inequalities.

This chapter also explains how cultural policymakers build the legitimacy of cultural policies based on concepts such as "normalisation" (or standardisation), an attempt to define Catalan culture as any other national culture. This process implies a certain idea of modernity, based firstly on a link between culture and national identity, and then including in the same framework notions as competitiveness or economic attraction. Identity, creativity and experience are merged and condensed in the economic dimension of culture. In the same vein, we have discussed the paradoxical implications of merging cultural heritage to local development and tourism, with its associated risk of packaging a culture to sell by the pound.

Having said that, our case study highlights several aspects that are usually underestimated in previous research. Firstly, the role of cultural assets that are normally understood as resources, that is inputs rather than outputs, open the way to further questioning. We have indeed called attention to the complex process of cultural production, which blurred the established frontiers between inputs and outputs. In this way, cultural assets can be both identified or produced, and must be critically analysed in further investigation.

Secondly, this chapter highlights how complex and paradoxical can be the relationships between culture, heritage and politics. In particular, new policies regarding cultural tourism can be an opportunity to foster more sustainable development processes (in cultural and economic terms), but at the same time an arena for new paths of the instrumentalisation of culture and status-quo reproduction. We also consider this issue as potentially relevant for future research.

Finally, the interdisciplinary approach adopted in this chapter has been a strength and a weakness at the same time. Applying concepts and methods coming from different disciplines can be considered puzzling and "time-consuming", but from our point of view it is essential to understand the complex relationship between sustainability and commodification in cultural and heritage policies. An epistemic and methodological effort for leaving behind the social science's partition into unconnected vessels is key to moving beyond partial portrayal of contemporary processes.

References

Abella, J., Alcalde, G., & Rojas, A. (2012). De la guadaña al forfait. Análisis del uso turístico de los museos etnológicos del Alto Pirineo catalán. *Pasos. Revista de Turismo y patrimonio cultural, 10*(5), 619–628. https://doi.org/10.25145/j.pasos.2012.10.076

Aikawa-Faure, N. (2009). From the proclamation of masterpieces to the convention for the safeguarding of intangible cultural heritage. In L. Smith & N. Akagawa (Eds.). *Intangible heritage* (pp. 13–44). Routledge.

Ajuntament de Barcelona. (1999). Pla estrategic del sector cultural de Barcelona, Ajuntament de Barcelona. https://bcnroc.ajuntament.barcelona.cat/jspui/handle/11703/84925

Alonso, P. (2014). From a given to a construct: Heritage as a commons. *Cultural Studies, 28*(3), 359–390. https://doi.org/10.1080/09502386.2013.789067

Alonso, P., & Macías, A. (2014). Etnografía de la financiación europea del desarrollo rural a través de un proyecto fallido. *Revista de Antropología Iberoamericana, 9*(3), 223–250. https://doi.org/10.11156/Aibr.090302

Anheier, H. K., & Isar, Y. R. (Eds.). (2008). *Cultures and globalization: The cultural economy.* Sage Publishing.

Baltà, J., & Dragićevic, M. (2017). Cultural rights and their contribution to sustainable development: Implications for cultural policy. *International Journal of Cultural Policy, 23*(2), 159–173. https://doi.org/10.1080/10286632.2017.1280787

Barbieri, N. (2015). A narrative-interactionist approach to policy change analysis. Lessons from a case study of the cultural policy domain in Catalonia. *Critical Policy Studies, 9*(4), 434–453. https://doi.org/10.1080/19460171.2015.1014519

Barbieri, N., & Fina, X. (2020). Polítiques culturals a Catalunya: tot buscant legitimitats. In R. Gomà & J. Subirats (Eds.), *Canvi d'època i de polítiques públiques a Catalunya* (pp. 404–421). Galaxia Gutenberg.

Belfiore, E. (2012). Defensive instrumentalism and the legacy of new labour's cultural policies. *Cultural Trends, 21*(2), 103–111. https://doi.org/10.1080/09548963.2012.674750

Bell, D., & Jayne, M. (2010). The creative countryside: Policy and practice in the UK rural cultural economy. *Journal of Rural Studies, 26*(3), 209–218. https://doi.org/10.1016/j.jrurstud.2010.01.001

Bendix, R. (2009). Heritage between economy and politics. An assessment from the perspective of cultural anthropology. In L. Smith & N. Akawaga (Eds.), *Intangible heritage* (pp. 253–269). Routledge.

Bendix, R., Eggert, A., & Peselmann, A. (2012). *Heritage regimes and the state*. Universitätsverlag Göttingen.

Bloomberg. (2013). *The rise of the intangible economy*. https://www.bloomberg.com/news/articles/2013-07-18/the-rise-of-the-intangible-economy-u-dot-s-dot-gdp-counts-r-and-d-artistic-creation

Bortolotto, C. (2011). Le trouble du patrimoine culturel immatériel. In C. Bortolotto (Ed.), *Le patrimoine culturel immatériel: enjeux d'une nouvelle catégorie* (pp. 21–46). Éditions de la Maison des sciences de l'homme.

Breglia, L. C. (2006). *Monumental ambivalence: The politics of heritage*. University of Texas.

Brumann, C., & Berliner, D. (Eds.). (2016). *World heritage on the ground. Ethnographic perspectives*. Berghahn Books.

Campillo, X., Ganyet, F., López, F., & Majorall, R. (1992). *L'Alt Urgell. Estructura territorial, recursos i activitat econòmica*. Caixa de Catalunya.

Collins, J. (2008). 'But what if I should need to defecate in your neighborhood, Madame?': Empire, redemption, and the 'tradition of the oppressed'. In a Brazilian world heritage site. *Cultural Anthropology, 23*(2), 279–328. https://doi.org/10.1111/j.1548-1360.2008.00010.x

Coombe, R. J., & Weiss, L. (2016). Neoliberalism, heritage regimes, and cultural rights. In L. Meskell (Ed.), *Global heritage: A reader* (pp. 43–69). Wiley-Blackwell.

Copertino, D. (2014). The tools of the trade: The materiality of architecture in the patrimonialization of 'Arab houses' in Damascus. *Journal of Material Culture, 19*(3), 327–335. https://doi.org/10.1177/1359183514540068

Cortés-Vázquez, J., Jiménez-Esquinas, G., & Sánchez-Carretero, C. (2017). Heritage and participatory governance: An analysis of political strategies and social fractures in Spain. *Anthropology Today, 33*(1), 15–18. https://doi.org/10.1111/1467-8322.12324

Cousin, S. (2008). L'Unesco et la doctrine du tourisme culturel. *Génealogie d'un bon tourisme, Civilisations, 57*, 41–56.

De Frantz, M. (2005). From cultural regeneration to discursive governance: Constructing the flagship of the 'Museumsquartier Vienna' as a plural symbol of change. *International Journal of Urban and Regional Research, 29*(1), 50–66. https://doi.org/10.1111/j.1468-2427.2005.00569.x

Del Mármol, C. (2012). *Pasados locales, políticas globales. Procesos de patrimonialización en un valle del Pirineo catalán* [Local Pasts, Global Politics. Heritagisation Processes in a Valley of the Catalonian Pyrenees], Germanias.

Del Mármol, C. (2017). The quest for a traditional style: Architecture and heritage processes in a Pyrenean Valley. *International Journal of Heritage Studies, 23*(10), 946–960. https://doi.org/10.1080/13527258.2017.1347888

Del Mármol, C., & Santamarina, B. (2019). Seeking authenticity: Heritage and value within the intangible economy. *Journal of Mediterranean Studies, 28*(2), 117–132.

Di Giovine, M. (2009). *The heritage-scape: UNESCO, World Heritage and Tourism*. Lexington Books.

Duro, J., & Rodrıguez, D. (2011). *Estimació del PIB turıstic per Catalunya, marques i comarques*. Technical report, Direcció General de Turisme, Generalitat de Catalunya.

European Commission. (1988). Future of the rural society. Commission communication transmitted to the Council and to the European Parliament on 29 July 1988. COM (88) 501 final 1988. Bulletin of the European Communities Supplement 4/88.

Franquesa, J. (2013). On keeping and selling: The political economy of heritage making in contemporary Spain. *Current Anthropology, 54*(3), 346–369. https://doi.org/10.1086/670620

Frigolé, J. (2005). *Dones que anaven pel món. Estudi etnogràfic de les trementinaires de la Vall de la Vansa i Tuixent* (Alt Urgell). Generalitat de Catalunya.

Frigolé, J. (2007). Producció cultural de lloc, memòria i terciarització de l'economia en una vall del Prepirineu. *Revista d'Etnologia de Catalunya, 30*, 70–80.

Fyall, A., & Rakic, T. (2006). The future market for world heritage sites. In A. Leask & A. Fyall (Eds.), *Managing world heritage sites* (pp. 159–175). Elsevier.

Garnham, N. (2005). From cultural to creative industries. *International Journal of Cultural Policy, 11*(1), 15–29. https://doi.org/10.1080/10286630500067606

Grätzer, M., Rengard, M., & Terlouw, F. (2015). The world heritage as a brand: Case study of world heritage brand usage by sites and their stakeholders. In *Context of Sweden and Denmark*. Linnaeus University. Retrieved June 3, 2022, from: https://www.diva-portal.org/smash/get/diva2:821916/FULLTEXT01.pdf

Goffman, E. (1975). *Frame analysis*. Penguin.

Hafstein, V. (2009). Intangible heritage as a list. From masterpieces to representation. In L. Smith & N. Akagawa (Eds.), *Intangible heritage* (pp. 93–111). Routledge.

Hafstein, V. (2014). Protection as dispossession: Government in the vernacular. In D. Kapchan (Ed.), *Cultural heritage in transit. Intangible rights as human rights* (pp. 22–57). University of Pennsylvania Press.

Halfacree, K. (1993). Locality and social representation: Space, discourse and alternative definitions of the rural. *Journal of Rural Studies, 9*(1), 23–37. https://doi.org/10.1016/0743-0167(93)90003-3

Haskel, J., & Westlake, S. (2017). *Capitalism without capital. The rise of the intangible economy*. Princeton University Press.

Harrison, R. (2013). *Heritage: Critical approaches*. Routledge.

Hawkes, J. (2001). *The fourth pillar of sustainability: Culture's essential role in public planning*. Common Ground.

Herzfeld, M. (1991). *A place in history. Social and monumental time in a Cretan town*. Princeton University.

Herzfeld, M. (2010). Engagement, gentrification, and the neoliberal hijacking of history. *Current Anthropology, 51*(2), 259–268. https://doi.org/10.1086/653420

Hottin, C. (2011). *De la théorisation à la mise en œuvre d'une nouvelle conception du patrimoine: le patrimoine immatériel. Patrimoine culturel immatériel. Premières expériences en France* [From the Theorisation to the Implementation of a New Concept of Heritage: Immaterial Heritage. Cultural Heritage. First Experiences in France]. Maison des Cultures du Monde – Actes Sud Babel.

Jiménez, S., & Prats, L. (2006). El turismo en Catalunya: evolución histórica y retos de futuro. *Pasos. Revista de Turismo y Patrimonio Cultural, 4*, 153–174. https://doi.org/10.25145/j.pasos.2006.04.011

Kiddey, R. (2017). From the ground up: Cultural heritage practices as tools for empowerment in the homeless heritage project. *International Journal of Heritage Studies, 24*(7), 694–708. https://doi.org/10.1080/13527258.2016.1274669

Kirshenblatt-Gimblett, B. (1998). *Destination culture. Tourism, museums, and heritage*. University of California Press.

Kirshenblatt-Gimblett, B. (2006). World heritage and cultural economics. In I. Karp, C. Kratz, L. Szwaja, & T. Ybarra-Fraust (Eds.), *Museum frictions: Public cultures/global transformations* (pp. 161–202). Duke University Press.

Korstanje, M. (2012). Reconsidering cultural tourism: An anthropologist's perspective. *Journal of Heritage Tourism, 7*(2), 179–184. https://doi.org/10.1080/1743873X.2011.639883

Korstanje, M. (2019). Lo No-Patrimoniable es Invisible a la Vista: La Obsesión Moderna por el Patrimonio Turístico. *El Periplo Sustentable, 36*, 432–446. https://doi.org/10.36677/elperiplo.v0i36.5068

Labadi, S. (2011). Intangible heritage and sustainable development: Realistic outcome or wishful thinking? *Heritage and Society, 4*(1), 115–118. https://doi.org/10.1179/hso.2011.4.1.115

Lai, S., & Can-Seng, O. (2015). Branded as a world heritage city: The politics afterwards. *Place Branding and Public Diplomacy, 11*(4), 276–292. https://doi.org/10.1057/pb.2015.12

Leblon, A. (2013). Les Paradoxes du Patrimoine. *Anthropologie et Sociétés, 37*(1), 177–193. https://doi.org/10.7202/1016153ar

López-i-Gelat, F., Tàbara, D., & Bartolomé, J. (2008). The rural in dispute: Discourses of rurality in the Pyrenees. *Geoforum, 40*, 602–612. https://doi.org/10.1016/j.geoforum.2009.04.008

Lowenthal, D. (1998). *The heritage crusade and the spoils of history*. Cambridge University Press.

Luzón, J., & Pi, C. (1999). La iniciativa comunitaria Leader de desarrollo rural en Europa: la experiencia de Catalunya. *Revista Econômica do Nordoste, 30*(2), 162–177. https://doi.org/10.29101/crcs.v28i0.16533

MacCannell, D. ([1976] 1999). *The tourist. A new theory of the leisure class.* University of California Press.

Mangset, P. (2009). The arm's length principle and the art funding system: A comparative approach. In M. Pyykkönen, N. Simainen, & S. Sokka (Eds.), *What about cultural policy?* (pp. 273–298). Jyväskyläm Minerva.

Mascarell, F. (2005). *La Cultura en l'era de la incerteza*. Roca.

Meskell, L. (2015). Transacting UNESCO world heritage: Gifts and exchanges on a global stage. *Social Anthropology, 23*, 3–21. https://doi.org/10.1111/1469-8676.12100

Meskell, L., & Brumann, C. (2015). UNESCO and new world orders. In L. Meskell (Ed.), *Global heritage: A reader* (pp. 22–42). Wiley.

Miller, D. (2008). The uses of value. *Geoforum, 39*, 1122–1132. https://doi.org/10.1016/j.geoforum.2006.03.009

Miller, T., & Yúdice, G. (2004). *Política Cultural. Gedisa.*

Moragas, M. (2008). *The cultural Olympiad of Barcelona in 1992: Good points and bad points. Lessons for the Future*. Centre d'Estudis Olímpics UAB, Barcelona.

Mulcahy, K. (2006). Cultural policy: Definitions and theoretical approaches. *The Journal of Arts Management, Law, and Society, 35*(4), 319–330. https://doi.org/10.3200/JAML.35.4.319-330

O'Brien, D. (2019). The cultural policy puzzle: Is cultural policy a problem or a solution for social inequality? *IPPR Progressive Review, 26*(2), 135–143. https://doi.org/10.1111/newe.12159

Parlament de Catalunya. (2000). DSPC 014/06 CPC 2000. http://www.parlament.cat/activitat/dspcc/06c014.pdf

Pascual, J. (2018). Cultural rights, local cultural policies and sustainable development: Constructing a coherent narrative. *Journal of Law, Social Justice & Global Development, 22*, 1–20. https://doi.org/10.31273/LGD.2018.2204

Porter, B., & Salazar, N. (2005). Heritage tourism conflict and the public interest: An introduction. *International Journal of Heritage Studies, 11*(5), 361–370. https://doi.org/10.1080/13527250500337397

Poulot, D. (2006). *Une histoire de patrimoine en Occident, XVIIIe –XXIe siècle* [A Heritage History in the West, 18th - 21st Centuries]. Presses Universitaires de France.

Prats, L. (2003). Turismo + patrimonio: ¿desarrollo? *Pasos, Revista de Turismo y Patrimonio cultural, 1*(2), 127–133. https://doi.org/10.25145/j.pasos.2003.01.012

Pratt, A. (2008). Locating the cultural economy. In H. K. Anheier & Y. R. Isar (Eds.), *Cultures and globalization: The cultural economy* (pp. 42–51). Sage.

Rius-Ulldemolins, J. & Klein, R. (2020). From a Barrio Chino urban stigma to the Raval cultural brand: Urban memory and cultural policies in the renewal of Central Barcelona. *Journal of Urban History*. December 2020. https://doi.org/10.1177/0096144220981117

Sánchez-Carretero, C., Múñoz-Albaladejo, J., Ruiz-Blanch, A., & Roura-Expósito, J. (Eds.) (2019). *El Imperativo de la Participación en la Gestión Patrimonial*. CSIC.

Santamarina, B. (2013). Los mapas geopolíticos de la Unesco: entre la distinción y la diferencia están las asimetrías. *Revista de Antropología Social, 22*, 263–286. https://doi.org/10.5209/rev_RASO.2013.v22.43191

Scott, A. (2008). Resurgent Metropolis: Economy, society and urbanization in an interconnected world. *International Journal of Urban and Regional Research, 32*(3), 548–564. https://doi.org/10.1111/j.1468-2427.2008.00795.x

Scott, A. (2014). Beyond the Creative City: Cognitive–cultural capitalism and the new urbanism. *Regional Studies, 48*(4), 565–578. https://doi.org/10.1080/00343404.2014.891010

Selwyn, T. (2007). The political economy of enchantment. Formations in the anthropology of tourism. *Suomen Antropologi: Journal of the Finnish Anthropological Society, 32*(2), 48–70.

Sivaramakrishnan, K., & Vaccaro, I. (2007). Postindustrial natures: Hyper-mobility and place-attachments. *Journal of Social Anthropology, 14*(3), 301–317. https://doi.org/10.1017/S0964028206002643

Smith, L. (2006). *Uses of heritage*. Routledge.

Smith, L., & Akagawa, N. (Eds.). (2009). *Intangible heritage*. Routledge.

Stylianou-Lambert, N., & Christodoulou-Yerali, M. (2014). Museums and cultural sustainability: Stakeholders, forces, and cultural policies. *International Journal of Cultural Policy, 20*(5), 566–587. https://doi.org/10.1080/10286632.2013.874420

Throsby, D. (2001). *Economics and culture*. Cambridge University Press.

Timothy, D. (2018). Making sense of heritage tourism: Research trends in a maturing field of study. *Tourism Management Perspectives, 25*, 177–180. https://doi.org/10.1016/j.tmp.2017.11.018

Tresserras, J. (2008). *L'economia de la identitat: el nostre valor afegit*. Fundació IRLA.

Tsing, A. (2015). *The mushroom at the end of the world. On the possibility of life in capitalist ruins*. Princeton University Press.

Tulla, A. (1994). *Procés de transformació agrària en àrees de muntanya. Les explotacions de producció lletera com a motor de canvi a les comarques de la Cerdanya, el Capcir, l'Alt Urgell i el Principat d'Andorra*. Institut Cartogràfic de Catalunya (col. Tesis doctorals), Barcelona.

UNESCO. (2003). *Convention for the safeguarding of the intangible cultural heritage*. UNESCO Publishing.

Urfalino, P. (1996). *L'invention de la politique culturelle*. La Documentation Française.

Urry, J. (1995). *Consuming places*. Routledge.

Vaccaro, I., & Beltran, O. (Eds.). (2010). *Social and ecological history of the Pyrenees: State, market, and landscape*. Left Coast Press.

Verloo, M. (2005). Displacement and empowerment: Reflections on the concept and practice of the Council of Europe Approach to gender mainstreaming and gender equality. *Social Politics: International Studies in Gender, State & Society, 12*(3), 344–365. https://doi.org/10.1093/sp/jxi019

Vlès, V. (2014). *Métastations: Mutations urbaines des stations de montagne—Un regardpyrénéen*. Presses Universitaires Bourdeaux.

Winter, T. (2011). The political economies of heritage. In H. Anheier & Y. Raj Isar (Eds.), *Heritage, memory & identity* (pp. 7–81). Sage Publications Ltd.

WORLD TOURISM ORGANIZATION (UNWTO) (2013). Sustainable tourism for development guidebook. .

WORLD TOURISM ORGANIZATION (UNWTO). (2018). Tourism and culture synergies. .

Wu, C. (2003). *Privatising culture*. Verso books.

Part IV
Cultural Leadership in Transition Tourism Beyond the Cultural Sector

Cultural Relationship Marketing Prospects and Potential

Oleksandr Savych and Tetiana Shkoda

Abstract The purpose of the research is to study the impact of cultural relationship marketing on business activity and substantiate the development prospects of its potential for a modern enterprise using the example of automotive components distributing companies in Ukraine.

The methodology of the research includes the following: structural and logic analysis and critical analysis for providing the literature review of the research, expert interviews for defining the characteristics of cultural relationship marketing, economic and mathematical modelling, and the graph-analytical methodology for substantiating the potential of cultural relationship marketing of the automotive components distributing companies in Ukraine.

The main findings of the research proved the importance of the cultural heritage of the company's residence country, its tourist attractiveness, religious and food specifics for building strong relations of automotive components distributors with their stakeholders. Management decisions are made by people, so it is important to build relationships between companies based on the relationships between people who represent the company, and on the other hand—to consider the characteristics of key managers, including their nature and behavior, formed by cultural characteristics of regions and countries.

The practical implications of the research related to the first-time research study's substantiation of the potential of cultural relationship marketing in the sphere of automotive components distribution.

The research limitations include: (1) expert interviews were conducted online because of the COVID-19 restrictions and (2) the research results are obtained before the war in Ukraine in 2022.

O. Savych (✉)
Department of Marketing named after A.F. Pavlenko, Kyiv National Economic University named after Vadym Hetman, Kyiv, Ukraine
e-mail: savych@kneu.edu.ua

T. Shkoda
Department of Business Economics and Entrepreneurship, Kyiv National Economic University named after Vadym Hetman, Kyiv, Ukraine
e-mail: shkoda-tetiana@kneu.edu.ua

E. Borin et al. (eds.), *Cultural Leadership in Transition Tourism*, Contributions to Management Science, https://doi.org/10.1007/978-3-031-14121-8_14

1 Introduction

Cultural relationship marketing is one of the new potential paths of cultural tourism development, tourism management and cultural destination management, governance and leadership, which is the thematic section of this monograph.

Studying the impact of cultural relationship marketing on business activity and substantiating the prospects of its potential for a modern enterprise using the example of automotive components distributing companies in Ukraine is the purpose of this research. Automotive components distributors perform as market makers in Ukraine, so they must satisfy all stakeholders. Nowadays, only rational benefits and advantages do not function independently of relationships and emotional structural elements. That is why, also in order to strengthen its position and power in the market, automotive components distributors need to empower and build their own brand, which is a strong emotional brand. Therefore, a smart company shapes its brand with the popularity and emotional heritage of culture, cuisine, and traditions of the country. All this shapes cultural relationship marketing in the business environment of the company. Our research is focused on such a group of stakeholders as suppliers, who are all European, Asian, and American companies, whose managers bring to bear different understandings of main lifestyle traditions and behavior. So, the automotive components distributors in Ukraine depend on mutual common familiarity and understanding, and toward that end, they use tourism to do it.

2 Literature Review

Previous studies considered the impact of culture on relationship marketing paying attention to critical importance of international relationships for business performance within cultural dimensions (Samaha et al., 2014), specifics of impact on international services in the banking sector (Schumann, 2009), cultural competence as the missing piece of relationship marketing (Norman & Ogbuehi, 2018), the predominance of the cultural perspective of fragmentation, which means a culture originating from relationships through intersections of cultures, a culture of boundaries (Larentis et al., 2018), etc. During business, tourist trips engender this intersection of different cultures, especially during business negotiations. And automotive components distributors try to attract suppliers to enter the specific market using cultural relationship marketing.

Some researchers (Nilson & Thorrell, 2018) consider cultural heritage as a vehicle for business purposes. Cultural heritage is also studied as a part of the past, which is also selected by people specifically for economic or business purposes (Khakzad, 2015). The authors believe that such an approach allows for using cultural heritage as an element of cultural relationship marketing and the whole research satisfies the extended concept of culture as the way of life according to T. Bennett (2015). In this context, the Hofstede's (1991) understanding of culture should be mentioned as the pattern of values, norms, and beliefs that impact the way people evaluate information. As a result, there is a difference in processing and evaluating environmental information. From the authors' point of view, this approach applies the extended concept of culture by Bennett (2015) at the organizational level.

Culture and heritage, as well as tourism, are considered by famous nation brand researchers (Fan, 2005; Anholt, 2006; Aronczyk, 2008; Melnychenko et al., 2021) as key elements of a nation brand. For example, M. Aronczyk (2008) considers "nation branding" as a concept and practice, which can be realized as a communications strategy and a practical initiative. In this case tourism and culture help to manage the image that is projected by the country to the world. Destination (tourism) branding and marketing initiatives are considered as a part of the nation's brand.

Scientists (Mazaraki et al., 2020) also consider the purpose of the tourism brand of Ukraine, which, from the authors' point of view, can be considered from a business perspective and applied to different business fields for building strong relations with stakeholders. In this case, the purpose of a national tourism brand is to form consistently an image of an extremely interesting country for doing business, which should be studied, with the help of appropriate tourist products and communications as with the help of attractive conditions for doing business, caring for the positive business and cultural experience and impressions of stakeholders during business trips. This point of view is supported by the earlier research of Evans (2003), who considers tourism as a cultural component of globalization that impacts nations as well as cities and businesses.

Cultural tourism is considered as an element of destination marketing by Ren and Blichfeldt (2011). However, these researchers admit the growing symbolic and intrinsic economic value brought by corporate brands to the cultural city and consider the issue through the example of Zakopane, a Polish ski resort town. They do not closely investigate the business context of tourism attractiveness and its usage in cultural relationship marketing.

Cultural relationship marketing's relationship with the automotive sector is studied within the framework of the concept of marketing efforts consolidation in the global automotive market (Savych & Shkoda, 2020, 2021), where distributors are considered as an important part of such processes as logistics, servicing, and final client decisions. Additionally, the authors of the concept highlight that the consolidation of marketing efforts is aimed at identifying the demand and satisfaction of the world automotive market, which includes also national, geographical, historical, and other conditions in each respective market. Based on this approach cultural relationship marketing is defined as establishing and developing successful relationships with stakeholders using culture as a marketing tool. The authors believe in the possibility of making widespread the concept of the marketing efforts consolidation on the automotive components distribution business, which is the interrelated business with the automotive sector previously studied in the authors' research.

So, *the proposed hypothesis* is the following one: if the nation brand, including tourism and cultural heritage, is attractive for businesses than they can better realize their potential of cultural relationship marketing.

3 Methodology

The methodology of the research includes the following: structural and logic analysis and critical analysis for the literature review, expert interviews for defining the characteristics of cultural relationship marketing, economic and mathematical modelling, and the graph-analytical methodology for substantiating the potential of cultural relationship marketing of the automotive components distributing companies in Ukraine.

The current vision and systemic limitations of cultural relationship marketing on business activity are formed through the proposition to assess the potential of cultural relationship marketing using the example of automotive components distributing companies in Ukraine. In the context of this study, the assessment of the potential of cultural relationship marketing is based on research in accordance with the elements of the graph-analytical method of analysis "Square of potential" (Fedonin et al., 2006). In particular, for the purpose of our study, the following groups of indicators are proposed, namely: the assessment of the characteristics of country's business tourist attractiveness (Vector I), the effectiveness of the organizational culture of management of the automotive components distributing companies (Vector II), the general effectiveness of marketing activities (Vector III), the cultural relationship marketing in business activity (Vector IV).

The group of indicators of the potential of cultural relationship marketing based on the example of automotive components distributing companies in Ukraine (Table 1) are measured on the following scale: weak—1 point, quite weak—2 points, not strong and not weak—3 points, quite strong—4 points, and strong—5 points.

The average estimation of the vectors of the potential of cultural relationship marketing at the example of automotive components distributing companies in Ukraine is conducted in accordance with formula (1).

$$K_m = \frac{\sum_{i=1}^{n}\sum_{j=1}^{k}\alpha_j\beta_{ij}}{n}, \tag{1}$$

where K_m—the average indicator of each vector, points;

$i = 1, 2, 3, \ldots, n$—number of experts;

$j = 1, 2, 3, \ldots, k$—the number of proposed characteristics of each vector;

α_i, β_j—weight of the jth characteristic of the vector and evaluation of the jth characteristic by the ith expert.

The expert interviews' method was applied to validate the proposed indicators according to formula (1). Such a method was used by Aronczyk (2008) for gathering her scientific research data on nation brand, in which there were a total of 30 interview respondents (10 main experts and 20 additional ones). In the authors' current research, the following criteria were taken for selecting the experts: (1) being the top manager in the automotive distributing company, (2) strong experience in attracting the suppliers in the network of automotive distributing association and

Table 1 Potential of cultural relationship marketing assessment

№	Vectors	Validity	Estimation
1	*2*	*3*	*4*
I vector	Characteristics of country's business tourist attractiveness		
1	The business image of products and services of a specific country	0.1	130
2	Easiness of communication with official authorities	0.15	150
3	The power to attract foreign companies to enter the country's business environment	0.15	138
4	The level of interest in visiting the country	0.15	148
5	The country's employee's reputation for competence, openness, and friendliness	0.1	155
6	The business image of cuisine and traditions of a specific country	0.1	149
7	The popularity and emotional heritage of the country's culture	0.1	145
8	The national business perception of a cultural heritage	0.15	151
I vector	The average estimation of characteristics		**3.95**
II vector	Evaluation of the effectiveness of the organizational culture of management of the automotive distributing companies' components		
1	The automotive components distributing companies work purposefully	0.1	155
2	The flexibility of the automotive components distributing companies is sufficient	0.15	165
3	The organizational culture is well visible	0.15	164
4	Power is delegated sufficiently	0.15	150
5	The management of the automotive components distributing companies have enough time to discuss strategic tasks	0.1	144
6	There is a spirit of cooperation in the automotive components distributing companies	0.15	153
7	Managers have enough information	0.1	147
8	The automotive components distributing companies have a strong corporate brand	0.1	155
II vector	The average estimation of indicators		**4.18**
III vector	Evaluation of the general effectiveness of marketing activities		
1	Corporate identity (developed areas)	0.15	153
2	Adaptation to conditions of unstable environment	0.15	164
3	Weak competitive pressure	0.1	119
4	Consumer commitment	0.15	156
5	Marketing strategies	0.15	157
6	Objectives of the enterprise	0.1	144
7	Opportunities to penetrate new markets	0.1	147
8	Opportunity to attract new employees	0.1	167
	The average estimation of indicators		4.11

(continued)

Table 1 (continued)

№	Vectors	Validity	Estimation
III vector			
IV vector	General assessment of the cultural relationship marketing in business activity		
1	Participation in conferences, exhibitions, and forums	0.15	130
2	Publications in tourist journals, corporate journals, own manuals, methodical materials	0.1	143
3	Internet projects on tourist, cultural subjects	0.1	154
4	Competitions of tourist and culturally innovative projects	0.1	143
5	Media	0.15	153
6	Social networks	0.15	151
7	Public hearings, round tables with government officials and the NGOs	0.1	146
8	CSR events	0.15	152
IV vector	The average estimation of indicators		3.96

(3) self-identification as the particular country's business representative with the help of such instruments as culture, tourism, cuisine, etc.

The expert interviews were made during September–October 2021 among managers of Nexus Automotive Ukraine (NAU, 2021), the automotive distributors association which unites five leading distributive companies of Ukraine, both cars and heavy-duty business fields. The number of experts is 37, which is equal to the number of top managers in Nexus Automotive Ukraine. This association is a member of the global Nexus Automotive International with a volume turnover per year of 30 billion euros (NAI, 2021). The Ukrainian specifics are chosen because the authors are Ukrainian researchers and consultants, and were given approval from Nexus Automotive Ukraine to interview their top managers.

4 Results

The obtained results of the experts' evaluations were calculated and represented in Table 1. The table contains four sections satisfying the appropriate vectors of the potential of cultural relationship marketing at the example of automotive components distributing companies in Ukraine. The indicators for each vector were chosen by the experts of Nexus Automotive Ukraine (NAU, 2021) during the interviews. The list of the proposed indicators was created by the authors based on previous related studies (Anholt, 2006; Fedonin et al., 2006; Gernego et al., 2021).

The schedule of vectors estimated by average estimations transformed into percentages (Tables 1–4) is represented within Fig. 2.

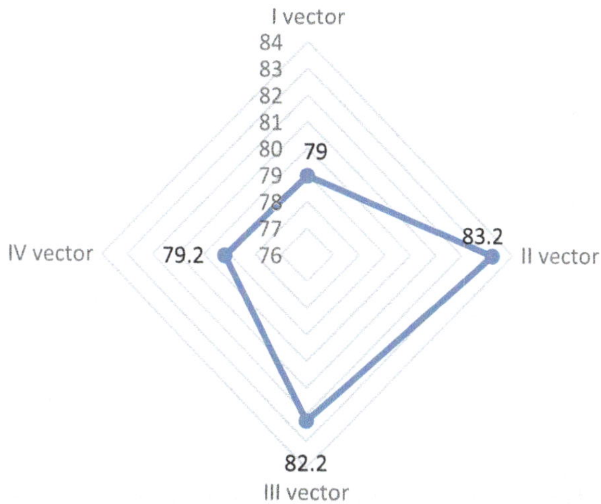

Fig. 1 The square of the potential of cultural relationship marketing at the example of the automotive components distributing companies in Ukraine (composed by the authors)

Using the obtained ratios of the components of the potential of the studied subjects makes it possible to determine the level of each component, and thus to identify the appropriate relationships between the levels of different vectors (Fig. 1) and decide which vectors to develop next.

Thus, the proposed method makes it possible to assess the prospects for the development of potential cultural relationship marketing based on the example of the automotive components distributing companies in Ukraine. According to the data of Fig. 1, such components of the potential of cultural relationship marketing as the country's business tourist attractiveness and the cultural relationship marketing in business activity are recommended as areas in need of more development by the automotive components distributing companies in Ukraine.

The results of the authors' research correspond to the data of Brand Finance (2021) for the nation brand of Ukraine (Fig. 2), components of which include tourism, culture and heritage, business, and trade.

The 55th position of Ukraine in the Brand Finance ranking for the nation's brand is not very high. It is located in the middle of rankings calculated for 100 countries. It demonstrates that the nation's brand of Ukraine is not very attractive. In the authors' research, the weakest vectors (Fig. 1) are the country's business tourist attractiveness and the cultural relationship marketing in business activity.

The comparison of these results with the authors' research results proves this research's hypothesis that if the nation's brand including tourism and cultural heritage is attractive to businesses then they can realize better their potential for cultural relationship marketing.

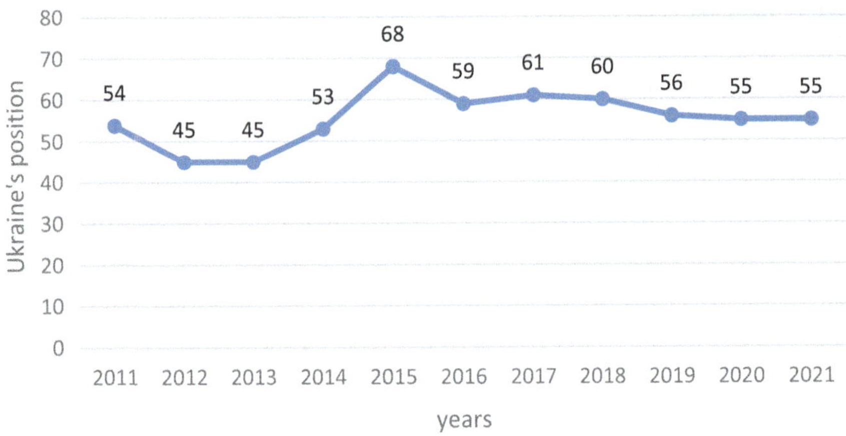

Fig. 2 The position of Ukraine in the Brand Finance ranking for the nation's brand, 2011–2021 (composed by the authors based on Brand Finance, 2021)

5 Conclusions

The main findings of the research proved the importance of the cultural heritage of the company's residence region, its tourist attractiveness, religious and food specifics for building strong relations of automotive components distributors with their stakeholders. The national business perception of cultural heritage is defined in the research as an element of such component of assessment of cultural relationship marketing as characteristics of country's business tourist attractiveness with the expert's estimated value of 151 and validity 0.15. The business image of cuisine and traditions of a specific country also belong to the elements of characteristics of a country's business tourist attractiveness with an estimated value of 149 and validity 0.1 according to the experts' opinion. So, it can be concluded that the importance of cultural heritage for building strong relationships with stakeholders of the business community of automotive distributing components in Ukraine is more important than cuisine and traditions, including religious ones.

Management decisions are made by people, so it is important to build relationships between companies based on relationships between people who represent the company on the one hand, and on the other hand that takes into account the characteristics of key managers, including their nature and behavior, formed by cultural characteristics of regions and countries.

The average estimation of the Vector II indicators of the effectiveness of the organizational culture of management of the automotive distributing companies' components is 4.18 (or 83.2%) that exceeds the average estimation of the Vector I indicators of the characteristics of country's business tourist attractiveness, which is equal to 3.95 (or 79%). The authors recommend to improve the relationship with such stakeholders as suppliers using those characteristics of country's business tourist attractiveness is under the real impact of the organizational culture of management of the automotive distributing companies' components.

The main research hypothesis is proved based on the improved research methodology and the obtained results of experts' evaluations. The experts represented the automotive distributors association which unites five leading distributive companies in Ukraine. The study is considered to become the research with the usage of the results of the Ukrainian experts' interviews.

The results of comparison of the research's results with the nation brand ranking Brand Finance also confirmed the research's hypothesis.

The research limitations are related to the availability of experts because of the COVID-19 restrictions in Ukraine and the necessity to use online applications for getting the appropriate research data.

Such vectors of the potential of cultural relationship marketing as the country's business tourist attractiveness and the cultural relationship marketing in business activity are recommended by the authors for further development by the automotive components distributing companies in Ukraine.

The research methodology is applicable in other industrial sectors and business segments, where it is possible to make expert interviews and process the obtained data according to this methodology. It is only recommended to use the Vector II indicators reflecting the specifics of the researched industrial sector or business segment.

The other possibility for applying the research methodology is to realize the research in different countries and make the appropriate comparison for understanding the specifics of the square of the potential of cultural relationship marketing and its components.

References

Anholt, S. (2006). The Anholt-GMI City brands index: How the world sees the world's cities. *Place Branding and Public Diplomacy, 2*(1), 18–31. https://doi.org/10.1057/palgrave.pb.5990042

Aronczyk, M. (2008). 'Living the brand': Nationality, globality and the identity strategies of nation branding consultants. *International Journal of Communication, 2*, 41–65.

Bennett, T. (2015). Cultural studies and the culture concept. *Cultural Studies, 29*(4), 546–568. https://doi.org/10.1080/09502386.2014.1000605

Brand Finance. (2021). *Nation brands 2021. The annual report on the most valuable and strongest nation brands*. Retrieved November 21, 2021 from https://brandirectory.com/rankings/nation-brands

Evans, G. (2003). Hard-branding the cultural city – From Prado to Prada. *International Journal of Urban and Regional Research, 27*(2), 417–440.

Fan, Y. (2005). Branding the nation: What is being branded. *Journal of Vacation Marketing, 12*(1), 5–14. https://doi.org/10.1177/1356766706056633

Fedonin, O. S., Riepina, O. I., & Oleksiuk, I. M. (2006). *Potential of the enterprise: Formation and evaluation*. Study Book.

Gernego, I., Shkoda, T., & Savych, O. (2021). Marketing tools in strategic management of young scientists' human capital. *Baltic Journal of Economic Studies, 7*(4), 43–53. https://doi.org/10.30525/2256-0742/2021-7-4-43-53

Hofstede, G., Hofstede, G. J., & Minkov, M. (1991). *Cultures and organizations: Software of the mind*. McGraw-Hill.

Khakzad, S. (2015). *Integrated approach in management of coastal cultural heritage*. Dissertation presented in partial fulfilment of the requirements for the degree of PhD in Engineering. Retrieved November 19, 2021 from https://www.academia.edu/13428696/Integrated_Approach_in_Management_of_Coastal_Cultural_Heritage

Larentis, F., Antonello, C. S., & Slongo, L. A. (2018). Organizational culture and relationship marketing: An interorganizational perspective. *Revista Brasilieira De Gestao De Negocios, 20*(1), 37–56. https://doi.org/10.7819/rbgn.v20i1.3688

Mazaraki, A., Mykhailichenko, H., Melnychenko, S., Melnyk, T., & Miroshnik, O. (2020). Development of tourism brand of Ukraine. *Financial and Credit Activity: Problems of Theory and Practice, 3*(34), 421–431. https://doi.org/10.18371/fcaptp.v3i34.215588

Melnychenko, S., Bosovska, M., & Okhrimenko, A. (2021). The formation of a nation tourism brand of Ukraine. *Baltic Journal of Economic Studies, 7*(2), 161–169. https://doi.org/10.30525/2256-0742/2021-7-2-161-169

Nexus Automotive Ukraine (NAU). (2021). Retrieved November, 20, 2021, from http://nexusua.com/en_GB/

Nexus Automotive International (NAI). (2021). Retrieved November, 20, 2021, from https://www.nexusautomotiveinternational.eu

Nilson, T., & Thorrell, K. (2018). *Cultural heritage preservation: The past, the present and the future*. Halmstad University Press. https://www.diva-portal.org/smash/get/diva2:1224014/FULLTEXT01.pdf

Norman, M., & Ogbuehi, A. (2018). *Cultural competence: The missing piece of relationship marketing*. Retrieved November, 21, 2021, from https://digitalcommons.kennesaw.edu/cgi/viewcontent.cgi?article=1296&context=ama_proceedings

Ren, C., & Blichfeldt, B. S. (2011). One clear image? Challenging simplicity in place branding. *Scandinavian Journal of Hospitality and Tourism, 11*(4), 416–434.

Samaha, S. A., Beck, J. T., & Palmatier, R. W. (2014). The role of culture in international relationship marketing. *Journal of Marketing, 78*(5), 78–98. https://doi.org/10.1509/jm.13.0185

Savych, O., & Shkoda, T. (2020). The concept of marketing efforts consolidation at the global car market. In J. Stankevičienė, T. Lankauskienė, I. Jokšienė, V. Davidavičienė, I. Skačkauskienė, R. Tamošiūnienė, V. Jurkevičius (Eds.), *Business and Management 2020: Selected papers of the 11th International Scientific Conference*, Vilnius, Lithuania, May 07–08, 2020. Vilnius, Technika, 456, 63–72. https://doi.org/10.3846/bm.2020.542

Savych, O., & Shkoda, T. (2021). Impact of key marketing tools on the global car market development. *Communications - Scientific Letters of the University of Žilina, 23*(4), A264–A276. https://doi.org/10.26552/com.C.2021.4.A264-A276

Schumann, J. H. (2009). *The impact of culture on relationship marketing in international services. A target group-specific analysis in the context of banking services*. Gabler Verlag. https://doi.org/10.1007/978-3-8349-8379-4

Inclusive Tourism, a Cultural Paradigm Shift in Approaching Tourism Development

Ingrid Molderez and Kavindya Perera

Abstract In this paper, the purpose is to highlight another concept of inclusive tourism than what is usually understood as "inclusiveness." Despite its very specific meaning: tourism is accessible to everybody by taking away the barriers that prevent a person from enjoying tourism, inclusiveness is much broader than only focussing on the demand side. The supply side of inclusive tourism stresses the underprivileged providers of tourism in the Base/Bottom of the Pyramid (BOP) areas. The paper is rather conceptual and describes the relevance of the BOP frameworks with respect to tourism. Research has shown that the most inspiring cases can be found where the two criteria of entrepreneurship and co-inventor meet. However, not many cases in the tourism sector can be found here. This approach of tourism is important because entrepreneurs are appreciated due to their capabilities. They develop tourism models from their perspective, in contradiction to models that focus on applying western business perspectives. The local community is involved because their skills and motivation for participation are central. Local people become a change agent and a developer for a greater understanding of their local culture which leads to positive contributions such as the conservation of the natural and cultural heritage. This approach of inclusive tourism is illustrated by the Hiriwaduna Village Tour in Sri Lanka. Despite the importance of this model, cases in the tourism sector are rare. This underlines the fact that the tourism sector is still too much seen from the demand side of the tourists and does not include enough input on how local people would like to develop their community into a flourishing and self-reliant one.

I. Molderez (✉)
KU Leuven, Faculty of Economics and Business, CEDON (Centre for Economics and Sustainable Entrepreneurship), Brussels, Belgium
e-mail: ingrid.molderez@kuleuven.be

K. Perera
KU Leuven, Faculty of Economics and Business, CEDON (Centre for Economics and Sustainable Entrepreneurship), Brussels, Belgium

Kyoto University, Kyoto, Japan

E. Borin et al. (eds.), *Cultural Leadership in Transition Tourism*, Contributions to Management Science, https://doi.org/10.1007/978-3-031-14121-8_15

1 Introduction

The United Nations (UN) is the backbone of sustainability and development of an ever-changing world and its issues related to the survival of people and the planet. Eight Millennium Development Goals (MDGs) were set up in 2000 by the UN for a period of 15 years with the aim of reducing extreme poverty. They were followed up in 2015 by 17 Sustainable Development Goals (SDGs) targeting peace and prosperity for people and the planet, now and into the future with a deadline of 2030 (United Nations, 2015). Everybody and every economic sector have a role to play, including tourism. Tourism is included in SDG 8, i.e., Decent Work and Economic Growth. The related target is 8.9.

> By 2030, devise and implement policies to promote sustainable tourism that creates jobs and promotes local culture and products.

Tourism provides one in every 11 jobs in the world (Rifai, 2017) and it has been a highlight for decades. Since 2020, it has been deeply affected by COVID-19 and associated economic recession. The world is changing and so is tourism. Travel industry could see an 80% decline in international arrivals for 2020 amid a crisis that threatens the livelihood of up to 120 million people. Hence, the contribution toward sustainability goals through tourism has a major impact on the global economic growth as well as on the society as a whole. However, it is unlikely that the past critics on tourism exclusion and poverty will offer much guidance since the low-income population has a wide range of different needs, among others as a result of the effect of COVID-19.

In order to achieve the target and overcome poverty, it is also required to create a better structure by engaging local people from deprived regions while considering them as entrepreneurs in inclusive tourism rather than just as sellers of low-quality souvenirs that have a negative impact on their culture and environment. Inspired by the literature related to the base of the pyramid, i.e., people living in deprived regions, this paper aims at emphasizing a tourism concept that centralizes local people with a low income and the belief that they know best what kind of tourism is affordable for the environment they are living in. A different kind of inclusive tourism is introduced than what is usually understood by the term inclusiveness. When striving for engagement through entrepreneurship and building local capacity from the bottom up, local people are given an opportunity to overcome their difficulties while contributing to tourism. We argue that considering a bottom-up approach and giving priority to the local community will add value to this inclusive tourism concept. In the second section of the paper different concepts of inclusive tourism are explained. In the third section, the base of the pyramid is described. How this concept relates to tourism is outlined in the fourth section. The final section illustrates this approach in the tourism sector with the Hiriwaduna Village Tour in Sri Lanka. The chapter ends with a conclusion.

2 Different Concepts of Inclusive Tourism

For a long time, inclusive tourism has had a very specific meaning within tourism research (Münch & Ulrich, 2011). Its synonyms accessible tourism and disabled tourism show that inclusiveness is mainly related to the demand side of this specific form of tourism. People's physical limitations, disabilities, age, or income should not represent barriers to them from enjoying tourist products and services. The idea is that every human being must have access to tourism, which makes it a social right. Although the concept is often related to specific categories, Münch and Ulrich (2011) recommend to interpret it in a more holistic way. Everything that is preventing a person from enjoying tourism, including poverty, must be eliminated. In this way, inclusive tourism is a fancier label than the much older concept of social tourism (Kakoudakis et al., 2017). Hunzicker (1951), cited in McCabe and Qiao (2020), was the first to use the term social tourism for economically weak or otherwise disadvantaged people who are participating in tourism. Since then, social tourism has become an umbrella concept for inclusive, accessible, disabled, equitable, and associative tourism. The benefits of social tourism are many: family bonding, quality time spent together, increased leisure time, better mental and physical health, opportunities for new experiences, relaxation, and recuperation (McCabe & Qiao, 2020). It also stimulates a fairer and more equal society when everyone has an opportunity to participate.

Elimination of inequalities and social exclusion play a critical role when fulfilling the inclusive tourism concept. Hence, it is important to acknowledge the local tensions, discriminatory practices, and the many barriers to social mobility. Social exclusion is generally defined as the denial of equal access to societal opportunities (Buvinic, 2004). It is rooted in inequalities between citizens in terms of the distribution of wealth, access to income sources, access to good quality social services and infrastructures, and social and political participation in civil society (Burchardt et al., 2002; Buvinic, 2004). Social Inclusion is a process that ensures that those at risk of poverty and social exclusion gain the opportunities and resources necessary to participate fully in economic, social, political, and cultural life and to enjoy a standard of living that is considered normal in society in which they live. It ensures that they have greater participation in decision-making which affects their lives and access to their fundamental rights (Commission of the European Communities, 2003: 9). Therefore, the obstacles related to social inclusion need to be eliminated in order to achieve sustainability.

Seeing tourism as a social right for everybody, irrespective of barriers, also has its side effects. When not managed properly it can become part of all-inclusive tourism, a dark side of tourism and well-known practice in the hospitality sector. Packages include unlimited food, drinks, entertainment, and activities at the booking price through the travel agency or the resort. For the last two to three decades, social scientists have critiqued all-inclusive resorts because they tend to result in enclaves that are out of bounds to the local population, limiting opportunities for local entrepreneurs to benefit by selling goods or services to tourists, and they result in high levels of leakage of tourist spending, with much going to foreign hotel chains and travel agents (Gibson, 2009; Scheyvens, 2011; Scheyvens & Biddulph, 2018).

In the past, especially governments, national health services, trade unions and other non-governmental organizations took the initiative of providing facilities for disadvantaged people to enjoy tourism. Encouraging people to participate in tourism is similar to initiatives of increasing cultural participation (Jancovich, 2017). However, as it becomes clear from Jancovich's (2017) research the greatest barriers are the resistance to change and the narrow range of people that are involved in the decision-making process. Also, Chinyowa (2015: 12) is rather critical about forcing people's participation:

> (. . .) strategies deployed in the name of participation, dialogue, giving voice and empowerment for the marginalized usually create the illusion of liberation while they reinforce, or even perpetuate the prevailing oppressive structures.

Therefore, participation should not be imposed, but led by the local people themselves. These kinds of bottom-up initiatives are, however, rather scarce (Jancovich, 2017) and their significance is often underrated by the government (Scheyvens, 2006). From a supply-side perspective, the same people from deprived positions should have the opportunity to participate as producers. This is especially the case for people living in economically constrained situations. The definition for inclusive tourism that includes both the demand and the supply side, is the one by Scheyvens and Biddulph (2018: 592):

> Inclusive tourism can be understood as transformative tourism in which marginalized groups are engaged in ethical production or consumption of tourism and the sharing of its benefits.

Seven elements are important in their concept of inclusive tourism: including marginalized people as tourism producers (1), as tourism consumers (2), enlarging the tourism map by places not conventionally frequented by tourists (3), widening participation in tourism decision-making (4), promotion of mutual understanding and respect (5), finding dignified and appropriate ways for self-representation (6), and transforming power relations (7) (Scheyvens & Biddulph, 2018).

In this chapter, marginalized is avoided as a word because of its negative connotation. Moreover, the focus should be on people living in or coming from deprived regions or situations. Focusing on people from deprived regions who are providing tourist products and services that are part of their local culture can result in a cultural change, a different mindset, or even a cultural paradigm shift. This includes locals engaging in delivering the type of cultures impacting tourists. One cannot become a cultural tourist (domestic or international) without cultural artefacts to experience, which in return requires a certain level of cultural capital or competence. The expansion of cultural tourism in the direction of intangible heritage and contemporary culture creates more attention for the increasing integration between tourism and the creative economy (United Nations, 2019). Creative experiences such as artistic creation, dance, and cookery, are now also being used to frame destination culture. Gastronomy—stimulated through the development of cultural routes linked to food and wine brands, including olive oil routes, cheese routes, and wine routes

(Richards, 2018). As a result of the multiplying effect of tourism on the economy, it also generates new jobs in other sectors, apart from valuing another culture, mutual learning, and boosting the pride of the local people (Scheyvens, 2006).

The tourism sector is already equipped with many options to integrate local culture from deprived regions, such as community-based tourism, ecotourism, homestays, rural tourism, and agrotourism (Scheyvens & Biddulph, 2018; Priatmoko et al., 2021). Especially community-based tourism and homestays are similar to the inclusive tourism model. Participation of the community toward planning and managing tourism is the central idea of community-based tourism (Schott & Nhem, 2018). Community-based tourism starts from the idea that tourists want to learn about local people's daily life. Priatmoko et al. (2021) give the example of residents preparing food that requires a slow-cooking fire or using banana leaves to wrap food. It is often thought that experts are the first to bring ideas to the local community who apply them. However, local people already integrate aspects of sustainability without using the concept of sustainability, such as the beach fale in Samoa, an indigenous, small-scale, locally owned initiative of open-sided huts built according to local traditions (Scheyvens, 2006). On many occasions, these activities were part of the rural village culture for a long time and even before sustainable tourism was conceptualized. Homestays are an important component of community-based tourism where local people provide tourists with a locally run experience such as accommodation, food, and cultural activities. Homestays are supplementary to the core livelihood activity which is mainly farming. These approaches focus on how the local traditions and cultures can add to the experience of the tourist, but do not take into account whether and how these experiences are beneficial from an entrepreneurial point of view. The participation of the community might not always offer enough challenges to individuals. Community-based tourism is managed by the village, with the advice and support of district tourism departments and NGOs and this can hinder the creativity of the entrepreneur (Pasanchay & Schott, 2021).

3 The Base of the Pyramid as an inclusive business model

Scheyvens and Biddulph (2018) distinguish their concept from the inclusive business approach because the latter supports a neoliberal model of growth. Golja and Požega (2012: 23) define an inclusive business model as a way to contribute to:

> poverty alleviation by including lower-income communities within its value chain while not losing sight of the ultimate goal of business

Socially inclusive businesses highlight the social goal. The Base/Bottom of the Pyramid (BOP) is a concept that is often used to refer to people who live in deprived regions. The concept was made popular by Prahalad (2005). His idea of selling to the poor was triggered by the BOP markets that are underserved by the global market system while the markets in industrialized countries were saturated and did not have enough growth opportunities. Although the income of the BOP people was very low, the number of potential customers was considered as being a profitable market (Prahalad, 2005). Prahalad (2005) was convinced of the combination of creating solutions to the problems people face living on less than US$2 a day and making money. Despite labeling it a radically different view and a potential contribution to sustainable development, the poor were primarily seen as potential customers, as an opportunity for multinationals in the north to adapt their products to the purchasing power of the south. When Prahalad (2005) and Hart (Prahalad & Hart, 1999, 2002) developed their concept of the BOP, it was mainly developed as a way to sell to the poor and therefore it is often labeled BOP 1.0. This approach got a lot of criticism and urged for a change. Especially Karnani (2007) was very critical and argued that most people do not have the skills, vision, and creativity of true entrepreneurs. According to Karnani (2007) the best way to eradicate poverty was employment at reasonable wages and lower prices, as well. It might indeed seem a bit naïve that big multinationals would have an aim like eradicating poverty. In the study by PwC (2015), the Sustainable Development Goal 1, i.e., No Poverty, was not ranked as a goal where businesses believed they could have a big impact and could represent a business opportunity in the future. Although the SDGs are more and more embraced by companies, they are still seen as being disconnected and invited for cherry picking (Leblanc, 2015; Molderez, 2021).

A follow-up for this rather narrow BOP concept was suggested by Simanis and Hart (2008) and was labeled BOP 2.0. The emphasis was now on business co-venturing that enabled to see the people at the BOP as business partners. The emphasis was on deep dialogue, expanding imagination, building shared commitment, and direct personal relationships facilitated by NGOs. This was a positive evolution because it started from the people, what they needed and how they could be part of the process that is creating the products or services. This second generation of BOP strategies was an opening toward two interpretations of the BOP. It is not only about people who live in circumstances of moderate poverty, but also about products or services that are created from the bottom up. Cases that fit into this approach are products and services that satisfy the needs of the poor. Philips' wood stove is an illustration of creating shared value, although still lacking co-creation in the real sense of the word (Hegarty, 2006). However, even when the product is co-created, as in the Solae Company, a BOP initiative launched in India by Du Pont (Simanis & Hart, 2008), the learning component is predominantly Western oriented. People living in deprived regions learn from multinationals how to create a product that satisfies a need in their market. This rather top-down approach triggered Chmielewski et al. (2020) to suggest BOP 3.0. In their concept, the link is made with a sustainable development model of poverty alleviation. The key characteristic is a community-centric approach with the first aim to create a self-reliant community. This third concept relates to Hunt et al.'s (2010) preference to use base instead of bottom which has a negative connotation. Furthermore, a base is also an entity on which something draws on or depends. While collaborating with people from deprived regions, they can become a change agent for their region. The contribution of the BOP concept is that poverty can be alleviated through financially profitable activities while in the past this was thought to be the responsibility of governments and NGOs only. Moreover, within BOP 3.0, poverty gets a much broader definition and moves from a focus on income only to an integration of social, political, emotional, and cultural aspects (Chmielewski et al., 2020, referring to Yurdakul et al., 2017, and to Cañeque & Hart, 2015).

Similar ideas of bottom-up creations can be found in Sacco's (2011) concept of Culture 3.0. Instead of cultural tourism driving culture-led development, the emphasis is now on projects that are created by the local people themselves, built on their own creativity, skills, and capabilities. This new dynamic can also create new entrepreneurship models for creative entrepreneurs (Sacco, 2011; Sacco et al., 2018), apart from innovation, well-being, sustainability, social cohesion, lifelong learning, soft power, and local identity. Also, Phelps' (2013) argument that nations prosper because of grassroots dynamism is relevant. His research on why some nations prospered between the 1820s and 1960s led to the important role of innovation driven by the bottom up by millions of people. Individual market participation was key, as also emphasized in the review by Godley (2014). However, Phelps' (2013) emphasis is not on the element of income that is earned thanks to market participation, but on the relationship between human psychology and market participation that makes people flourish, i.e., meaningful work, self-expression, personal growth because of the experience of the new. Phelps (2013: vii) refers to:

> (...) new problems, new insights, and new ideas to develop and share (...) mass flourishing comes from broad involvement of people in the process of innovation.

instead of thinking that innovation was driven by a few isolated visionaries. The source of this flourishing were the right values, such as the desire to create, explore, and meet challenges. Unfortunately, according to Phelps (2013), these values are under threat from institutions that put the state and the community over the individual. Although Phelps (2013) can only be read from a neoliberal perspective, his work is an invitation to emphasize the role individuals can play toward a sustainable transition. Participation is seen on the individual level without any enforcement from the top, as is often the case in development interventions, which has led to negative connotations such as the repressive myth (Chinyowa, 2015), the participation myth (Jancovich, 2017), orthodoxy, dogma, norm, rationalized myth, institutionalized environmental expectation, and symbolic implementation (Schindler, 2022: 37–64, referring to different authors). Apart from describing the phenomenon, Schindler (2022) examines how organizations govern it. Decoupling, i.e., between policies and practices, and between means and ends, is an often used strategy. Giving ample space to Phelps (2013), mass flourishing can be seen as a way to avoid the fallacy.

4 Inclusive Tourism from the BOP Approach

Chmielewski et al. (2020) emphasized the importance of a self-reliant community while working on poverty alleviation from a market-oriented perspective. This next phase in the BOP literature was not yet included in Kolk et al.'s (2014) review of BOP articles. Nevertheless, their allocation of different articles on BOP was based on two similar criteria: mode of engagement of the poor, i.e., either as the recipient of existing products or as a co-inventor, and the position of the poor in the value network, i.e., either as a consumer or as an entrepreneur. While the concept of BOP is most promising when the poor are seen as co-inventors and as entrepreneurs, the majority of the reviewed articles were located in the combination recipient/consumer. The most challenging quadrant, where poor people are valued for their capabilities remained a niche. It is in the combination of co-inventor and entrepreneur that inclusiveness is best represented. Combining the co-creation and the entrepreneur is important because co-creation is often seen as the participation of the customer in creating the product. Rahman et al. (2019) accentuate that co-creation should go further, i.e., interactions between customers, the firm, and other actors. A diversity of BOP-business models therefore occurs.

In Kolk et al.'s (2014) literature review, only one article is related to tourism. This can be explained by the original emphasis of the BOP concept on multinationals. The concept is, however, more and more applied by small and domestic companies and not-for-profit organizations (Kolk et al., 2014). Another explanation might be that poverty can be researched from similar concepts such as underdog entrepreneurship (Morgan, 2020), disadvantaged entrepreneurship (Maalaoui et al., 2020), frugal innovation (Rosca et al., 2017; Lange et al., 2021), or social entrepreneurship (Biddulph, 2018). Despite the very few BOP cases in tourism, the global market has already discovered BOP areas with a lot of negative economic, environmental, and social impacts caused by large-scale foreign-owned businesses (Hall et al., 2012).

The tourism sector is accused of providing opportunities for the privileged middle and upper classes to travel and enjoy leisure activities in other places, creating profits, particularly for large companies and creating exclusive enclaves for the rich, while development opportunities associated with tourism are not open to those who are poor because of deprived areas and situations (Harrison, 1992; Gibson, 2009; Jamal & Camargo, 2014; Scheyvens & Biddulph, 2018). These people are still victims of considerable exploitation at the hands of both local tourism agencies hiring them and the entire system they embrace by taking up opportunities within global tourism (Arellano, 2011). If tourism does not respect local cultures and is uncontrolled, it will have a negative impact on local communities, their heritage, and environment, and can exacerbate inequalities according to the International Labor Organization (2016). This means new dynamics are needed in destinations to guarantee inclusive tourism opportunities and benefits to the local people in deprived regions.

The responsibility of the stakeholders is to develop tourism which utilizes local cultural resources and benefits local community, contributing to the cultural values and improving the local livelihood (Sacco, 2011; Sacco et al., 2018). In order to create awareness and overcome exploiting the poor and growing unemployment issues in the tourism industry, especially in the most vulnerable areas, it is necessary to engage the travelers (global or domestic) in meaningful connections with the local people who can become a change agent while co-creating a product or service that is of benefit for both. At the same time, creating a greater understanding of the local culture drives positive contributions, such as the conservation of the natural and cultural heritage and building the local pride and confidence by enriching the value proposition of the destination (Scheyvens, 2006).

Maximizing positive impact and minimizing negative impact could be achieved by enhancing the local community to actively participate in the tourism decision-making process. For example, decentralizing decision-making power by empowering people locally on the ground can lead to more effective development outcomes (Timothy, 2007). Local communities are likely to give more practical recommendations for stimulating events that will allow more organic growth to inclusive tourism. Today, a large proportion of tourism demand, especially the millennial segment, expresses interest in visiting destinations and consuming products that have a clear environmental and cultural value, and in which the presence of the local population is almost essential (WTO, 2018).

When considering the responsibilities of the stakeholders, the economies of developing countries, where 83% of the world's population lives, are growing fast. However, this growth comes with increasing inequality which undermines prospects for progress for millions of people. Thus, economic growth in itself is insufficient for fighting poverty and inequality unless it is both sustainable and inclusive (WTO, 2018). Hence, it is important to implement responsible production of tourism by existing tourism businesses and by focusing on inclusiveness. Mainstream operators can take an inclusive approach to tourism production by transforming their core activities, such as providing decision-making roles and ownership opportunities for staff; mentoring local people in relation to starting their own small businesses associated with tourism; introducing inclusive procurement strategies; and offering dignified work, good training, and fair remuneration (Ashley et al., 2008; Walmsley, 2012; Hughes & Scheyvens, 2016).

In this process toward a more inclusive approach, commitment of the public authorities is relevant, at least when they contribute to stimulating the values that are necessary for flourishing (Phelps, 2013). The local government can encourage the local social agencies and indigenous organizations to develop simplified and transparent administration of social welfare registration for local people. Official authorities can deliver one set of cash benefits and standardize the eligibilities of applicants to increase the accessibility and take up the costs (Wang & Wang, 2019). In order to do so and to identify what is best for the people, immersing in the rhythms of local customers or producers or their needs is required when improving the possibilities for vulnerable groups to participate in the tourism system. People from vulnerable groups can provide a different value to the destination through their traditional skills (craftworkers, cooks, farmers, etc.), if needed training and advice can be provided to facilitate their entry into the market and place them in contact with end consumers. In the case of unskilled people belonging to vulnerable groups, training in professional tourism skills can improve their employability and their contacts with the tourism sector (WTO, 2018).

The advanced concept of BOP, i.e., BOP 3.0, might offer a different view on how to provide tourist services to people living in deprived regions. Instead of trying to apply a Western business perspective, the focus is on the development of self-reliant communities and their businesses (Chmielewski et al., 2020). The result of

Chmielewski et al. (2020) analysis of 21 organizations in the Philippines, among them one in the tourism sector, is a flipped model of the first and second BOP framework. Within these organizations, existing skills, products, and resources of the community form the starting point to develop the business. Their entrepreneurs believe in their own capabilities, rather than being pushed by a foreign company. Adjusting the business to the community is what they call a bottom-up approach. Making use of what is present in the community makes the community more self-reliant.

There are limited cases that focus on co-creation where the learning goes from the bottom to the top, i.e., learning from the competencies of people from the base of the pyramid and co-creating a product or service that is of benefit for both. Miss Arlene's oven, analyzed in Hunt et al. (2010) being a rare example of persistence of local people who start with something small within the tourism sector and dream of generating a positive impact for the whole community. João Pessoa, one of the oldest cities in Brazil that mainly attracts domestic tourists, is another example that goes in the direction of the combination of co-inventor and entrepreneur (Hall et al., 2012). The Federal Government Agency opted for a tourism policy based on local cultural attractions and nature tourism. One key product that had become central in the policy was the locally produced and naturally colored cotton. The farmers could not afford chemical herbicides and pesticides, but this disadvantage became an advantage. The eco-friendly yarn was used by women from impoverished communities to produce diverse textile products and garments. The entrepreneurial opportunity that was offered by the yarn, was a slower process in comparison with mass tourism, but in the long-term people from these BOP areas participated equally (Hall et al., 2012). Tourism was not seen as a separate sector, but other sectors, like agriculture and the textile sector, were involved. Local people were included on the basis of their capabilities. They did not feel restrained to participate because the entry barriers, like fluency in foreign languages, were absent. The local community was involved in a positive, appreciative way because their skills and motivation for participation were central. In this way, there is a link with community-based entrepreneurship (Peredo & Chrisman, 2006) that often has poverty alleviation as an aim (Ohe, 2019). In an inclusive tourism model, it is important that the tourism sector is not seen as "an island"—only there for serving tourists. Such a structure is very vulnerable to external factors such as disruptions in the economic system and environmental disasters (Pascariu et al., 2022). Inclusive tourism does not only take care of a decent income but also for one that can sustain. Biddulph (2018) is rather pessimistic whether small-medium scale initiatives will be able to significantly transform overall tourism. However, these small enterprises are very important as enablers for Phelps' (2013) mass flourishing.

5 Hiriwaduna Village Tour in Sri Lanka as an Illustrative Case for a Flourishing Concept of Inclusive Tourism

The Hiriwaduna Village Tour in Sri Lanka is selected as an illustrative case for a flourishing concept of inclusive tourism. Three qualitative methods were used: document analysis, participant observation, and semi-structured interviews with the creator of the tour and with the manager and two employees from a Destination Management Organization that makes use of the tour. The Hiriwaduna Village Tour was created by Nilantha Kodithuwakku, a naturalist from Cinnamon Lodge, in 2006. An interview with him forms the basis of this exemplary case of inclusive tourism. The purpose of this case study is to show the tourism potential of the Hiriwaduna Village Tour through the participation of the members of the community, as well as the impact of Hiriwaduna Village Experience on the local community.

Mainly because of its history, culture, and landscape, Sri Lanka is an interesting country to illustrate this concept. For three decades, Sri Lanka was a country in war. Only since 2009 the country stepped into a period of hope for peace and prosperity (Jayasuriya & Weerakoon, 2019). The end of the conflict has offered new opportunities, especially in the tourism sector (Bandara, 2019), but sometimes also at the expense of local communities and their culture (Ranasinghe & Cheng, 2018). Despite Sri Lanka's relatively young position as a tourist destination, researchers urge for sustainable management of its resources, like in the cases of the deleterious impacts on whale watching (Buultjens et al., 2016a), or the precarious impact of tourism on the Vedda Community (Ranasinghe & Cheng, 2018) and on the coastal zones (Marasinghe et al., 2021). Tourism often offers new opportunities for countries that have suffered from civil wars and local conflicts. This is also the case for Sri Lanka where the government has positioned tourism as a central pillar of its economy (Bandara, 2019). The reason for this is the positive impact of tourism on employment and export earnings. However, because of its natural assets and historical heritage, the positive effects on Sri Lanka can be larger than only at the macroeconomic level. In line with Sacco et al. (2018) arguments, emphasizing the cultural sites will bring other effects, too. Especially culture as a form of soft power can be helpful to increase the visibility and improve the image of Sri Lanka that has to change to a country where different ethnic people can live in harmony.

One of the main goals of the tourism policy of the government is

> to promote inclusive economic growth through tourism
>
> but according to Bandara's (2019: 188) research,
>
> (...) a large part of conflict areas is yet to benefit (...)

The region of the case is not part of the former conflict area. Being located in the North Central province, Hiriwaduna might be a telling case for a bottom-up approach to tourism. As is also argued by Buultjens et al. (2016b), it is important to stimulate small, local tourism businesses and to recognize the informal sector because both contribute to the resilience of the community. Sri Lanka is primarily an agricultural economy. The main crops that are produced in this plantation economy are tea, rubber, coconut, and rice. Cropland is widely distributed in every province of Sri Lanka, but the North Central province has the largest part, i.e., almost 22% of the total cropland area (Wei, 2021: 25). Hence, this agricultural aspect is central in the tour.

Hiriwaduna is a village in the Dambulla area and has been developed through the years by the locals with the support of Destination Marketing Organizations who see the potential of tourism in this village without any government intervention (Weerasinghe, 2018). Hiriwaduna is a village of approximately 2500 people. The villagers, who are mainly farmers, consider themselves as very rich since they have everything they need. However, this is not considering the city standards, but they are able to produce their own food and barter their crops with the neighbors. The tour shows an insight into the village life in Sri Lanka. It all starts with a walk stopping to look at the butterflies and row the tourists across the reservoir with a local boat rider. After reaching the village house through the paddy fields and vegetable cultivations the tourist can relax and taste a Sri Lankan lunch meal prepared by a village woman using the farmers' handpicked vegetables from the land.

Nilantha Kodithuwakku decided to create a village experience for the guests who arrive at the Cinneman Lodge Hotel and this was never been done before. He wanted to search for a clay house with straw roofs in Hiriwaduna village and the first house he could get in touch with was owned by Mr. Siril who is a farmer. The hotel inspected the house to see if the amenities were in order and requested their permission to bring tourists in, to share their life experience with the tourists. The house was located beside a paddy field close to the lake and the jungle. There were no bullock cart rides in the beginning. After seeing how people commuted within the village and that most of the carts were used to pull the tree barks, the founder wanted to include this experience in the tourist's excursion list. This indirectly resulted in stoppage of the cutting of trees in this village and ultimately broadened the scope of revenue and employment opportunities for the community in Hiriwaduna.

In the development of the Hiriwaduna Village Tour, the local community plays a very important role, such as being food providers, coconut pluckers, women who know how to cook, and boat riders. In 2007, the lake only had one boat but with the introduction of this new service to the village, the number of boats increased to around 40. With the introduction of such special services, this village was also supported by the local corporate events to build its service levels and train the locals for betterment. Lately, the locals were willing to supply services by themselves and created their own small businesses around this concept. Some excursions included the tourists taken to the schools. Afterward, the tourists bought books for children and engaged with their studies. After several years of introducing Hiriwaduna Village Tour to the tourists, the other neighboring villagers also started following similar excursions to create employment opportunities for themselves and they are still in this business up to date.

It is part of the grassroots program organized by Cinnamon Nature Trails to benefit the surrounding communities by providing self-employment and partnership opportunities. Participants of the tour get the occasion to be emerged in the nature and culture of a real Sri Lankan village. Tour operators in Hiriwaduna, Habarana, promote the village trek as a tour package. It is not only the heritage and nature but also the authentic culture that is shared. This diversified tourism product offers a unique community-based experience for the tourists compared to other natural and cultural attractions at the periphery. The tour is a combination of culture, nature, tradition, community, and other simple but traditional ways of living with a lot of people benefiting from it, like the farmers who sell their crops within the village, the farmer's wives who are independent while providing hospitality and cooking for the guests, the boat riders who earn extra income since their normal work was to pluck flowers, the bullock cart riders who have a new way of earning money without exhausting their cows, i.e., by pulling the tree trunks.

The above outcomes show that Hiriwaduna village Tour had the potential of offering a unique experience enriched with natural rice fields and plantations as well as the preservation of the local culture and traditions. In addition, local communities participated as owners, employees, and also as investors in the development of this tourism experience, thereby emphasizing the possibility for many people in the community to flourish. As a community booster and to promote the advanced BOP concept, i.e., BOP 3.0 as an inclusive tourism approach, it is important to cater more toward rural entrepreneurship in the cultural tourism sector while highlighting how well a rural community needs to market its hospitality and other natural resources toward a sustainable tourism.

6 Conclusion

The inclusive tourism concept that was central in this chapter is about filling the needs of the local community through empowerment where they can aim higher than meeting their basic needs and entrepreneurship opportunities to cater to domestic and international responsible travelers. It is important to bridge the gap between the travelers who would want to make a change and the people from deprived regions who wish to share their culture and local pride with these travelers.

In contradiction to other models of inclusive tourism, the emphasis is on the people from deprived regions who invent the product and at the same time set it in the market. As analyzed, this is the most difficult segment in the framework of different BOP approaches, but also the most challenging and rewarding one. These entrepreneurs participate in the tourism project not as simply bottom, but as the basis to create a tourism sector that is founded in and by the community so that everybody can flourish. Tourism is then not seen as an island alongside other sectors, but is fully integrated. In this respect, local communities will be less vulnerable and less dependent on foreign demand.

Providing authentic cultural experience to the travelers through inclusive tourism can result in an increase of demand for this specific concept of inclusive tourism in the future. It is important that inclusive tourism benefits an entire community, destination, and country and is a valuable tool in striving toward the achievement of sustainable development. Hence, this is not only about creating jobs for the local community from deprived regions but also about enabling them to take pride in their community through engagement in tourism where options are given for everybody to flourish.

Acknowledgments We would like to thank the reviewers for their very supportive comments and Prof. Dr. Carina Branzila, University of Iasi, Romania, for her valuable feedback.

References

Arellano, A. (2011). Tourism in poor regions and social inclusion: The porters of the Inca Trail to Machu Picchu. *World Leisure Journal, 53*(2), 104–118. https://doi.org/10.1080/04419057.2011.580551

Ashley, C., Haysom, G., & Spenceley, A. (2008). The development impacts of tourism supply chains: Increasing impact on poverty and decreasing our ignorance. In A. Spenceley (Ed.), *Responsible tourism: Critical issues for conservation and development* (pp. 129–156). Routledge.

Bandara, J. S. (2019). Tourism in post-conflict development: Making use of new opportunities in Sri Lanka. In D. Weerakoon & S. Jayasuriya (Eds.), *Managing domestic and international challenges and opportunities in post-conflict development: Lessons from Sri Lanka* (pp. 173–194). Springer.

Biddulph, R. (2018). Social Enterprise and inclusive tourism. Five cases in Siem Reap, Cambodia. *Tourism Geographies, 20*(4), 610–629. https://doi.org/10.1080/14616688.2017.1417471

Burchardt, T., Le Grand, J., & Piachaud, D. (2002). Introduction. In J. Hills, J. Le Grand, & D. Piachaud (Eds.), *Understanding social exclusion*. Oxford University Press.

Buultjens, J., Ratnayake, I., & Gnanapala, A. (2016a). Whale watching in Sri Lanka: Perceptions of sustainability. *Tourism Management Perspectives, 18*, 125–133. https://doi.org/10.1016/j.tmp.2016.02.003

Buultjens, J., Ratnayake, I., & Gnanapala, A. (2016b). Post-conflict tourism development in Sri Lanka: Implications for building resilience. *Current Issues in Tourism, 19*(4), 355–372.

Buvinic, M. (2004). Introduction: Social inclusion in Latin America. In M. Buvinic, J. Mazza, & R. Deutsch (Eds.), *Social inclusion and economic development in Latin America* (pp. 3–31). InterAmerican Development Bank.

Cañeque, F. C., & Hart, S. (Eds.). (2015). *Base of the pyramid 3.0: Sustainable development through innovation and entrepreneurship*. Greenleaf Publishing Limited.

Chmielewski, D. A., Dembek, K., & Beckett, J. R. (2020). Business unusual: Building BoP 3.0. *Journal of Business Ethics, 161*(1), 211–229. https://doi.org/10.1007/s10551-018-3938-7

Chinyowa, K. C. (2015). Participation as 'repressive myth': A case study of the interactive Themba theatre organisation in South Africa. *Research in Drama Education: The Journal of Applied Theatre and Performance, 20*(1), 12–23. https://doi.org/10.1080/13569783.2014.975109

Commission of the European Communities. (2003). *Joint report on social inclusion: Social security and social integration*. European Communities. Retrieved from https://ec.europa.eu/employment_social/social_inclusion/docs/final_joint_inclusion_report_2003_en.pdf

Godley, A. C. (2014). Creativity and growth. In Edmund Phelps (2013), Mass flourishing. How grassroots innovation created jobs, challenge and change. *International Journal of the Economics of Business, 21*(2), 255–260. https://doi.org/10.1080/13571516.2014.919737

Golja, T., & Požega, S. (2012). Inclusive business - What it is all about? Managing inclusive companies. *International Review of Management and Marketing, 2*(1), 22–42.

Gibson, C. (2009). Geographies of tourism: Critical research on capitalism and local livelihoods. *Progress in Human Geography, 33*(4), 527–534. https://doi.org/10.1177/0309132508099797

Hall, J., Matos, S., Sheehan, L., & Silvestre, B. (2012). Entrepreneurship and innovation at the base of the pyramid: A recipe for inclusive growth or social exclusion? *Journal of Management Studies, 49*(4), 785–812. https://doi.org/10.1111/j.1467-6486.2012.01044.x

Harrison, D. (Ed.). (1992). *Tourism and the less developed countries*. Belhaven Press.

Hegarty, D. (2006). Satisfying a burning need. *Password Phillips Research Technology Magazine, 28*, 28–31.

Hughes, E., & Scheyvens, R. (2016). Corporate social responsibility in tourism post-2015: A development first approach. *Tourism Geographies, 18*(5), 469–482. https://doi.org/10.1080/14616688.2016.1208678

Hunt, J., Lane, P., & Farris, J. (2010). The lesson of Miss Arlene's oven: Entrepreneurship at the base of the economic pyramid. *International Journal of Entrepreneurship, 14*, 85–110.

Hunzicker, W. (1951). *Social tourism: Its nature and problems*. International Tourists Alliance Scientific Commission.

International Labor Organization (ILO). (2016, May 17). *Sustainable tourism: A driving force of job creation, economic growth and development*. Retrieved from https://www.ilo.org/global/about-the-ilo/newsroom/news/WCMS_480824/lang%2D%2Den/index.htm

Jamal, T., & Camargo, B. A. (2014). Sustainable tourism, justice and an ethic of care: Toward the just destination. *Journal of Sustainable Tourism, 22*(1), 11–30.

Jancovich, L. (2017). The participation myth. *International Journal of Cultural Policy, 23*(1), 107–121. https://doi.org/10.1080/10286632.2015.1027698

Jayasuriya, S., & Weerakoon, D. (2019). Introduction: Challenges and opportunities of Sri Lanka's post-conflict economic development overview. In D. Weerakoon & S. Jayasuriya (Eds.), *Managing domestic and international challenges and opportunities in post-conflict development. South Asia economic and policy studies* (pp. 1–12). Springer.

Karnani, A. (2007). The mirage of marketing to the bottom of the pyramid. How the private sector can help alleviate poverty. *California Management Review, 49*(4), 90–111. https://doi.org/10.2307/41166407

Kakoudakis, K. I., McCabe, S., & Story, V. (2017). Social tourism and self-efficacy: Exploring links between tourism participation, job-seeking and unemployment. *Annals of Tourism Research, 65*, 108–121. https://doi.org/10.1016/j.annals.2017.05.005

Kolk, A., Rivera-Santos, M., & Rufin, C. (2014). Reviewing a decade of research on the "base/bottom of the pyramid" (BOP) concept. *Business & Society, 53*(3), 338–377. https://doi.org/10.1177/0007650312474928

Lange, A., Hüsig, S., & Albert, M. (2021). How frugal innovation and inclusive business are linked to tackle low-income markets. *Journal of Small Business Management*, 1–34. https://doi.org/10.1080/00472778.2021.1924380.

Leblanc, D. (2015). Towards integration at last? The sustainable development goals as a network of targets. *Sustainable Development, 23*(3), 176–187. https://doi.org/10.1002/sd.1582

Maalaoui, A., Ratten, V., Heilbrunn, S., Brannback, M., & Kraus, S. (2020). Disadvantage entrepreneurship: Decoding a new area of research. *European Management Review, 17*(3), 663–668. https://doi.org/10.1111/emre.12424

Marasinghe, S., Perera, P., Simpson, G. D., & Newsome, D. (2021). Nature-based tourism development. In Coastal Wetlands of Sri Lanka: An importance–performance analysis at Maduganga Mangrove Estuary. *Journal of Outdoor Recreation and Tourism, 33*, 100345. https://doi.org/10.1016/j.jort.2020.100345.

McCabe, S., & Qiao, G. (2020). A review of research into social tourism: Launching the annals of tourism research curated collection on social tourism. *Annals of Tourism Research, 85*, 103103. https://doi.org/10.1016/j.annals.2020.103103

Molderez, I. (2021). Corporate social responsibility and the sustainable development goals (SDGs). In W. L. Filho, A. M. Azul, L. Brandli, A. Lange Salvia, & T. Wall (Eds.), *Decent work and economic growth. Encyclopedia of the UN sustainable development goals* (pp. 116–128). Springer.

Morgan, H. M. (2020). *Underdog entrepreneurs: A framework of success for marginalized and minority innovators*. Springer International Publishing.

Münch, H., & Ulrich, R. (2011). Inclusive tourism. In A. Papathanassis (Ed.), *The long tail of tourism* (pp. 159–169). Springer Nature.

Ohe, Y. (2019). *Community-based rural tourism and entrepreneurship*. Springer Pte. Limited.

Pasanchay, K., & Schott, C. (2021). Community-based tourism Homestays' capacity to advance the sustainable development goals: A holistic sustainable livelihood perspective. *Tourism Management Perspectives, 37*, 100784. https://doi.org/10.1016/j.tmp.2020.100784

Pascariu, G. C., Cehan, A., & Branzila, C. (2022). International tourism regionalization: Major trends and shock disruption. *Transylvania Revue*, special issue 2022 (forthcoming).

Peredo, A. M., & Chrisman, J. J. (2006). Toward a theory of community-based enterprise. *Academy of Management Review, 31*(2), 309–328. https://doi.org/10.5465/amr.2006.20208683

Phelps, E. (2013). *Mass flourishing: How grassroots innovation created jobs, challenge, and change*. Princeton University Press.

Prahalad, C. K. (2005). *The fortune at the bottom of the pyramid: Eradicating poverty through profits*. Wharton School Publishing.

Prahalad, C., & Hart, S. (1999). *Strategies for the bottom of the pyramid: Creating sustainable development* (1–26). Ann Arbor.

Prahalad, C. K., & Hart, S. (2002). The fortune at the bottom of the pyramid. *Strategy+Business, 26*, 54–67.

Priatmoko, S., Kabil, M., Purwoko, Y., & Dávid, L. D. (2021). Rethinking sustainable community-based tourism: A villager's point of view and case study in Pampang Village, Indonesia. *Sustainability, 13*(6), 3245.

PwC. (2015). *Make it your business: Engaging with the sustainable development goals*. Retrieved from http://www.pwc.com/gx/en/sustainability/SDG/SDG%20Research_FINAL.pdf

Ranasinghe, R., & Cheng, L. (2018). Tourism-induced mobilities and transformation of indigenous cultures: Where is the Vedda Community in Sri Lanka Heading to? *Journal of Tourism and Cultural Change, 16*(5), 521–538. https://doi.org/10.1080/14766825.2017.1393081

Rahman, M., Bose, S., Babu, M. M., Dey, B. L., Roy, S. K., & Binsardi, B. (2019). Value co-creation as a dialectical process: Study in Bangladesh and Indian Province of West Bengal. *Information Systems Frontiers, 21*(3), 527–545. https://doi.org/10.1007/s10796-019-09902-4

Richards, G. (2018). Cultural tourism: A review of recent research and trends. *Journal of Hospitality and Tourism Management, 36*, 12–21. https://doi.org/10.1016/j.jhtm.2018.03.005

Rifai, T. (2017). *Travel and tourism competitiveness report 2017: Tourism and job creation – Advancing the 2030 agenda*. World economic forum. Retrieved from https://reports.weforum.org/travel-and-tourism-competitiveness-report-2017/tourism-and-job-creation-advancing-the-2030-development-agenda/

Rosca, E., Reedy, J., & Bendul, J. C. (2017). Does frugal innovation enable sustainable development? A systematic literature review. *European Journal of Development Research, 30*(1), 136–157. https://doi.org/10.1057/s41287-017-0106-3

Sacco, P. L. (2011). *Culture 3.0: A new perspective for the EU 2014-2020 structural funds programming*. Report for the OMC working group on cultural and creative industries. Retrieved from http://archive.interarts.net/descargas/interarts2577.pdf

Sacco, P. L., Ferilli, G., & Blessi, G. T. (2018). From culture 1.0 to culture 3.0: Three socio-technical regimes of social and economic value creation through culture, and their impact on European cohesion policies. *Sustainability, 10*(11), 3923. https://doi.org/10.3390/su10113923.

Scheyvens, R. (2006). Sun, sand, and Beach Fale: Benefiting from backpackers-the Samoan way. *Tourism Recreation Research, 31*(3), 75–86. https://doi.org/10.1080/02508281.2006.11081507

Scheyvens, R. (2011). *Tourism and poverty*. Routledge.

Scheyvens, R., & Biddulph, R. (2018). Inclusive tourism development. *Tourism Geographies, 20*(4), 589–609. https://doi.org/10.1080/14616688.2017.1381985

Schindler, E. M. (2022). *Structuring people: The myth of participation and the organisation of civil society in development*. Springer Fachmedien.

Schott, C., & Nhem, S. (2018). Paths to the market: Analysing tourism distribution channels for community-based tourism. *Tourism Recreation Research, 43*(3), 356–371. https://doi.org/10.1080/02508281.2018.1447837

Simanis, E., & Hart, S. (2008). *The base of the pyramid Protocol: Toward Next Generation BoP Strategy*. https://doi.org/10.13140/2.1.5097.0402.

Timothy, D. J. (2007). Empowerment and stakeholder participation in tourism destination communities. In T. C. Andrew Church (Ed.), *Power and space* (pp. 203–216). Routledge.

United Nations (2015). *Transforming our world: The 2030 agenda for sustainable development*. United Nations Publications. Retrieved from https://www.unep.org/resources/report/transforming-our-world-2030-agenda-sustainable-development

United Nations (2019). *International Year of Creative Economy for Sustainable Development, 2021*. Retrieved from https://digitallibrary.un.org/record/3835223?ln=en#record-files-collapse-header

Walmsley, A. (2012). Decent work and tourism wages: An international comparison. *Progress in Responsible Tourism, 2*(1), 90–99.

Wang, J.-H., & Wang, S.-Y. (2019). Indigenous social policy and social inclusion in Taiwan. *Sustainability, 11*(12), 3458. https://doi.org/10.3390/su11123458

Weerasinghe, I. (2018). Tour Operators' role in product diversification for enhancing visitor satisfaction: a case study of Hiriwadunna Village Tour, Dambulla, Sri Lanka. Conference paper for Tourism Leaders' Summit (TLS) and International Tourism Research Conference (ITRC) - 2018, At: BMICH, Colombo. *Journal of Tourism Economics and Applied Research, 1*(1). Retrieved from http://hdl.handle.net/70130/4679. https://www.researchgate.net/publication/328161673_TOUR_OPERATORS'_ROLE_IN_PRODUCT_DIVERSIFICATION_FOR_ENHANCING_VISITOR_SATISFACTION_A_CASE_STUDY_OF_HIRIWADUNNA_VILLAGE_TOUR_DAMBULLA_SRI_LANKA

Wei, X. (2021). *The geography and remote sensing analysis of Sri Lanka*. Springer. https://doi.org/10.1007/978-981-33-4627-7.

World Tourism Organization. (2018). *Global report on inclusive tourism destinations – Model and success stories*. UNWTO. https://doi.org/10.18111/9789284420094.

Yurdakul, D., Atik, D., & Dholakia, N. (2017). Redefining the bottom of the pyramid from a marketing perspective. *Marketing Theory, 17*(3), 289–303. https://doi.org/10.1177/1470593117704265

Toward a Demand Profile of WWOOFers in Portugal: Ecological, Rural, and Voluntary Principles in Tourism

Carla Nogueira, Michelle Lins de Moraes, João Filipe Marques, and Antónia Correia

Abstract Environmental crisis worldwide was the critical driver to Worldwide Opportunities on Organic Farm (WWOOF) tourists. These tourists, like many others, decide their holiday experiences based on preferences, time, and money constraints. With the increase in the perception of ecological crises, new concerns have arisen in tourism practices. This chapter aims to characterize the WWOOFers tourists by elaborating a typology from a sociological and economic perspective. These tourists' profiles were obtained through the analysis of reviews submitted on the WWOOF website. This website offers cultural experiences based on trust and nonmonetary exchange. The reviews function as a kind of application prior to the experiences, which include dimensions such as age group, class, education, level of commitment to the project, length of stay, activities developed, and perceptions about the experience. The investigation focused on two main phases. On the one hand, to establish some sociodemographic characteristics of these tourists through quantitative analysis and on the other hand, it identifies possible profiles through qualitative content analysis. The results corroborate the trends identified in other studies and advance with three exploratory profiles: sustainable gappers, sustainable escapers, and sustainable lifers.

C. Nogueira (✉) · J. F. Marques
Research Centre for Tourism, Sustainability and Well-being CinTurs, University of Algarve, Algarve, Portugal
e-mail: cfnogueira@ualg.pt; jfmarq@ualg.pt

M. L. de Moraes
Center for Research in Organizations, Markets and Industrial Management COMEGI, Lusiada University, Lisbon, Portugal
e-mail: lnsmhl@unife.it

A. Correia
Center for Advanced Studies in Management and Economics CEFAGE, University of Algarve, Algarve, Portugal
e-mail: acorreia@ualg.pt

E. Borin et al. (eds.), *Cultural Leadership in Transition Tourism*, Contributions to Management Science, https://doi.org/10.1007/978-3-031-14121-8_16

1 Introduction

The environmental movement emerged in the 1960s to respond to the ecological consequences introduced by the increase in industrialization and globalization (McCormick, 1992). In the late 1980s and early 1990s, environmental concerns began to take on more transversal characteristics, both in the institutional and political sphere and in the dimension of individual choices (Spindler, 2013). This environmental concern has also echoed in the tourism sector as ecotourism is one of the segments with the highest growth rate in recent years (Rajović & Bulatović, 2015). Ecotourism is based on a logic that comprises tourists' environmental concerns, both in terms of destination choices and the practices they develop (Blamey, 1997).

This chapter analyzes the profiles of tourists seeking experiences through the WWOOF platform—Worldwide Opportunities on Organic Farms, in Portugal, mobilizing the idea of ecotourism and relating it to the rural dimension of agrotourism and the principle of volunteering. WWOOF can be defined as "an international movement in which host farms provide room and board in exchange for labor by tourist-volunteers" (Yamamoto & Engelsted, 2014).

In order to contribute to the expansion of knowledge inherent to these dynamics, the present chapter features the profiles of the tourists visiting the WWOOFs elaborated through a sociological and economic perspective. This study can be considered relevant since a better understanding of this demand profile is essential to comprehend its future behavior (Kim & Park, 2017). Also, the fact that managers related to this niche have limited skills, resources (Sharpley, 2002), and strategic knowledge to achieve their goals through tourism activities (Prince & Ioannides, 2017) are also important factors to reflect and deconstruct.

In this sense, this chapter is organized into four main parts. First, the theoretical foundations on which it is based are presented, placing WWOOFF tourists' (or WOOFers) experiences at the center of an articulation between ecotourism, agrotourism, and voluntary tourism. Subsequently, the methodological approach is explained, and the results are presented. The results are divided between a sociodemographic presentation of these tourists and the identification of their characteristics and main motivations. This allowed the construction of a typology and the identification of three theoretical and exploratory categories. The chapter ends with the discussion of the results, with some concluding lines and with an outline of the main strategic axes for the future study of this emerging tourism market trend.

2 Toward a Definition of WWOOFers Tourists

This chapter analyses ecotourism practices within the specificities of organic farms. These places focus on organic agriculture and can represent an innovative way to contribute to sustainable rural development and rural tourism (Yamamoto &

Engelsted, 2014). The farms under analysis are part of the WWOOF platform. WWOOF is a network that brings together a group of organizations looking for volunteer workers, allowing them to experience life on an organic farm (McIntosh & Bonnemann, 2006). This network emerged in the mid-1970s, supported by the environmental social movement and manifesting a countercurrent to industrial agriculture and promoting sustainability (Kosnik, 2014). At the microlevel, this promotion occurs through sincere encounters between host and volunteer (Prince, 2019). Thus, aligned with previous studies (Ord & Amer, 2010), it is possible to consider that the tourism experiences that take place on these farms embody characteristics of agrotourism, voluntary tourism (Yamamoto & Engelsted, 2014), and ecotourism (Deville et al., 2016).

In the case of agrotourism, it can be understood as tourism in rural areas (Pulina et al., 2006), composed of small-scale enterprises (Sharpley, 2002), and tourist experiences with working farms direct authentic interaction (Arroyo et al., 2013). In the WWOOF farms, this configuration is confirmed since it is also characterized by small independent farmers and authentic tourist experiences that privilege direct contact between tourists and agriculture tasks developed in loco (McIntosh & Campbell, 2001). Sin (2009) defines volunteer tourism as:

> a form of tourism where the tourists volunteer in local communities as part of his or her travel.

Some of the primary tourist motivators in this segment are cultural immersion and the desire to make a difference (Brown, 2005). According to Dernoi (1988), this type of tourism implies establishing direct personal/cultural intercommunication and understanding between host and guest.

In this sense, the cultural dimension represents one of the main motivations of volunteer tourists, which is materialized through collaborative goal-orientated projects inside a community's own space, established through cross-cultural relationships (Wearing & Grabowski, 2011). According to Deville et al. (2016), among WWOOFer's main reasons for joining WWOOF are: meeting local people, experiencing local life/culture, and learning about organic farming.

In this work, the "organic farming" dimension present in the Ord and Amer (2010) study was replaced by ecotourism since it can be considered a more comprehensive category, as contemplate nature-based products with a fundamental concern about the sustainability of the practices developed (Blamey, 2001). Despite the close relationship between traditional ecotourism and WWOOF activities (Deville & Wearing, 2013), there are also some differences. WWOOF presents the nature of volunteering (McIntosh & Bonnemann, 2006), the philosophy around sharing associated with the developed practices (Lai et al., 2020), the horizontal principles of power distribution of these farms' management (McIntosh & Campbell, 2001), and the integration of "free work" (of the volunteer tourist) in work and life of WWOOF hosts (Yamamoto & Engelsted, 2014). According to Deville et al. (2016), WWOOFing can be considered a different kind of tourism as it

> requires a tourist to shed that role or identity and become more like a person (. . .) in conventional tourism forms, the tourist is often engaged in a diversionary mode seeking

> distraction and escapes from their boredom and routinized daily lives and various degrees of alienation.

The WWOOFers, more specifically (Nordbø et al., 2020).

> see traveling as an important part of their lives. They have adventurous travel spirits, seek new adventures, visit new destinations, and spend time with locals.

According to Deville et al. (2016), the main reasons of WWOOFers for joining WWOOF are: meeting local people, experiencing local life/culture, saving money/cheap traveling, life experience, and learning about organic farming.

Tourist experiences through WWOOF aggregate agrotourism characteristics due to the locus where they take place, ecotourism, due to the concerns associated with organic production and environmental practices, and voluntary tourism due to the fact that the motivations of the visitors are based on personal and meaningful experiences (Pearce, 1990). Tourists are no longer passive spectators, and the WWOOFs visitors are active protagonists as they co-create their own travel experiences (Campos et al., 2015). This search for authentic experiences and personal growth (McIntosh & Bonnemann, 2006; Miller & Mair, 2015) implies, on the one hand, a set of previous individual concerns and aspirations and, on the other hand, the materialization of these manifestations in travel processes.

In terms of sociodemographic profile, most of the WWOOFers are foreigners, well-educated, female, young (Deville et al., 2016; McIntosh & Bonnemann, 2006; Nordbø et al., 2020; Ord & Amer, 2010), single, without children, and from cities (Deville, 2011; Nordbø et al., 2020). Furthermore, these tourists generally have WWOOFed before (Nordbø et al., 2020) and present a short length of stay (Ord & Amer, 2010). It can explain that WWOOFers are also positively influenced by the reduced cost of this kind of travel (Deville et al., 2016). In this context, another motivator is the possibility to escape from mass tourism (Ateljevic & Doorne, 2001). In terms of post-WWOOFing, previous studies highlighted the high levels of satisfaction of these tourists with the experience and their intention to recommend it to other travelers (Deville et al., 2016). It is an essential aspect since word of mouth can be considered the main WWOOF informal information channel (Nordbø et al., 2020; Deville et al., 2016).

3 Methodology

3.1 Collection of Textual Information

To identify the profiles of the tourists, this study contemplated reviews from the WWOOF platform that offers experiences in more than 100 countries and has about 12,000 hosts and 100,000 volunteers (https://wwoof.net/). In Portugal, the number of hosts and volunteers is, respectively, 108 and 1400 (https://wwoof.pt/).

On this platform, tourists (volunteers) can complete a set of data to provide some sociodemographic characteristics and their preferences and motivations. Tourists

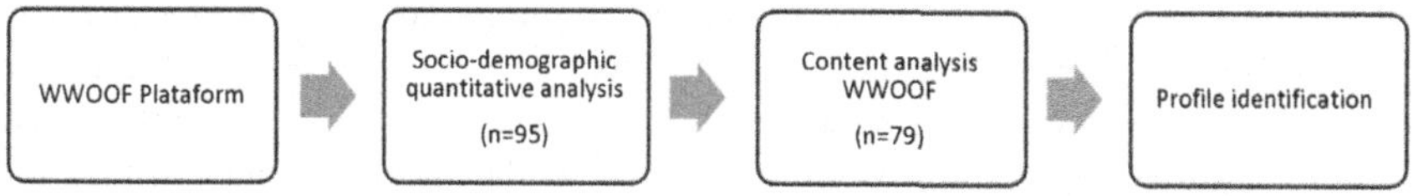

Fig. 1 Data collection process and analysis

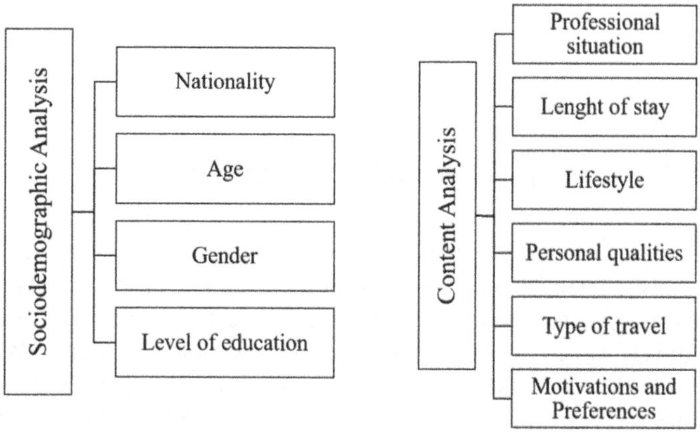

Fig. 2 Dimensions under analysis

have punctual participation in the projects and visit the communities for a certain period, usually between 1 week and 6 months.

The sample of this study is composed of all the reviews written on WWOOF Portugal until 07/02/2020. These reviews were analyzed considering two main objectives: on the one hand, to identify the sociodemographic characteristics of tourists seeking this type of experience; and, on the other hand, to analyze their profiles through an in-depth analysis of the content provided in the descriptions they deliver.

In the first stage, the sociodemographic characterization, the totality of reviews available on the platform were analyzed ($n = 95$). The number of reviews considered in this study represents 0.1% of the volunteers enrolled on this platform worldwide. Each country has an individualized platform where the farms available in that country can be found. WWOOF Portugal® lists 165 biological farms across the country. For this study, only the WWOOF Portugal® platform was used, and in this case, the 95 available reviews were analyzed for the first phase. Of these 95, 79 provided more detailed information by describing their goals, concerns, and motivations and underlining some personal characteristics that they consider essential for this type of experience (Fig. 1). The content analysis fell on these 79 and was developed using the *Nvivo* software (Fig. 1).

Figure 2 outlines the dimensions analyzed for the development of this exploration. The sociodemographic analysis focused on dimensions such as nationality, gender, age, and education level. Content analysis duplicates some of these dimensions, such as education, to enable the framing and to define the different profiles. Besides, other dimensions were included, such as the professional situation, the type of trip they tend to favor, and their motivations to apply for this experience.

After identifying the dimensions addressed in the literature and creating the categories, the reference information under analysis was handled using the *Nvivo* software, respecting the coding process. The direct transcriptions of the information provided by tourists (volunteers) were analyzed, clustering the profiles.

4 Results

4.1 Sociodemographic Quantitative Analysis

Regarding the sociodemographic quantitative analysis, it was possible to verify that most tourists are international, only 11% were Portuguese (Table 1). This configuration confirms previous studies' findings regarding the tendency of this kind of tourist to travel away from home to participate in nature-based activities (Wolf-Watz et al., 2011). The nationalities with a higher level of recurrences were Germany (21%), France (16%), and the USA (12%). The category "others" includes the Czech Republic, Switzerland, and Denmark with 2% each and Finland, Ireland, Japan, Russia, South Africa, and Sweden with 1% each. Given that distance generally influences tourism demand negatively (McKercher & Mak, 2019), the configuration presented by the demand considered in this study can indicate its deep predisposition to know more about different cultures (McKercher et al., 2008).

Furthermore, among these tourists, the prevalence was people under 31 years (66%), and women (69%) are highlighted. These results are aligned with previous studies, such as that developed by McIntosh and Bonnemann (2006). However, this high participation of women needs further research in the future.

A count was made of the most frequent words for the dimensions of occupation and the experience's objectives. In terms of occupancy, the most recurrent was the

Table 1 Tourists by nationality ($N = 95$)

Nationality	%
Germany	21
France	16
USA	12
Portugal	11
Belgium	6
Spain	6
Netherlands	5
Canada	4
Italy	4
UK	4
Others	12

category "student," followed by "teacher" and "manager." Concerning the self-description, some of the words with the highest frequencies were "want," "new," "experiences," organic, and permaculture. These results are aligned with Adalilar et al. (2015), according to which this kind of tourist is interested in authentic experiences of a sustainable lifestyle (Adalilar et al., 2015).

4.2 Content Analysis

Tourists staying in WWOOFs are heterogeneous, making it difficult to create standards, although there are some common characteristics. Often, these actors are students in periods of interregnum or in transition between qualification and entry into the labor market: *"Before I start looking for a job opportunity in Germany, I want to take the time to do wwoofing"* (P19, 22 years). Other studies had already noted the predominance of younger visitors who may be more interested in socio-cultural learning experiences, be more willing to work as volunteers (McIntosh & Bonnemann, 2006).

In other cases, they are individuals who already have professional activities related to the work they will develop in the farms: *"I have lived and worked as an HIV/AIDS Training Volunteer in a rural village in southern Botswana for the past two years"* (P8, 35 years). In this case, we face visitors whose primary motivation is learning and developing practices consistent with their ecological concerns (Lai et al., 2020). This aspect is directly related to the lifestyle that these tourists pursue. Tourists looking for the WWOOF experience seek to maintain a lifestyle based on sustainability, with ecological convictions, a connection with nature and with a spiritual focus; they are also looking for an experience based on an educational dimension (Yamamoto & Engelsted, 2014): *"since I was a child I have been very concerned with the environment and for several years I would like to learn and try alternative and more sustainable lifestyles (...)"* (P12, 31 years).

There is also a set of descriptions that show the need for a change in the professional trajectory of this person. As this reviewer states, *"I left my career as a neuroscientist in 2015 to find a job that would make me happier. I ended up participating in a sustainable agriculture internship and then went on to work as a full-time farmer"* (P3, 45 years). In these cases, volunteer tourists are more motivated by sociocultural dimensions. The experience is seen as a "means to an end," which is the permanent adoption of this lifestyle (Lai et al., 2020). While some studies identify that the main motivations behind volunteering were cultural immersion, making a difference, and seeking camaraderie (Brown, 2005), others add precisely this personal development dimension (Broad & Jenkins, 2008).

In general, these actors favor long trips, using public transport with the least possible environmental impact. Their main objective is an immersion in the local culture, often being solo travelers. This view is consistent with backpackers' travel behavior and long-term budget travelers (Riley, 1988) who prefer to travel alone. They perceive WWOOF as a type of tourism that is consistent with these values. The following statements confirm this idea: *"(...) I don't just want to travel. I want to create value and meet local people."* (P24, 26 years old), *"We spent the last two years living on a boat on the Avon River. We are now traveling through Portugal in our Motorhome, made by ourselves"* (P56, 34 years old).

Regarding the dimension of the objectives, the agents essentially seek to apply or deepen the knowledge and skills they already have: *"(...) I have been interested in making woofs for some time now to be able to learn, especially about permaculture and energy systems"* (P5, 26 years). Also, as highlighted by previous studies (Deville et al., 2016), many introduced a lot of information into the descriptions that were categorized as a desire to meet new people and new cultures: *"I am also very interested in meeting new people, in alternative social structures, community life, seeing how humans can live together in a different system (...)"* (P58, 34 yeas), as well as a need to (re)connect with nature: *"I intend to have more contact with the farm life and nature as I live in the city"* (P15, 22 years). Within the information analyzed, it was also possible to see that some people are looking for this type of experience as a starting point for a change of country, mainly to adapt to the mores, food and language: *"After visiting Portugal on my last summer vacation, I look forward to returning. It was a country that I would like to choose to live in"* (P11, 32 years).

4.3 Profiles

The previous analysis made it possible to verify some general trends. However, some aspects distinguish the different tourists (volunteers) present at WWOOF. These differentiating characteristics allowed the creation of a typology comprising three theoretical and exploratory categories.

The "Sustainable Gappers" are younger participants, aged between 18 and 23 years old, who are essentially looking for a temporary experience to connect with nature, meet new people and cultures and experience a different type of tourism. These young tourists are significantly influenced by financial costs (Song et al., 2018). This enhances the importance of travel as an immersive cultural experience and the idea of adventure. Usually, the time available for travel is between 2 weeks and 2 months. The general tourism literature has pointed out that younger tourists prefer shorter stays (Santos et al., 2015; Song et al., 2018), including in WWOOFing experiences (Ord & Amer, 2010). Therefore, these individuals are primarily students and develop this type of travel between breaks in the school terms, the transition between study cycles, or an interruption between the end of qualification and the entry into the labor market. In this sense, they are young people who attend higher education degrees or who finished them less than a year ago. They are not yet inserted in the labor market or have temporary jobs that accumulate with their student situation.

The second category, the "Sustainable Escapers" are individuals aged between 25 and 33 years who seek the experience of connecting with nature in the farms as a way to escape their day-to-day routine and to reach the state of "existential authenticity" as theorized by the sociology of tourism (Steiner & Reisinger, 2005). This will of escape emerges because they live in large cities, with jobs that are often, in their perception, exhausting. The main objective of these individuals is the "transformative travel" process (Sampaio et al., 2014). They value the possibility of "*knowing the world*" and being in constant "*travel*" rather than professional stability. They indicate the availability to travel between 1 and 6 months, dividing their year between a few months of work and a few months of travel. Most of them have secondary education qualifications. At the professional level, they tend to have seasonal jobs, with low wages or are unemployed.

The last type, "Sustainable Lifers," comprises individuals who are over 35 years old and looking for structural change or even a rupture with their former way of life. The search for WWOOF appears as a transition stage where they intend to accumulate knowledge and experience to create their own community and/or organic farm or even to stay there indefinitely. In this sense, they do not perceive this experience as much as a travel, but as a moment of a life-changing process. According to Deville et al. (2016), WWOOFers "often staying in specific places for long periods as part of an effort made to remedy western anomie/alienation." For the most part, these are people who have left their previous jobs and are looking for a simpler lifestyle, closer to nature, where they can see their values reflected. They indicate a willingness to travel for more than 6 months and sometimes do not indicate any limit. This type is composed of people with higher education qualifications. They claim to have had competitive jobs with high levels of responsibility and leadership, and with high remunerations.

5 Discussion of Results and Conclusions

This chapter identified some general traits of the tourists seeking new experiences through the WWOOF platform. These experiences can be identified as agrotourism because they take place on small farms in rural areas whose existence contributes, directly or indirectly, to the territories' development (Yamamoto & Engelsted, 2014).

The common characteristics of all the WWOOF tourists include solid environmental and sustainable concerns, both about the activities developed and about the way the farms are managed. These characteristics are also directly aligned with the dimensions that support ecotourism practices: nature-based products and a sustainable management that comprise environmental education elements that contribute to conservation (Wearing, 2004).

The experiences offered usually involve accommodation and food in exchange for the farm work, thus assuming voluntary tourism characteristics. In addition, many of the analyzed WWOOFers manifest motivations such as personal development, learning, and the opportunity to materialize their values. Volunteer tourists seek vacations that address these needs, namely self-discovery and the chance to re-evaluate their values (Weiler & Richins, 1995). However, this type of tourist practice also implies the desire to make a difference in the world and contribute, in a positive way, to the natural or social environment (Wearing, 2004). WWOOFers value the sense of place, community, and cultural exchange. The focus of these experiences is the interaction with the members from different cultures along with several altruistic and self-serving motivations (Wearing & Grabowski, 2011).

On the one hand, this cultural dimension was evident in the analysis, since several textual passages of the observed reviews corroborate the vision of this desire to contact and learn about other cultures. On the other hand, it was not possible to perceive any statements that could be directly categorized as a manifestation of an altruistic motivation. However, it should be noted that the descriptions analyzed have limitations since they function as a presentation of WWOOFer to the farms. In this sense, it is not possible to say that these tourists do not express this desire to contribute selflessly to these places. This gap could be filled with more detailed and targeted interviews.

Three theoretical and exploratory categories were identified. While the first two types, although different, have characteristics similar to those of backpackers and long-term budget travelers (Riley, 1988), "Sustainable Lifers" deviate somewhat from the traditional definition of tourist, in the sense that this experience is perceived as an opportunity to permanently break with their lifestyle. For them, it is seen as a long-term solution and not as a temporary experience.

The characteristics of these categories are aligned with previous studies that have highlighted, for example, that WWOOFers can remain for long periods in the farms as a way to live in different socio-economic contexts (Deville et al., 2016). Furthermore, the literature indicates that younger tourists prefer shorter stays (Santos et al., 2015; Song et al., 2018) and are highly influenced by travel costs (Song et al., 2018).

Regardless of the dissimilarities of these three categories, some characteristics are transversal. The analysis shows that these tourism experiences are frequently seen as a means of escape from modern life's alienation. Still, they are also increasingly regarded as opportunities for personal growth and individual learning. Such behavior has been described as Transformative Tourism (Reisinger, 2013). A type of tourism that can lead travelers to a positive change in their attitudes and values fosters cross-cultural tolerance and may also result in a lifestyle switch toward more sustainable life options and global citizenship awareness.

Some limitations of the present investigation are also identified. On the one hand, there is a significant participation of female tourists seeking this type of experience, which requires a more in-depth analysis to determine the causal relationships between gender and this kind of tourism. On the other hand, concerning the profiles identified, it should be noted that these tourists' characteristics vary considerably considering the country where they come from (McIntosh & Bonnemann, 2006). In this sense, they cannot be inferred for populations other than the one directly analyzed.

Acknowledgments Carla Nogueira acknowledges the support of FCT (SFRH/BD /117398/2016) for her PhD grant. João Filipe Marques is financed by National Funds provided by FCT—Foundation for Science and Technology, through project UIDB/04020/2020.

References

Adalilar, S. N., Alkibay, S., & Eser, Z. (2015). Ecovillages as a destination and a study of consumer approaches to ecovillages. In A. I. Iacob (Ed.), *2nd Global Conference on Business, Economics, Management and Tourism* (23, pp. 539–546).

Arroyo, C. G., Barbieri, C., & Rich, S. R. (2013). Defining agritourism: A comparative study of stakeholders' perceptions in Missouri and North Carolina. *Tourism Management, 37*, 39–47. https://doi.org/10.1016/j.tourman.2012.12.007

Ateljevic I, Doorne S (2001) Nowhere left to run: A study of value boundaries and segmentation within the backpacker market of New Zealand. In: JA Mazanec, , AG Woodside, GI Crouch (eds), Consumer psychology of tourism - hospitality and leisure (vol 2, pp 169–186). CAB International.

Blamey, R. K. (1997). Ecotourism: The search for an operational definition. *Journal of Sustainable Tourism, 5*(2), 109–130. https://doi.org/10.1080/09669589708667280

Blamey, R. K. (2001). Principles of ecotourism. In D. Weaver (Ed.), *The encyclopedia of ecotourism* (pp. 5–22). CABI.

Broad, S., & Jenkins, J. (2008). Gibbons in their midst? Conservation volunteers' motivations at the gibbon rehabilitation project, Phuket, Thailand. In K. Lyons & S. Wearing (Eds.), *Journeys of discovery in volunteer tourism: International case study perspectives* (pp. 72–85). CABI.

Brown, S. (2005). Travelling with a purpose: Understanding the motives and benefits of volunteer vacationers. *Current Issues in Tourism, 8*(6), 479–496. https://doi.org/10.1080/13683500508668232

Campos, A. C., Mendes, J., Valle, P., & Scott, N. (2015). Co-creation of tourist experiences: A literature review. *Current Issues in Tourism*, 369–400. https://doi.org/10.1080/13683500.2015.1081158

Dernoi, L. A. (1988). Alternative or community-based tourism. In L. D'Amore & J. Jafari (Eds.), *Tourism: A vital force for peace* (pp. 89–94). D'Amore and Associates.

Deville, A. (2011). *Alice in WWOOFerLand: Exploring symbiotic worlds beyond tourism.* Unpublished doctoral dissertation, University of Technology - Sydney.

Deville, A. W., Wearing, S., & McDonald, M. (2016). WWOOFing in Australia: Ideas and lessons for a decommodified sustainability tourism. *Journal of Sustainable Tourism, 24*(1), 91–113. https://doi.org/10.1080/09669582.2015.1049607

Deville, A., & Wearing, S. (2013). WWOOFing tourists, beaten tracks and transformational paths. In Y. Reisinger (Ed.), *Transformational tourism: Tourist perspectives* (pp. 151–168). CAB International.

Kim, K.-H., & Park, D.-B. (2017). Relationships among perceived value, satisfaction, and loyalty: Community-based ecotourism in Korea. *Journal of Travel & Tourism Marketing, 34*(2), 171–191. https://doi.org/10.1080/10548408.2016.1156609

Kosnik, E. (2014). Work for food and accommodation: Negotiating socio-economic relationships in non-commercial work-exchange encounters. *Hospitality & Society, 4*(3), 275–291. https://doi.org/10.1386/hosp.4.3.275_1

Lai, P.-H., Chuang, S.-T., Zhang, M.-C., & Sanjay, Z. N. (2020). The non-profit sharing economy from a social exchange theory perspective: A case from world wide opportunities on organic farms in Taiwan. *Journal of Sustainable Tourism, 28*(12), 1970–1987. https://doi.org/10.1080/09669582.2020.1778709

McCormick, J. (1992). *Rumo ao Paraíso: A história do movimento ambientalista*. Relume Dumará.

McIntosh, A. J., & Bonnemann, S. M. (2006). Willing workers on organic farms (WWOOF): The alternative farm stay experience? *Journal of Sustainable Tourism, 14*(1), 82–99. https://doi.org/10.1080/09669580608668593

McIntosh, A. J., & Campbell, T. (2001). Willing workers on organic farms (WWOOF): A neglected aspect of farm tourism in New Zealand. *Journal of Sustainable Tourism, 9*(2), 111–127.

McKercher, B., & Mak, B. (2019). The impact of distance on international tourism demand. *Tourism Management Perspectives Journal, 31*, 340–347.

McKercher, B., Chan, A., & Lam, C. (2008). The impact of distance on international tourist movements. *Journal of Travel Research, 47*(2), 208–224.

Miller, M. C., & Mair, H. (2015). Volunteer experiences on organic farms: A phenomenological exploration. *Tourism Analysis, 20*(1), 69–80. https://doi.org/10.3727/108354215X14205687167662

Nordbø, I., Segovia-Pérez, M., & Mykletun, R. J. (2020). WWOOFers in Norway – Who are they? Scandinavian. *Journal of Hospitality and Tourism, 20*(5), 419–437. https://doi.org/10.1080/15022250.2020.1766559

Ord, C., & Amer, J. (2010). Contribution of volunteer tourism to organic farms: An analysis of the WWOOF exchange in Canada. *International Ecotourism Club*. Retrieved February 15, 2021, from http://ecoclub.com/education/articles/502-volunteer-tourism-organic-farms

Pearce, P. L. (1990). Farm tourism in New Zealand: A social situation analysis. *Annals of Tourism Research, 17*, 337–352.

Prince, S. (2019). Volunteer tourism and the eco-village: Finding the host in the pedagogic experience. *Hospitality & Society, 9*(1), 71–89. https://doi.org/10.1386/hosp.9.1.71_1

Prince, S., & Ioannides, D. (2017). Contextualizing the complexities of managing alternative tourism at the community-level: A case study of a Nordic eco-village. *Tourism Management, 60*, 348–356. https://doi.org/10.1016/j.tourman.2016.12.015

Pulina, M., Dettori, D. G., & Paba, A. (2006). Life cycle of agrotouristic firms in Sardinia. *Tourism Management, 27*(5). https://doi.org/10.1016/j.tourman.2005.10.023

Rajović, G., & Bulatović, J. (2015). Eco tourism with special review on eco-village "Štavna". *Scientific Electronic Archives, 8*(1). https://doi.org/10.36560/812015157

Reisinger, Y. (2013). *Transformational tourism: Tourist perspectives*. CABI.

Riley, P. J. (1988). Road culture of international long-travel budget travellers. *Annals of Tourism Research, 15*, 313–328.

Sampaio, S., Simoni, V., & Isnart, C. (2014). Tourism and transformation: Negotiating metaphors, experiencing change. *Journal of Tourism and Cultural Change, 12*(2), 93–101.

Santos, G. E. O., Ramos, V., & Rey-Maquieira, J. (2015). Length of stay at multiple destinations of tourism trips in Brazil. *Journal of Travel Research, 54*(6), 788–800. https://doi.org/10.1177/0047287514532370

Sharpley, R. (2002). Rural tourism and the challenge of tourism diversification: The case of Cyprus. *Tourism Management, 23*(3), 233–244. https://doi.org/10.1016/S0261-5177(01)00078-4

Sin, H. L. (2009). Volunteer tourism: "Involve me and I will learn?". *Annals of Tourism Research, 36*(3), 480–501. https://doi.org/10.1016/j.annals.2009.03.001

Song, H., Wang, Y., & Sparks, B. A. (2018). How do young Chinese friendship groups make travel decisions? A content and interaction process analysis. *Journal of Travel & Tourism Marketing, 35*(6), 772–785. https://doi.org/10.1080/10548408.2017.1421117

Spindler, E. A. (2013). The history of sustainability: The origins and effects of a popular concept in Jenkins. In I. Jenkins & A. Schröder (Eds.), *Sustainability in tourism: A multidisciplinary approach* (pp. 9–32). Springer.

Steiner, C. J., & Reisinger, Y. (2005). Understanding existential authenticity. *Annals of Tourism Research, 2*(33), 299–218.

Wearing, S. (2004). Examining best practice in volunteer tourism. In R. A. Stebbins & R. T. Graham (Eds.), *Volunteering as leisure/Leisure as volunteering*. Wallingford.

Wearing, S., & Grabowski, S. (2011). Volunteer tourism and intercultural exchange- Exploring the 'Other' in the experience. In A. Benson (Ed.), *Volunteer tourism - Theoretical frameworks and practical applications*. Routledge.

Weiler, B., & Richins, H. (1995). Extreme, extravagant and elite: A profile of ecotourists on earthwatch expeditions. *Tourism Recreation Research, 20*, 29–36.

Wolf-Watz, D., Sandell, K., & Fredman, P. (2011). Environmentalism and tourism preferences: A study of outdoor recreationists in Sweden, Scandinavia. *Journal of Hospitality and Tourism, 11*(2), 190–204. https://doi.org/10.1080/15022250.2011.583066

Yamamoto, D., & Engelsted, A. K. (2014). World wide opportunities on organic farms (WWOOF) in the United States: Locations and motivations of volunteer tourism host farms. *Journal of Sustainable Tourism, 22*(6), 964–982. https://doi.org/10.1080/09669582.2014.894519

The manufacturer's authorised representative in the EU is Springer Nature Customer Service Centre GmbH, Europaplatz 3, 69115 Heidelberg, Germany. If you have any concerns regarding our products, please contact ProductSafety@springernature.com

Printed and bound by CPI Group (UK) Ltd, Croydon, CR0 4YY
28/09/2026
02224093-0008